THE MULTIDIMENSIONALITY OF ELECTRICITY INDUSTRY REFORM

A STRATEGIC ANALYSIS FOR
THE CASE OF ASIAN COUNTRIES AND VIETNAM

NGUYEN THI NGOC MAI

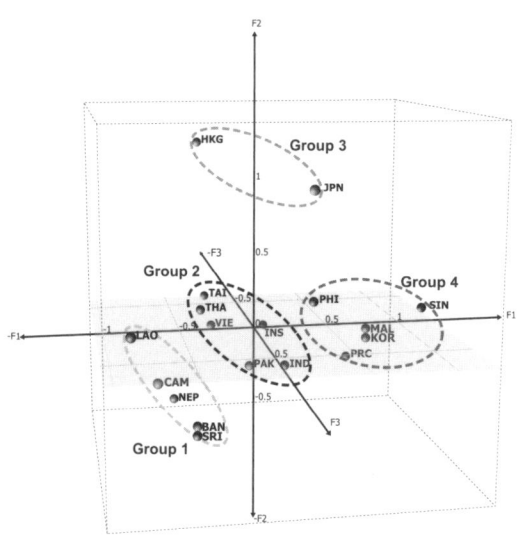

Kwansei Gakuin University Press

Copyright © 2007 by NGUYEN THI NGOC MAI

All right reserved.

No part of this book may be reproduced in any form or by any means without permission in writing from the author.

Kwansei Gakuin University Press
1-1-155 Uegahara, Nishinomiya, Hyogo, 662-0891, Japan
ISBN: 978-4-86283-007-4

ACKNOWLEDGEMENTS

This book owes a great deal to many people.

First, I am exceedingly indebted to my academic advisor, Prof. Munenori Nomura, who thoroughly guided my work during 'periods of suffering and doom' that are typical for research and completing this book. He also has been instrumental in transforming every bit of discouragement into efficacious motivational impulses for the entire period of my two years of study at the School of Economics at Kwansei Gakuin University.

I would like to express my gratitude and sincere thanks to Prof. Noriyuki Doi and Prof. Shoichi Ito who willingly served as co-advisors and kindly answered my queries. I owe a debt of gratitude to both professors for the numerous and valuable suggestions I received. The author of the book would like to express her great gratitude to Prof. Shoichi Ito for his kindness in the provision of computer software essential for carrying out empirical analysis.

I would like to extend sincere thanks and gratitude to all the professors at the School of Economics for their encouragement and valuable support. Especially, I would like to express special thanks to Prof. Norihiko Toyohara and Prof. Toshihiro Okada for their assistance that enabled me to advance my knowledge on microeconomics and econometrics and helped me to complete my research successfully. The author of this book also cannot forget the valuable instruction and support kindly provided by Prof. Toshihiro Matsuda of Otemon Gakuin University.

During my study in Japan, I was supported by a scholarship from the Japanese Government and Ministry of Education, Culture, Sports, Science and Technology. I am extremely grateful for their generous support. As luck would have it, this book was funded by the Federation of Electric Power Companies of Japan. Grateful acknowledgement is also made to the strong support from Tokyo Electric Power Company in checking the whole book. The author is sure that without such valuable support, of both finance and expertise, the book would not have been finished.

I would like to express my gratitude to my Japanese colleagues for their support and encouragement. I am especially grateful to Mr. Hideki Yukimura, who kindly gave me a guidance to pursue my studies in Japan, as well as taking care of me throughout this period. I would like to thank also Mr. Masahiro Sado, Mr. Toshiaki Toyama, and Mr. Shoji Kusuda for their enthusiasm and constant help.

I would like to thank to the leaders and staff of Ministry of Industry, Electricity of Vietnam, Institute of Energy, faculties of Hanoi University of Technology for their fruitful assistance in the data collection. Thanks were made also to Prof. Nguyen Minh Due, Prof. Tran Van Binh, Prof. Dam Xuan Hiep, Dr. Tran Thanh Lien, Dr. Duong Quang Thanh, Dr. Nguyen Anh Tuan, Mr. Dang Hoang An, Mrs. Pham Thi Ly, and Mrs. Le Thi Minh Thu for their assistance in conducting this study.

Sincere thanks are due to Mr. Naoya Tanaka of Kwansei Gakuin University Press, all faculty staff, secretaries, and classmates from the School of Economics at Kwansei Gakuin University for their teaching and direct and indirect assistance.

Acknowledgement is also given to my younger sister for her special help throughout the entire period of this study.

Last, but not least, I sincerely thank my parents, my parents-in-law, and other family members for their support and encouragement throughout the period of my studies in Japan.

I also want to thank my husband, Mr. Pham Le Phu for his tremendous inspiration, encouragement, moral and academic support throughout my period of study. I thank my beloved five-year old daughter for her dedication and loving care.

Thanks to all friends, especially my best friends who shared a bowl of rice in Nada dormitory and those who have made my stay in Japan a memorable one.

TABLE OF CONTENTS

Acknowledgements
Table of Contents
List of Tables
List of Figures
List of Abbreviations

INTRODUCTION······ 15

PART I: THE ESSENTIAL ASPECTS OF ELECTRICITY INDUSTRY REFORM······ 21

Chapter 1: Background and Introduction of Electricity Industry Reform······ 22
1. 1. Functions of Electric Industry
1. 2. Specific Characteristics of Electricity Industry and Traditional Organization of the Industry
1. 3. The Motivation behind EI Reform
1. 4. The Basic Architectures of Reform

Chapter 2: Restructuring and Competition in Electricity Industry······ 32
2. 1. Structural Options for the Electricity Industry
2. 2. Components of Restructured Systems
2. 3. Electricity Industry Unbundling
2. 4. Open Network Access
2. 5. Competitive Electricity Market

Chapter 3: Electricity Industry Privatization······ 54
3. 1. Definition of Privatization
3. 2. Rationales for Privatization
3. 3. Managing Privatization
3. 4. Main Trends in Privatization
3. 5. Specific Issues of the Privatization Process in the Electricity Industry
3. 6. Partial Sales – the Rationales
3. 7. Golden Shares and Special Rights
3. 8. Investment Vehicles and the Role of IPPs

Chapter 4: Electricity Regulation······ 73
4. 1. Definition of Regulation
4. 2. Regulatory Governance in Liberalized Environment

4. 3. Mechanisms to Regulate Prices
4. 4. Regulation of Entry
4. 5. Regulation and the Mechanisms to Address Social and Environmental Issues

Chapter 5: Worldwide Analysis of Electricity Industry Reform – the Development Overview and the New Trend of Reform······ 91
5. 1. Electricity Industry Reform Overview
5. 2. The New Trend of Electricity Industry Reform
5. 3. Reform from the Viewpoint of Several Main Stakeholders

PART II: ELECTRICITY INDUSTRY REFORM MODELING FOR ASIAN COUNTRIES······ 111

Chapter 6: The Evaluation of Electricity Industry Reform in Asian Countries······ 112
6. 1. Overview
6. 2. Build Criteria to Measure Regulatory Environment, Ownership and Competition Outcomes
6. 3. Comparison of EI Status in Asian Countries
6. 4. Crucial Issues in Electricity Industry Reform of Asian Countries
Appendix 6_1: A Brief Summary (one-page overview) of Electricity Industry Reform in Asian Countries

Chapter 7: Analysis of Effects of Competition, Ownership and Regulation on Electricity Performance in Asian Countries and Predictions about the Shape of the Industry······ 185
7. 1. Reviews on Empirical Studies of Electricity Sector Reform
7. 2. Rationales of the Study
7. 3. Data and Modeling
7. 4. The Research Hypotheses
7. 5. Results and Discussion
7. 6. Predictions about the Probability and Shape of Change in Electricity Industry Reform: Group Modeling

Chapter 8: Verification Analysis: the Case of Four Asian Countries, Group Representatives······ 207
8. 1. Policy Implications for the Reform in Asian Electricity Industries
8. 2. The Case of Japan
8. 3. The Case of Singapore
8. 4. The Case of Thailand
8. 5. The Case of Bangladesh
8. 6. Generalization Issues for Asian Modeling in Electricity Industry Reform

Chapter 9: Nuclear Power Policy under the Process of EI Reform in Asian Countries······ 234

9. 1. Current Status and Future Potential of Nuclear Power in Asian Countries
9. 2. Effects of Electricity Reform on the Development of Nuclear Power
9. 3. British Energy's Experience regarding Nuclear Power Issues in Liberalized Electricity Industry
9. 4. Impacts of Industry's Reform and Mechanisms for the Promotion of Nuclear Power in Asian Countries

PART III: MULTIDIMENSIONAL ELECTRICITY INDUSTRY REFORM – A STRATEGIC ANALYSIS FOR THE CASE OF VIETNAM······ 261

Chapter 10: Analysis of Electricity Industry Reform to Date in Vietnam······ 262
10. 1. Historical Development of Vietnam's Electricity Industry
10. 2. The Need for Electricity Industry Reform
10. 3. Reform to Date in Vietnam's Electricity Industry

Chapter 11: A Proposed Design for EI Reform in Vietnam – Overall Design······ 272
11. 1. Considerations for Policy Selection of Electricity Reform
11. 2. Expected Benefits and Potential Risks
11. 3. Overall Design for Electricity Industry Reform

Chapter 12: Regulation Reform in Vietnam's Electricity Industry······ 280
12. 1. Building an Effective Regulatory Framework
12. 2. Mechanism to Regulate Price
12. 3. Mechanism for Long-term Planning and Investment Incentives
12. 4. Mechanisms to Address Social and Environmental Issues

Chapter 13: Restructuring and Market Development in Vietnam's Electricity Industry······ 295
13. 1. Market Development Standpoints in Vietnam's Electricity Industry
13. 2. Competitive Electricity Generation Market
13. 3. Competitive Electricity Wholesale Market
13. 4. Competitive Electricity Retail Market

Chapter 14: Implementing Partial Privatization in Vietnam's Electricity Industry······ 311
14. 1. Private Investment Promotion in Electricity Industry
14. 2. Equitization and Electricity Industry Reform
14. 3. Proposed Equitization Plan
14. 4. Applying Cost-benefit Analysis into the Proposed Equitization Plan

CONCLUSIONS, RECOMMENDATIONS AND FURTHER RESEARCH······ 325
REFERENCES······ 329
INDEX······ 340

LIST OF TABLES

Table 2. 1   Characteristics of theoretical electricity industry models
Table 4. 1   Features of rate of return and price cap regulation
Table 5. 1   Electricity reform by regions (1998) (% of countries where the reform has occurred)
Table 5. 2   OECD multinational electricity companies active in Asia-Pacific
Table 6. 1   List of indicators for measuring reform outcome
Table 6. 2   Reform process – correspondent factors
Table 7. 1   The results from regression analysis
Table 8. 1   Proposals for future industry structure
Table 9. 1   Recent nuclear reactor additions and reactors under construction in Asia, as of April 2004
Table 9. 2   List of nuclear power reactors in Asian countries 4 January 2006
Table 9. 3   Nuclear power data summary, Japan
Table 9. 4   Nuclear power data summary, Korea
Table 9. 5   Nuclear power data summary, China
Table 9. 6   Nuclear power data summary, Chinese Taipei
Table 14. 1 Investment requirements for electricity industry development in the period of 2006–2025

LIST OF FIGURES

Figure 2. 1 Vertical integration monopoly model
Figure 2. 2 Single buyer model
Figure 2. 3 Wholesale competition model
Figure 2. 4 Full customer choice model
Figure 2. 5 Transmission companies vs. independent system operator
Figure 2. 6 Pricing of network services
Figure 2. 7 Trading in the power pool
Figure 2. 8 Trading with pool and bilateral contracts
Figure 2. 9 Trading with bilateral/multilateral contracts
Figure 2. 10 How an ideal spot electricity market would work
Figure 3. 1 Range of service models from fully public to fully private
Figure 3. 2 Choosing the appropriate privatization method
Figure 3. 3 Sectoral distribution of privatization proceeds: 1990–2003
Figure 3. 4 The difference between new greenfield investments and privatization of existing infrastructure assets
Figure 3. 5 Number of private participation transactions in developing countries
Figure 4. 1 Regulation in vertically integrated monopoly
Figure 4. 2 Regulation in network-opening model
Figure 4. 3 Cost-plus mechanism
Figure 4. 4 Price cap mechanism
Figure 4. 5 Hybrid regulatory mechanisms
Figure 5. 1 The post-privatization labor productivity levels in electricity distribution in Argentina, Chile and the UK
Figure 5. 2 Energy losses among Argentina's distribution companies, at privatization in 1999
Figure 5. 3 EU average real price (2004 euro/MWh)
Figure 5. 4 Allocation of ADB loan for energy sector 2002
Figure 6. 1 Principal coordinates and scree plot obtained from MCA
Figure 6. 2 Eigenvalue obtained from MCA
Figure 6. 3 Interpreting axis meaning
Figure 6. 4 Two-dimensional visualization of country groups
Figure 6. 5 Three-dimensional visualization of country groups
Figure 6. 6 Asian electricity industry reform in a group pattern

Figure 6. 7	Private investment trends in electricity in developing countries 1990–2002
Figure 7. 1	Impact of reform options on net electricity generation per capita
Figure 7. 2	Impact of reform options on installed capacity per capita
Figure 7. 3	Impact of reform options on the capacity utilization rate
Figure 7. 4	Impact of reform options on electricity loss
Figure 7. 5	Impact of reform options on the share of nuclear power generation
Figure 8. 1	Electric power utilities in Japan
Figure 8. 2	Structural changes in Japanese electric power sector
Figure 8. 3	Key points in the design of institutional arrangement
Figure 8. 4	Electricity industry structure from 1 Oct 1995 to 31 Mar 2001
Figure 8. 5	Electricity industry structure from 1 Apr 2001 to 31 Dec 2002
Figure 8. 6	Electricity industry structure since 1 Jan 2003
Figure 8. 7	Structure of wholesale market before and after 1 Jan 2003
Figure 8. 8	Thailand's electric power sector in short-, medium- and long-term plan
Figure 8. 9	Present structure of the electricity industry of Bangladesh
Figure 8. 10	Electricity industry structure in Bangladesh after reforms
Figure 9. 1	Nuclear share in total power generation 2004
Figure 9. 2	Nuclear generation projection for Asia
Figure 9. 3	Japan's organization chart in nuclear power
Figure 9. 4	Evolution of primary energy supply structure in Japan, 1975–2002
Figure 9. 5	R & D fund distribution by sector (2001)
Figure 9. 6	The share of nuclear power generation in Asian countries
Figure 9. 7	Scheme of back-end funding
Figure 10. 1	Power sector structure in the period of 1954–1995
Figure 10. 2	Power sector structure in the period of 1995–2004
Figure 10. 3	Legal framework of electricity industry reform
Figure 10. 4	EVN restructuring activities in 2005
Figure 11. 1	Imaginable multidimensional reform strategy
Figure 11. 2	Relation between reform aspects
Figure 12. 1	An institutional framework for separation of functions
Figure 12. 2	The distinction between policy making, regulation and ownership
Figure 12. 3	The main policy instruments for regulators
Figure 12. 4	The scope of regulating price in different phases of market development
Figure 12. 5	The current planning process
Figure 12. 6	System planning under the single buyer market
Figure 12. 7	System planning under wholesale and retail competition

Figure 12. 8   Public benefit consideration in EI reform
Figure 12. 9   Policy, regulatory, financing and institutional mechanisms to advance public benefits
Figure 13. 1   Market transformation phases
Figure 13. 2 a Competitive generation market step 1: an internal EVN market
Figure 13. 2 b Competitive generation market step 2: a fully competitive generation market
Figure 13. 3 a Wholesale electricity market: trial step
Figure 13. 3 b Fully competitive wholesale market
Figure 13. 4 a Retail electricity market: trial step
Figure 13. 4 b Fully competitive retail market
Figure 14. 1   A flow chart of the cost-benefit model
Figure 14. 2   Self-financing ratio of EVN in the period of 2004–2010
Figure 14. 3   Debt service coverage ratio of EVN in the period of 2004–2010
Figure 14. 4   Debt/Equity ratio of EVN in the period of 2004–2010
Figure 14. 5   Cash balance of EVN in the period of 2004–2010

LIST OF ABBREVIATION

ABAC	APEC Business Advisory Council
ABWR	Advanced Boiling Water Reactor
ADB	Asian Development Bank
AEC	Atomic Energy Commission
AGC	Automatic Generation Control
APEC	Asia–Pacific Economic Cooperation
BE	British Energy
BG&E	Baltimore Gas and Electric
BOO	Build, Operate and Own
BOT	Build, Operate and Transfer
BPDB	Bangladesh Power Development Board
CA	Correspondence Analysis
CAEA	China National Atomic Energy Authority
CCGT	Combined Cycle Gas Turbines
CEB	Ceylon Electricity Board
CERC	Central Electricity Regulatory Commission
CfD	Contract for Differences
CLP	China Light & Power
CNEC	China Nuclear Engineering and Construction Group Corporation
CNEIC	China Nuclear Energy Industry Corporation
CNEPP	Comprehensive Nuclear Energy Promotion Plan
CNNC	China National Nuclear Corporation
COS	Cost of service
CPI	Consumer Price Index
CTU	Central Transmission Utility
DAE	Department of Atomic Energy
DCF	Discounted Cash Flow
DESA	Dhaka Electricity Supply Authority
DESCO	Dhaka Electric Supply Company Ltd.
Discos	Distribution companies
DMCs	Developing Member Countries
DSM	Demand Side Management
E&W	England and Wales

EAC	Electricity Authority of Cambodia
EBRD	European Bank for Reconstruction and Development
EDC	Electricite du Cambodge
EdL	Electricite du Lao
EGAT	Energy Generating Authority of Thailand
EHV/HV	The share in electricity consumption of eligible
EI	Electricity Industry
EL	Electricity Law
EMA	Energy Market Authority
EMC	Energy Market Company
EPDC	Electric Power Development Co Ltd
EPIRA	Electric Power Industry Reform Act
ERC	Energy Regulatory Commission
ESCJ	Electric Power System Council of Japan
EUIC	Electric Utility Industry Council
EVN	Electricity of Vietnam
FBR	Fast Breeder Reactor
FCC	Full Customer Choice Model
FERC	Federal Energy Regulator Commission
Gencos	Generation companies
HEC	Hong Kong Electric Company
IAEA	International Atomic Energy Agency
IEA	International Energy Agency
IMF	International Monetary Fund
IPO	Initial Public Offering
IPPs	Independent Power Producers
ISO	Independent System/Market Operator
JAERI	Japan Atomic Energy Research Institute
JAPCO	Japanese Atomic Power Company
JEPX	Japan Electric Power Exchange
JBIC	Japanese Bank of International Cooperation
JNC	Japan Nuclear Cycle Development Institute
JPOWER	Japan Electric Power Development Co., Ltd.
KESC	Karachi Electricity Supply Corporation
KHNP	Korea Hydro and Nuclear Power Company
KLSE	Kuala Lumpur Stock Exchange
KPX	Korea Power Exchange
LECO	Lanka Electricity Company (Pvt.) Ltd.
LILW	Low and Intermediate Level radioactive Wastes

LNG	Liquid Natural Gas
MBO	Management Buy–out
MCA	Multiple Correspondence Analysis
MEA	Metropolitan Electricity Authority
MEBO	Management/Employee Buy–out
METI	Ministry of Economy, Trade and Industry
MEXT	Ministry of Education, Culture, Sports, Science and Technology
MIME	Ministry of Industry, Mines and Energy
MITI	Ministry of Industry and International Trade
MOCIE	Ministry of Commerce, Industry and Energy
MOEP	Ministry of Electric Power
MOI	Ministry of Industry
MOST	Ministry of Science and Technology
NAV	Net Asset Valuation
NDRC	National Development and Reform Commission
NEA	Nepal Electricity Authority
NEPC	National Energy Policy Council
NEPRA	National Electric Power Regulatory Authority
NGC	National Grid Company
NHPC	National Hydroelectric Power Corporation
NISA	Nuclear and Industrial Safety Agency
NPCIL	Nuclear Power Corporation of India Ltd.
NSC	Nuclear Safety Commission
NSO	Neutral transmission System Organization
NTPC	National Thermal Power Corporation
NUR	Northern Utility Resources Sdn Bhd
ODA	Official Development Assistance
OECD	Organisation for Economic Co–operation and Development
OPF	Optimal Power Flow
PAEC	Pakistan Atomic Energy Commission
PBR	Performance Based Regulation
PCA	Principle Component Analysis
PEA	Provincial Electricity Authority
PEMC	Philippine Electricity Market Corporation
PEPCO	Potomaic Electric Power Corporation
PGCB	Power Grid Company of Bangladesh Ltd.
PHWR	Pressurised Heavy Water Reactor
PLN	Perusahaan Listrik Negara–Djakarta
PPAs	Power Purchase Agreements

PPS	Pakistan Power Sector
PPSs	Power Producers and Suppliers
PUB	Public Utilities Board
PURPA	Public Utilities Regulatory Policy Act
PX	Power Exchange
R&D	Research and Development
RLDC	Regional Load Dispatch Centre
ROR	Rate of Return
RPC	Rural Power Company
SBM	Single Buyer Model
SBU	Strategic Business Unit
SCAs	Scheme of Control Agreements
SCSTI	The State Commission on Science, Technology and Industry for National Defence
SDRC	State Development and Reform Commission
SDPC	State Development Planning Commission
SEP	Singapore Electricity Pool
SERC	State Electricity Regulatory Commission
SESB	Sabah Electricity Sdn. Bhd.
SESCo	Sarawak Electricity Supply Corporation
SFR	Self Financial Ratio
SO	System/Market Operator
SOEs	State Owned Enterprises
SP	Singapore Power
SPCC	State Power Corporation of China
STUs	State Transmission Utilities
T&D	Transmission and Distribution
TEPCO	Tokyo Electric
TOs	Transmission Owners
TPA	Third Party Access
VIM	Vertical Integration Monopoly Model
WB	World Bank
WCM	Wholesale Competition Model

INTRODUCTION

For much of the 20th century, electric utility companies have traditionally been owned and operated by the state. In the power sector there is usually little competition and public monopolies tend to dominate the business. The commonly held belief was that, like all public utility or infrastructure services, electricity production and distribution were natural monopolies. In many countries, electric utility companies are still vertically integrated as their responsibilities cover generation, transmission, and distribution in order to take advantage of economies of scale and to mobilize capital investment for the large systems of generation and transmission. Moreover, as they provide a basic social service, electric utilities have been under heavy government regulation, particularly with respect to pricing and the quality of service.

This thinking, however, radically changed in 1980s. The first remarkable change was the approval of the Public Utilities Regulatory Policy Act (PURPA) in the United State in 1978. The next two decades, the 1980's and 1990's, saw dramatic changes in views about how the electricity industry (EI) should be owned, organized, and regulated. The new model for the electricity industry is reliance on the private sector to improve efficiency, promote innovation, and enhance service. Electricity industry reform, at different levels, has spread throughout most of the world, from developed to developing countries.

In contrast, after a period known as the *"liberalization tide"*, at the end of the 1990's and especially the beginning of the 2000's, the view regarding reform of the EI changed again. After a series of financial crises, corporate scandals, the California electricity crisis, and a series of major blackouts in North America and other areas of the world, the need to find how EI reform should be carried out arose. Several countries re-examined their reform strategies, while others even temporarily delayed the privatization process.

"There is no method of easily solving the problem of annoying the power industry" [Joskow and Schmalensee, 1985].

This book, to some extent, aims to contribute to policymakers' efforts of addressing the question of how EI reform should be carried out. The book endeavors to introduce a multidimensional and strategic analysis of electricity industry reform. Two paths of logic are followed in this book. First, the multidimensionality of electricity reform is addressed. Second is systematic analysis.

Multidimensionality, what does it mean? Traditionally, electricity industry reform considers two fundamental dimensions: ownership and competition. The dimension of competition covers commercial arrangements for selling energy: separating or "unbundling" integrated industry structures and introducing competition and choice. It includes the following four models: (i) monopolization at all levels, (ii) purchasing agency or single buyer, (iii) wholesale competition, and (iv) retail competition. Regarding the dimension of ownership, there is the process of changing from government to private ownership, which is the reduction of governmental control in the power sector. The three most common forms of ownership/ management are: (i) direct government ownership, (ii) a government owned corporation, and (iii) a privately owned corporation. Recently, more attention has been placed on a third dimension of the sector reform − regulation, which involves developing a regulatory framework for the electric industry. Developing a proper regulatory framework is a precondition that must be met for creating a reform policy. Regulation is defined here as the creation and enforcement of rules that promote the efficiency and optimal operation of markets, as well as devising a mechanism to address specific social and environmental issues. The role of the regulator is to maximize social benefits by taking into account social values and the need to maintain financial viability and promote economic efficiency of the sector in the long run.

All three dimensions mentioned above are three aspects of the problem that should be examined through comprehensive analysis. While plenty of literature reviewing electricity reform exists regarding the combination of competition and ownership, it is true that very few analyses cover the full story of the reform process, especially regulation, even though regulation is the linchpin of reform success. Moreover, evidence is emerging that reform has been mainly designed to address economic concerns, in particular financial concerns, with insufficient consideration for social and environmental issues [Wamukonya, 2003 a, 2003 b]. Considering the multidimensionality of reform, this book tries to give policy makers a comprehensive view of industry reform. The multidimensionality of reform is consistently shown in all parts of the book, from the standard prescription to empirical analysis and policy implications of reform modeling for Asian countries and Vietnam.

The second path of logic followed by the book is systematic. First, electricity industry reform is addressed from the widest scope, a worldwide analysis; however, the level of analysis is limited to a prescribed standard or general overview of the status of reform. In the second stage of the analysis, reform is examined from a more concentrated scale of the electricity industry in Asian countries. The analysis covers empirical examination based on econometric modeling, which allows policymakers to see the impact of each reform policy option on EI performance, deriving predictions about the probability and shape of change in electricity industry reform in several representative countries. In the final stage of the book there is a pointed analysis of reform policy implications for the electricity industry in Vietnam. Thus, through the entire analysis, the scope of reform is gradually narrowed from a worldwide perspective to a singular one; in contrast, the depth of analysis increases stage by stage.

The book is divided into three parts, reflecting the three stages of analysis. It starts with Part I – The standard prescription of electricity industry reform, a broad and extensive analysis about reform in the electricity industry, which can be applied to any country. Except for the background chapter, all issues related to reform are grouped into three chapters, from Chapter 2 to Chapter 4, which respectively cover the three sides of reform: ownership, competition, and regulation. Moreover, Chapter 5 carries out a worldwide comparison of electricity industry reform. Valuable lessons should be drawn from the successes and failures. Finally, several main stakeholders, who have an important role in the reform process, talk about electricity industry reform from several points of view.

Part II of the book endeavors to carry out a comprehensive analysis of the EI reform in Asian countries. The twenty-year development of worldwide EI reform is recognized in Part I, demonstrating plenty of successes, as well as failures. Moreover, it raises the urgent need to analyze the successes and failures associated with past reforms and to identify the instruments and policies that should guide ongoing and future efforts. Assessing reform's effects on performance will require systematically collecting cross-country data, defining and constructing basic economic measures for various aspects of reform and industry performance, and determining appropriate techniques for econometric estimation. Several investigations have been carried out on the economic and social impacts of electricity industry reform. However, up-to-date empirical analyses have largely been carried for OECD countries in Latin America and Eastern Europe, where outcomes have been systematically monitored. Meanwhile, only limited research of EI reform in Asian countries, such as ADB reviews, has been carried out. The second part of this book, therefore, is devoted to an open and comprehensive analysis of industry reform in Asian countries.

Chapter 6 includes details and evaluation of EI reform. In this chapter, the writer tries to build a set of core indicators for EI reform. Such indicators, of course, follow the main line of this book and will be divided into three categories: regulation, competition, and privatization. The results of this part are essential for empirical analysis of industry performance in the next chapter. Chapter 6 also extends to classify country groups based on the level of advances in electricity industry reform and on the political and economic conditions within the three-dimension matrix of restructuring, privatization, and regulation. Based on this classification, a representative from each country group will be selected for further analysis in Chapter 8. Chapter 7 carries out an econometrics analysis, which allows policymakers to see the impact of each reform policy option on EI performance. Based on that, this book intends to derive some predictions about the probability and shape of change in electricity industry reform of several representative countries. Furthermore, the scope of this analysis includes the consideration of nuclear power policy in the process of EI reform in Asian countries.

Part III of the book presents the deepest and most pointed analysis of policy implications of multidimensional reform in Vietnam's electricity industry. First, this part endeavors to examine up-to-date industry reform, and then applies policy implication findings from Part I and Part II to establish a comprehensive strategy for electricity industry reform in Vietnam. This strategy covers most aspects of reform, from developing an efficient regulatory framework, restructuring of the Electricity of Vietnam, introducing a competitive market, to implementing privatization to address regulation mechanisms for several of the most sensitive social-benefit issues. These challenging tasks will be discussed and resolved respectively in chapters 10 to 14.

During the 1990's, Vietnam has not stood out the present-day trend to many of developing countries engaged in attracting private participation in the power sector. In view of the Vietnamese government's own shortage of capital, private international power companies have been identified as key problem-solvers in the country's efforts to meet the skyrocketing demand for electricity. The annual growth rate of demand is 13–15% (1995–2010).

The first milestone of Vietnam's power sector reform was marked in 1995 with the foundation of Electricity of Vietnam, which serves as a for-profit cooperative, regulated by the Law of State Enterprise. Electricity of Vietnam (EVN) is responsible for generation, transmission, and distribution of electricity for the whole country. Since 1999, independent power producers (IPPs) has started to involve in power generation. As of 2000, EVN's share in power generation is 93.8% and 91.2% in energy and power, respectively. Independent power producers make-up the remaining share. As the number of new IPPs in operation has gradually increased, such as Phumy 2. 2, Phumy 3, and Caongan, Camau, the share of IPPs is expected to continuously increase.

EVN itself has made many improvements in financial, accounting and business activities in an effort to corporatize and commercialize itself. A bulk supply tariff has been applied to power companies. Since January 1, 2005, EVN has launched an internal trial market for power plants within the EVN. Initially, that market was based on a production cost bidding mechanism to gain market operation experience.

EVN has implemented several stages of organizational restructuring, such as drafting an Electric Power Group project or conducting several trial establishments of one-member limited companies, which is transferred from a dependent accounting subsidiary.

It can be said that the electricity industry in Vietnam is in a preparation period of reform, moving from a central planning mechanism to a market force mechanism. The industry now is waiting to develop reform policy. However, which strategy is appropriate for setting up a reform program during this slow period of worldwide electricity reform? This question will be addressed in Part III of the book.

This part, to some extent, uses information and results from [Mai and Nomura, 2005]. However, focus is put on the details and comprehensive analysis of multidimensional reform in Vietnam's electricity industry.

PART I.
THE ESSENTIAL ASPECTS OF ELECTRICITY INDUSTRY REFORM

Electric utility companies were traditionally owned and operated by the state until the end of 1970's. During the 1980's and 1990's, a period known as the "liberalization tide" in electricity industry reforms occurred worldwide. The new model for electricity industry is reliance on the private sector to improve efficiency, promote innovation, and enhance service. Electricity industry reform, at different levels, has spread worldwide, from developed to developing countries, and includes three dimensions: ownership, competition, and regulation.

All three dimensions of the reform process should be examined through comprehensive analysis. While plenty literature reviewing electricity reform exists regarding the combination of competition and ownership, it is true that very few analyses cover the full story of the reform process; especially regulation, even though regulation is the linchpin of reform success. The objective of this book is to introduce a multidimensional and strategic analysis of electricity industry reform. Multidimensional analysis is a consistent idea throughout this book.

The book starts with Part I – The standard prescription of the electricity industry reform, a broad and extensive analysis of reform in the electricity industry, which can be applied to any country. Except for the background chapter, various issues related to the reform process are grouped into three chapters, from Chapter 2 to Chapter 4, which respectively covers the three sides of reform: ownership, competition, and regulation.

Moreover, Chapter 5 carries out a worldwide comparison of electricity industry reform. Valuable lessons should be drawn from the successes and failures. Finally, several main stakeholders, who have an important role in the reform process, talk about electricity industry reform from several points of view.

CHAPTER 1
BACKGROUND AND INTRODUCTION OF ELECTRICITY INDUSTRY REFORM

This chapter outlines basic concepts of the electricity industry, which may be well-known with senior experts in the power sector. However, it will give readers a comprehensive understanding of the history of electricity industry reform.

1.1 Functions of Electric Industry

The chain of producing and supplying electricity traditionally includes such major components as: generation (production), transmission (delivery of bulk power, a wires business), distribution (delivery to wholesale and retail customers, a wire business), system planning and system operations, bulk power markets (sales at wholesale), and retail sales (energy sales to retail customers). The first four components are physical functions and the others are merchant functions.

Each function will be described in this section, respectively.

- **Generation (Production)**

Electricity is generated from different sources including coal, oil, gas, water, nuclear power, and renewable resources. Those multiple fuel resources are turned into exactly the same product. It all has to be totally standardized; otherwise electric appliances do not run properly.

The electricity generation technology consists of two main parts: (i) production of motive force and (ii) conversion of motive force to magnetic electricity. The later component is almost the same for all types of power plants. Different generation technologies are used for the former part – motive force preparation. Motive force is conventionally produced by via steam technology which uses coal, oil, gas, solar energy, or even heat radiated from a nuclear reactor, to boil water. Potential energy also comes from falling water, tidal energy, turning windmills, or by direct combustion of fuel, as in a jet engine.

Each kind of electric generation technology has different advantages and disadvantages. Nuclear power plants are economically competitive, but its disadvantages include nuclear safety and radioactive waste disposal, which receives strong opposition from the public and the media. Hydro power is clean energy;

however problems include flooding (inundation) and ecological changes. Conventional thermal power plants including coal, oil, or gas-fired plants, produce significant amounts of greenhouse gas pollution. Renewable energy, such as solar energy, wind power, etc has coped with ungathered and low efficiency.

Technical advances in the 1980s, which combined steam generators with direct fuel combustion, increased the maximum efficiency of new power plants to over 50%. The maximum efficiency of a conventional power plant is normally only 30–35%. A new technology, called combined cycle gas turbines (CCGT), used natural gas as fuel and it came in much smaller packages. This technology is one of the forces shaping the electricity industry worldwide because economies of scale are not an essential part of the electricity production process; therefore it opened the door to competition in generation.

- **Transmission**

Transmission is the business of delivering bulk power from a generator to the distribution network via a high voltage wire. The transmission grid is a set of nodes connected by high-voltage lines and substations throughout the country and recently even across borders in a regional network.

The transmission system is quite fragile. If overloaded, it becomes unstable and can cause widespread blackouts[1]. Electricity flows have to be managed on a continuous, real-time basis; no traffic jams, no busy signals are allowed. For those reasons, the transmission system requires the constant attention of a system operator to integrate the operation of generating power plants with a transmission system on a second by second basis.

The grid is considered a natural monopoly because it is costly to duplicate existing power lines.

- **Distribution**

This is also a wire business delivering electricity from the transmission network to wholesale and retail customers via a low voltage wire. Transmission and distribution together are the transport systems; however, as for the transport element of distribution, transmission is like a major highway while distribution is like a local road.

However, on a practical level, it is not is easy to separate between transmission from distribution due to the unclear range of voltage levels. Certain voltage levels are considered to be equivalent to transmission in some areas, but are classified as distribution in other regions.

---

1   One of the most famous blackouts happened in Northern United State in 2003.

The distribution network, similar to the transmission system, is considered a natural monopoly for the same reason, to avoid the wasting of social resources.

- **System Planning and System Operations**

System planning is the process of identifying the need for the new facilities or demand-side-management (DSM) which is required to accommodate load demand and electricity growth. System operation is the process of planning, scheduling, and dispatching the resources, both supply-side and demand-side, to meet reliability and economic functions. The system functions are intended to ensure both minimum cost expansion of the system, considering capital and operating cost of generation, transmission, and DSM, and minimum cost for reliable operation of the system, considering incremental costs.

Long-term planning may be performed independently by generation and by transmission sections or other special organizations. However, a check is needed to protect consumers from development of plans that are financially favorable to a single entity while being detrimental to the ultimate consumer of services provided by the entity.

The main role of system operation is to ensure the reliability of the transmission grid and to keep supply and demand in balance at any times. This requires scheduling and centralization of the information coming from all the participants using the grid and authority to dispatch plants to react to imbalances. For this reason, it is generally thought to be the job for a single operator.

System operation and transmission can sometimes be merged into one entity to facilitate unique instruction.

- **Bulk Power Markets**

Bulk power markets function as the process by which the buying and selling of bulk power products occurs among trading entities. The buyer in the bulk power market is not the final customer, except in the case of large customers. Typically, the bulk power buying entity resells electricity to the final or end-use customers.

It is important to note that the bulk power market's function must coordinate closely with the system operations function. The system operation function initiates generation control actions that are required to implement delivery of a bulk power product from one trading partner to another.

Basic products sold in the bulk power market are capacity and energy. Capacity sales usually entitle the purchaser to energy purchases at an agreed price. If capacity transactions do not include must-take energy provisions they may be viewed as optional contracts (that is, the energy is taken only when it is economical). Moreover, a number of services that have traditionally been bundled exist, however they are increasingly being viewed as separate services that may be sold independently or as part of the basic product package, such as: frequency regulation service, spinning reserves, quick-start reserves, reactive power capacity and energy, black-start capacity, etc.

Electricity products may be traded through various market mechanisms, including forms of brokering, power pooling arrangements, bilateral transactions between the producer and the buyer, and bilateral transactions between buyers. It is not unusual for multiple market mechanisms to operate in parallel, either for the same or for different products.

- **Retail Sales**

Retail sale is the function of selling electricity to final customers. It involves a series of commercial functions – procuring, pricing, and selling electricity and also metering its use, billing, and collection of payment.

Retailing is a merchant function, traditionally it is always a part of the distribution process, but is increasingly being viewed as an independent entity in the competitive electricity market.

1.2 Specific Characteristics of Electricity Industry and Traditional Organization of the Industry

For much of the 20th century, the electricity industry has had a monolithic structure, with a single entity owning generation and transmission capacity and performing all system operations. This entity transmits power to one or more distribution companies that hold exclusive rights to serve households and businesses in specific regions. This vertical integration of functions was the traditional structural organization of the electricity industry worldwide. Vertical integration was almost always accompanied by a legal monopoly within its served area – only one company could provide electricity to customers in that area. In most countries, the government or government-owned companies monopolized the industry. In several other countries[2], private companies dominate the industry, but still follow the rule of vertical integration. In this section, the reasons for vertical integration and monopolization will be discussed.

The reasons for vertical integration and monopolized integration in the electricity industry derive from characteristics of transmission networks and from operating and investment relationships between generations, transmission, and supplying electricity to customers. It could be outlined four elements relevant to the hundred-year history of the monopolistic electricity industry as follows:

---

2   Those countries include Germany, Japan, Spain, and the United States.

- Natural monopoly aspects of transmission and distribution: Transmission and distribution are quintessential natural monopolies because they entail substantial, largely sunk fixed costs. Competition would lead to the wasteful duplication of network resources. Thus, in most countries a single entity governs transmission network for all or most of the country.
- Requirement of a short-term balance: Electric generation and consumption need to be balanced continuously and almost instantly for a network to meet specific physical parameters (frequency, voltage, stability).

While:
- Electricity cannot be stored economically because storage technologies, such as batteries and hydro-electric pumps, are extremely inefficient. Thus, electricity is the ultimate real-time product, with production and consumption occurring at essentially the same time.
- The demand is almost completely inelastic in the short run. Little or no supply and demand balancing can be conducted on the demand side.
- Due to physical constraints on production and transmission, achieving real-time balancing of supply and demand is difficult and requires intensive system coordination. Network congestion constrains the ability of remote generators to respond to the supply needs of a given area. Moreover, generating units have capacity constraints that cannot be breached without risking costly damage. As a result the amount of electricity that can be delivered in an area at a given time is limited and supply is highly inelastic — especially at peak times [Borenstein, 2000].
- Unlike other switched networks, such as railroads or telecommunications, where routing in the physical delivery of products can be specified, flows in electricity networks are hard to control. This physical property of electricity transmission implies that an imbalance of demand and supply at any location on the grid can affect the stability of the entire system. Thus, matching between a supplier and a customer is effectively part of the overall balancing of the system. Mismatching could disrupt the delivery of electricity for all suppliers and consumers.
- Requirement of long-term balance: In the long-term, of course it is possible to adjust the supply capacity. However, time requirements for construction in the electricity industry are usually long, especially for power plants. Proper forecasts should be made and preparation for investment and construction made in advance. Moreover, allocation investments need to be appropriately allocated between generation, transmission, and distribution. All such duties are facilitated by vertically integrated entities.

- Economy of scale: historically, bigger and bigger power plants means had lower and lower investment as well as operation cost resulting that generation also was a natural monopoly.

Of these four elements, It is undoubted that the first element resists forever-transmission and distribution are still considered natural monopolies. The fourth element is no longer a constraint. Thanks to the advent of CCGT, economies of scale are no longer considered sufficient to warrant monopolies in generation. The optimum output of CCGT is 250–400 Megawatt (MW) compared to the optimum output of a nuclear plant at 900–1,000 MW or optimum output of a coal-fired plant at 500–600 MW.

In many electricity markets, the complexities of short and long-term coordination of generation and transmission have tried to solve. For example, the short-term balance could be resolved by market mechanisms with standby generators providing ancillary services in response to changing demand and supply conditions. However, many difficulties and problems still exist and it is likely lead to failures, similar to the expensive experience of California's electricity market.

1.3 The Motivation behind EI Reform

What are the benefits of reform? At first, we will look at the major theoretical differences between regulation and competition. The difference lies in who takes the responsibility for various risks, which involve market demand and prices, technological changes, management decisions about maintenance, investment, and credit risk.

Under regulation, customers bear most of the risks and all the costs, including the costs of bad management decisions. Under regulation, if new technology is invented, the customers continue to pay for the old technology, whereas under reform the old technology producers find their assets are worth less. Under regulation, if demand turns out to be less than forecasted, prices have to rise to cover the cost of excess capacity, whereas under competition excess capacity causes prices to fall.

Under competition, owners initially bear the risks[3]. They pay for mistakes and mismanagement, but also profit from good decisions and management. Under competition, generators also bear the risk of technological changes, so the incentive to choose the best technology and avoid costly mistakes is strong. They also bear the risk of changes in demand and prices, so they need to be flexible in their building plans, and watch the market constantly. They need to adjust their maintenance schedules to get plants back on line when prices are high. The profit incentive works

---

3 The producers may be able to shift risk through insurance and hedging.

everywhere to make producers sharpen their pencil.

Forces driving structural changes in the electricity industry differ between countries—especially between industrial and developing countries. In developed countries, operating efficiency is low due to a lack of competition resulting from monopolization of the power sector, both public and private, and excess capacity of the electricity system. While in developing and transitional countries reforms have been driven by power shortages, debt and lack of public funds for badly needed investments, poor operation and financial performance of state-owned electricity systems (with low labor productivity, poor service quality, and high system losses) and the government's desire to raise revenue through privatization[4]. The following paragraphs will describe the motivation behind the forces driving structural change.

- **Excess Capacity in Industrial Countries**

For about 30 years after World War II, industrial countries experienced remarkably high and steady growth in demand for electricity. However, in the 1970s this growth was interrupted and has never returned to its previous level. This resulted in significant excesses in capacity in developed countries. Excess capacity made competition between generating companies feasible and attractive.

- **Need for Investment in Developing Countries**

Many developing countries are at a stage of economic development where demand for electricity increases rapidly, requiring enormous investment. Between 1999 and 2020, global electricity consumption is projected to increase by 2.7 percent a year, but in the developing world the increase is projected to increase by 4.2 percent a year [Kessides, 2004].

Electricity systems are under stress in many developing countries. The balance between demand and supply is tight, and the lack of spare capacity often leads to blackouts. Thus, significant investment is needed in generation, transmission, and distribution. On the other hand, the governance structure in the sector – typically vertically integrated, state-owned, and centrally planned – is poorly suited for mobilizing the long-term capital needed for an adequate and reliable electricity supply.

- **Technological Innovation**

Recent technological advances have dramatically altered the cost structure of electricity generation. From the start of the 20th century until the early 1980s, technological developments led to larger and more efficient fossil-fuel power plants

---

4 According to [Bacon and Besant-Jones, 2001], [IEA, 1999], [Joskow, 2003].

built farther and farther from cities and factories. However, in recent years, technological improvements in gas turbines and the development of CCGT have recast economies of scale in electricity, reversing a 50-year trend toward large, centralized power stations[5]. Combined-cycle gas turbines can be brought online faster (within 2 years) and at a more modest scale (50–500 megawatts). Aero-derivative gas turbines can be efficient from as little as 10 megawatts [Balzhiser, 1996].

The advantages of CCGT are further strengthened by the strong growth of the natural gas industry, both in capacity exploitation and pipeline technology, which allows for competition with electricity transmission lines.

Moreover, small-scale generation technology, such as photovoltaic, fuel cells, etc., has significantly improved. Although until now the competitiveness of such technologies was still lower than central electricity systems, the potential of is bright, promising to end natural transmission and distribution networks monopolies.

- **The Problem of State-owned Industry**

The performance of state-owned electricity companies varies considerably across countries. In many developing and transition economies these entities suffered from low labor productivity, deteriorating fixed facilities and equipment, poor service quality, chronic revenue shortages and inadequate investment, and serious theft and nonpayment problems [Kessides, 2004].

Infrastructure performance was generally much better in advanced industrial countries. Still, high construction costs (caused by delays and changing environmental and safety requirements) and expensive, politically driven programs led to problems in the electricity sector.

In developing and transitional economies, the main cause of deteriorating infrastructure performance was underinvestment, which was largely due to the failure of governments to prescribe cost-reflective tariffs, especially during periods of high inflation. Under state ownership, managers and politicians favor under-pricing to stimulate demand and secure political support. The under-pricing policy creates in transparency and discourages foreign investment. Due inefficient pricing policies, most electricity companies were burdened by debt that resulted in the growth of government budget deficits. A 1989 survey of 360 electricity utilities in 57 developing and transitional economies found that the average annual return on revalued net fixed assets was less than 4 percent, well below the 10 percent target normally set by donors. In 1991, these utilities financed just 12 percent of their investment requirements and revenue covered only 60 percent of power sector costs.

---

5   Cited from [Bayless, 1994], [Casten, 1995].

Under-priced electricity posed a fiscal burden of $90 billion a year in the early 1990s, or 7 percent of governmental revenue in developing countries—larger than required power investments of about $80 billion a year[6]. Debt and fiscal crises, combined with extraordinarily weak performance, stimulated strong pressure for electricity reform.

1. 4   The Basic Architectures of Reform

The basic architectures of reform will be presented and classified in this part. While a number of variations are potentially available, it can be divided into three dimensions: competition, ownership, and regulation.
- The dimension of competition covers commercial arrangements for selling energy. It includes:
  - Vertical separation of competitive segments (e.g. generation and retail supply) from regulated segments (transmission, system operations, and distribution) either structurally (through divestiture) or functionally (with internal walls separating affiliates within the same corporation).
  - The creation of wholesale spot energy and operating reserve market institutions to support requirements for the real time balancing of supply and demand, to ensure network voltage, frequency and stability parameters within narrow limits, and to facilitate economical trading opportunities among suppliers and between buyers and sellers.
- The dimension of ownership is the process of changing from government to private ownership. Governmental control over the power sector is reduced and involvement of the private sector increases in new power projects or through the privatization of state-owned utilities.
- The third dimension, regulation, covers regulatory framework. It includes the establishment and application the legal and regulatory rules and the creation of a regulatory authority.

Privatization, restructuring, and deregulation are separate phenomena, however, they often come together and in the case of the electricity industry, create a multi-dimension reform process, in which the success or failure of each dimension is often strongly influenced by the kind and degree of the others. The study of reform, therefore, requires comprehensive analysis. While plenty of literature reviewing electricity reform exists regarding the combination of competition and ownership, it is true that very few analyses cover the full story of the reform process, especially regulation, even though regulation is the linchpin of reform success. The objective of this book is to introduce a multidimensional and strategic analysis of electricity

---

6   Cited form [Besant-Jones, 1993], [Newbery, 2001 b].

industry reform. To do that, at first, the next three chapters, from Chapter 2 to Chapter 4 will summarize essential issues related to each of three reform aspects.

CHAPTER 2
RESTRUCTURING AND COMPETITION IN ELECTRICITY INDUSTRY

In Chapter 1, we discussed the functions of the electricity industry, as well as the competitive potential for each function. This chapter will present more details regarding the modeling applied for such functions, the scope of restructuring and competition in electricity industry, as well as many related issues, including: (i) structural options for industry, (ii) industry unbundling, (iii) open network access, and (iv) competitive markets.

In the electricity production chain, two functions can be considered as candidates for competition, they are generation and retail supply. The generation function is a major candidate for competition. Old arguments from economies of scale are no longer hold and although not entirely solved, problems with the coordination of generation and transmission have been eased by advanced computer systems. The retail function can be competitive as far as the bidding, pricing, and selling of electricity. There are many possible models of electricity competition. The following section will discuss four basic models applied to the electricity industry.

The other functions of the electricity supply chain cannot to be competitive, including transmission, system operations, and distribution. Reform in these areas is based on the introduction of Third Party Access (TPA) to networks. TPA establishes the obligation of an electrical network operator to provide access to users of that network.

2. 1 Structural Options for the Electricity Industry

- **Models for Industry Structure**

Although there are many variations of operating models for EI structure, four basic models can be constructed. These models have been largely adapted from the excellent study on competition and choice in electricity markets [Hunt and Shuttleworth, 1996], including: (i) Vertically-integrated monopolies, (ii) Single buyers, (iii) Wholesale competition, and (iv) Full customer choice.

Vertical Integration Monopoly Model (VIM)
- Characteristics

In this model, there is no competition and no consumer choice. Typically this model

```
┌─────────────────────────┐
│  ┌───────────────────┐  │
│  │    GENERATION     │  │
│  └─────────┬─────────┘  │
│            ▼            │
│  ┌───────────────────┐  │
│  │   TRANSMISSION    │  │
│  └─────────┬─────────┘  │
│            ▼            │
│  ┌───────────────────┐  │
│  │   DISTRIBUTION    │  │
│  └─────────┬─────────┘  │
│            ▼            │
│  ┌───────────────────┐  │
│  │    CONSUMERS      │  │
│  └───────────────────┘  │
└─────────────────────────┘
```

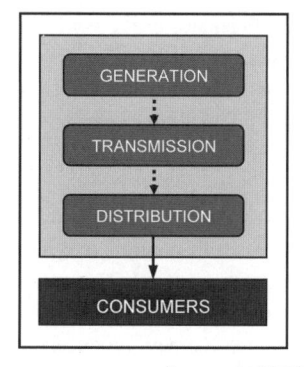

Source: [APERC, 2000]
Figure 2. 1 Vertical integration monopoly model

is characterized by the existence of one vertically and horizontally integrated system (Figure 2. 1). In some cases, a number of vertically-integrated monopolies may exist, but each has their own separate franchised area of operation, usually mandated by law. The monopoly utility owns and operates all generation plants, and transmission and distribution networks. It may be obligated to supply consumers with electricity, but consumers for their part are captive and are not free to choose their supplier. The utility is usually tightly regulated through such mechanisms as price controls.

− Advantages

Historically, this model developed as a logical way of dealing efficiently with a number of the rather unique characteristics of the electricity supply industry at a time when rapid industrialization required rapid growth in supporting the infrastructure. Electricity is physically more complicated to deliver to end-use consumers than most other goods. Balance must be constantly ensured between supply and demand at all times. In the early days of electricity infrastructure development, it was easy to imagine and argue that generation, transmission, and distribution were intimately inter-related, were natural components of the monopoly in the overall supply system, and best handled by a single monopolized structure. This approach also allowed for the construction of large-scale generation plants and transmission systems at a time when economies of scale were important in the industry. This model allows subsidies and cross-subsidies since no competitive market exists, as well as investment in public goods, such as rural electrification and the distribution of power to poor communities.

− Disadvantages

The shortfalls of a vertically-integrated monopoly structure have really only become evident in recent years, and in economies with relatively large and mature electricity supply industries. The major deficiencies of the vertical monopoly structure that have been identified include: the lack of incentives to improve services and lower costs; the lack of transparency; poor investment decision-making; "gold-plating" of infrastructural components; and political interference.

Single Buyer Model (SBM)
- Characteristics

This model can be considered as the first step towards deregulation. One or several vertically-integrated monopolies still control the sector; however some private investment is made possible by licensing Independent Power Producers (IPPs) to build generation capacity (see Figure 2. 2). IPP's can be created from existing utilities by divestiture, or they may be new producers who enter the market when a new plant is needed. With this model, it is possible to have competition in generation, with a single buyer purchasing wholesale electricity. In reality, this may not happen; instead each IPP might negotiate a separate long-term Power Purchasing Agreement (PPA) with the respective government, often on terms that are quite favorable to the IPP.

- Advantages

The major advantage of this model is that it allows for direct inward investment by private investors, and allows investment risks to be shared. This model still allows governments to use the electricity industry to meet social policy obligations and create public goods such as (i) "obligation to supply". Although consumers cannot

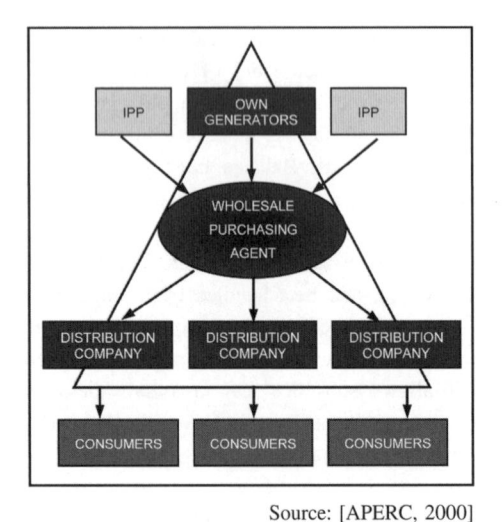

Source: [APERC, 2000]
Figure 2. 2 Single buyer model

choose their supplier, it is common for the utility to be obligated to supply all consumers, even those in remote areas, (ii) maintain uniform tariff pricing to all consumers, regardless of their remoteness from major centers of supply or the transmission grid.

The single buyer model can represent a good transitional structure, where sophisticated arrangements needed for a more complete market structure are not in place and would be hard to establish.

− Disadvantages

This model suffers the same problems as the previous model; there is no competitive market, so it is difficult to maintain economic efficiency. Other potential problems can arise if the single buyer discriminates unfairly between generators.

Wholesale Competition Model (WCM)

− Characteristics

This model (Figure 2. 3) allows a distribution company retailing electricity to consumers to choose their supplier. This creates competition in the generation and wholesale supply. In this model, separate distribution companies purchase electricity from any competing IPP generator. The distribution companies maintain a monopoly over energy sales to final customers (usually within franchise areas). In this model, an existing generation company has to compete against new entrants. The ability of governments to direct the choice of new generation technology is no longer desirable or necessary.

− Advantages

As the choice of generation assets (in terms of capacity additions and fuel type) is left to the market, economic efficiency can be improved, and risks transferred from

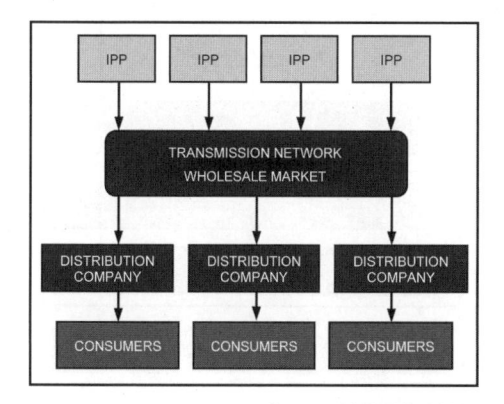

Source: [APERC, 2000]
Figure 2. 3 Wholesale competition model

the government to private investors. Although investors may seek a long-term contract before building generation capacity, the existence of a wholesale electricity market (which normally includes a spot market) means that such contracts are not essential.

The importance of this model lies in the fact that policy-makers have taken the important philosophical step of rejecting heavy handed regulation as an adequate tool to manage the sector, and believe that competition can be introduced into the electricity supply industry, and that social and economic benefits will flow from this decision.

– Disadvantages

The ability of generators to accommodate social policy obligations connected with generation virtually disappears under this model. This model represents only partial movement towards the introduction of competition in the electricity supply industry. Consumers are still captive, so the full economic benefits of a fully deregulated market are not achieved. Additionally, because generation assets are being valued by the wholesale market, the issue of stranded costs begins to arise.

Full Customer Choice Model (FCC)

– Characteristics

In this model (Figure 2. 4), competition has been introduced into all levels of the industry, ideally from wholesaling down to individual domestic consumers. A key

Source: [APERC, 2000]

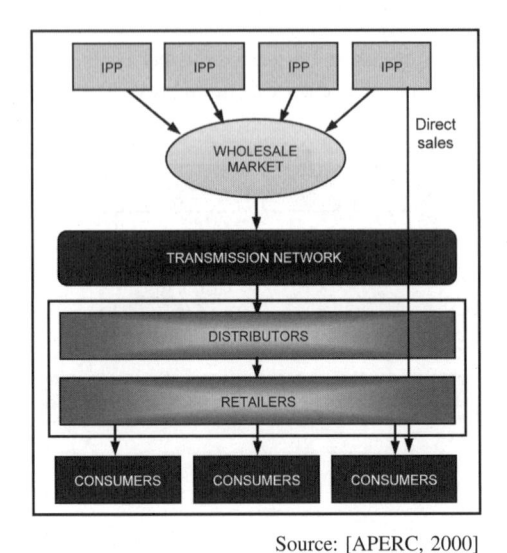

Figure 2. 4 Full customer choice model

component of this model is direct (or third party) access to transmission and distribution networks. With the right regulatory structure in place, any electricity consumer should theoretically be able to purchase from any retail supplier, who in turn is purchasing electricity from a competitive wholesale market.

Ideally, the network functions of transmission and distribution (natural monopolies) are completely separated from the functions of generation and retailing. Dispatch can be handled by the transmission network owner, but where more than one transmission network exists, this is best handled by an ISO.

With this structure, firms should be free to enter into the competitive functions of generation and retailing. There are no permitting or licensing requirements. This allows anyone, including householders, to build their own generation plants.

– Advantages

The advantage of this model is that it optimizes economic efficiency. Costs in the generation sector are driven down substantially, and this is usually expressed in lower retail tariffs. At the retail level, evidence of competition is emerging, with the range of goods and services being offered by electricity retailers increasing in both quantity and quality.

– Disadvantages

The ability to use the electricity sector as a tool to deliver social policy obligations disappears under this model. If there is a need to provide power to poor and/or isolated communities, or to promote a certain sector of the economy, this has to be handled using other social policy tools.

- **The Choice of a Model**

Table 2. 1 below describes the four structural models in terms of their

Table 2. 1 Characteristics of theoretical electricity industry models

Model of Power Sector Structure	VIM Model	SBM Model	WCM Model	FCC Model
Definition	Monopoly At every level	Monopoly Single Buyer	Competition among power generators Plus choice for distributors	Competition among power generators Plus choice for consumers
Competition in Generation?	No	Yes	Yes	Yes
Retailer choice?	No	No	Yes	Yes
Customer choice?	No	No	No	Yes

Source: Adapted from [Hunt and Shuttleworth, 1996]

characteristics, from the most regulated (VIM model) to the least (FCC model). A distinguishing feature among the different models is the degree of market openness, that is, the extent to which the different functions are open to competition. Opening of the market may effect different functions such as generation and supply to the end-user, and within each function, can reach all market players or a subset of them (e.g. all end users or only some of them).

The choice of models is up to the policy makers. However, to ensure the right decision is made, they have to first understand the existing arrangements, and then foresee the potential implications for structural change and the need for new institutions. Different models require different levels of structural change and rearrangement of functions in the industry, new institutions and new roles for regulation.

2. 2   Components of Restructured Systems

In 1. 1 of Chapter 1, we discussed six functions of the electricity system. Under restructuring, depending on the level of restructuring and the regulatory framework, some of these functions may be consolidated or further unbundled. In this part, we consider the structural components of the electricity market.

- Generation companies (Gencos). Gencos are responsible for operating and maintaining generation plants in the generation sector and in most cases are the owners of the plant. Open transmission access allows Gencos to access the transmission network without distinction and to compete.
- BOT plants operators and contracted IPPs. Build, operate and transfer (BOT), and build, operate and own (BOO) or independent power (IPPs) plants who have long-term contracts with surrounding, usually national utilities, play an important role in providing additional generation in many fast-growing systems. Take-or-pay power purchase agreements are often in force as an economic incentive to investors.
- Distribution companies (Discos) and retailers. Discos assume the same responsibility on the distribution side as in a traditional supply utility. However, as a result of liberalization, Discos can be restricted to maintaining the distribution network and providing facilities for electricity delivery. Meanwhile, retailers are separated from Discos and provide electricity sales to end-users. Another trend in developing countries is to sell to an investor, or to corporatize portions of the distribution system so that investment for reinforcement can be raised and better operating practices implemented.
- Transmission owners (TOs). In transmission networks that were previously state-owned before restructuring, the integrity will obviously be retained and distinguishing between the owner and the operator is redundant. In the case of

private electricity companies, such as those in the US, former electric utilities may sell off their other assets and become regulated transmission and distribution (T&D) companies. The basic premise of open transmission access is that transmission operators treat all users on a non-discriminatory basis with respect to access and use of services. This requirement cannot be ensured if transmission owners have financial interests in energy generation or supply. A requirement, therefore, is to designate an independent system operator to operate the transmission system.

- Independent system operator (ISO). The ISO is the supreme entity in the control of the transmission system. The basic requirement of an ISO is disassociation from all market participants and absence from any financial interest in the generation and distribution business. However, there is no requirement, in the context of open access, to separate transmission ownership from operation. For example, the National Grid Company (NGC) in England and Wales (E&W) is both a transmission owner and an operator. The roles and responsibilities of ISO varies widely. This issue will be discussed in the next section. In countries such as India and China where regional grids are owned by regional or state governments and where grid system interconnection is beginning to grow rapidly towards a nationwide status, the protocols of future ownership and operation are still evolving.
- Power exchange (PX). The PX handles the electric power pool, which provides a forum to match the electric energy supply and demand is based on bid prices. The timeline of the pool market may range from half an hour to a week or longer. The day-ahead market is typically used to facilitate energy trading one day before each operating day. An hour-ahead market is also useful, since it provides additional opportunities for energy trading to redress short-term imbalances. In the E&W system, ISO and PX functions both exist within the NGC.
- Scheduling coordinators (SCs). SCs aggregate participants in energy trade and are free to use protocols that may differ from pool rules. In other words, market participants may enter an SC's market under the SC's rules and this could give rise to different market strategies. In some markets, such as E&W, SCs are not allowed to operate. In many new situations such as California, SCs are an integral component of the market.
- Related services
Supplying electricity involves other activities as well. Construction and maintenance services provide and maintain generating plants and grid assets needed to supply electricity. There are also many new financial services, such as those offered by a growing number of power exchanges that facilitate trade in

electricity. Financial instruments (e.g. electricity futures) are also being developed to enable better risk management.

The disaggregation of functions and the introduction of competition is reflected in the changing shape of the market and market participants. Under competition a vertically integrated utility gives way to a number of different and more specialized market players including generation, transmission, and distribution and supply companies. The unbundling of various functions, often mandated or facilitated by regulations, is a key factor in the development of these new markets.

2.3 Electricity Industry Unbundling

- **Why the Industry should be Unbundled?**

The main reason to unbundle is to avoid discrimination by vertically separating monopolies from competitive activities. Transmission and distribution remain monopolized in a liberalized industry. If these monopolies are vertically integrated with the competitive activities of generation and end-user supply, they have an incentive to use their monopoly power against competitors. Competition law makes discrimination illegal in most countries, and discriminatory behavior is punishable by competition authorities. However, this may not be a completely effective and requires the affected parties to engage in lengthy antitrust procedures with no guarantee that they will win.

A second reason to introduce unbundling is to improve the effectiveness of regulation. Some degree of separation between regulated and competitive activities is needed to regulate effectively. For instance, the regulation of transmission revenues requires separate and transparent transmission accounting. Stronger separation facilitates more effective ring-fencing of regulated activities and, therefore, is a more cost reflective pricing method for grid services.

Vertical separation aims to limit the ability as well as the incentive of grid monopolies to distort competition. In particular, electricity industries must deal with the separation of: (i) generation and transmission/system operator, (ii) generation and distribution, and (iii) distribution and end-user supply. Such unbundling can be implemented through the forms described below.

- **Forms of Separation**

There are three basic approaches to vertical separation, including accounting and functional separation, operational separation, and ownership separation.

Functional and Accounting Separation

Accounting separation means keeping separate accounts for generation and transmission activities within the same vertically integrated entity. On this basis, a vertically integrated entity charges itself the same price for transmission services as it does others, while also offering separate charges for generation services. Functional separation is accounting separation plus (i) relying on the same information about its transmission system as the other market players when buying and selling power, and (2) separating employees involved in transmission from those involved in power sales.

These separations are relatively easy to implement. However, this is a weaker separation form and it requires a large and costly involvement of regulatory and competition authorities and may fail to prevent discrimination.

Operational Separation
Under this form of separation, operation of and decisions about investment in the transmission/distribution grid are the responsibility of an entity that is independent of generation owner(s); however ownership of the transmission grid remains with the generation owner(s). Operational separation may be effective in preventing discrimination if there are many transmission/distribution owners. However, in order to promote efficiency, operational separation requires the development of sophisticated and still largely untested governance structures.

Ownership Separation
Ownership separation solves nearly all concerns about discrimination because it eliminates the incentive as well as the ability to discriminate. Imposing ownership separation of transmission on vertically integrated companies may be difficult due to legal obstacles or opposition from utilities. This issue is particularly significant when

VERTICAL INTEGRATION	TRANSMISSION COMPANY APPROACH	ISO APPROACH
• SYSTEM OPERATOR • PLANNER • TRANSMISSION OWNER • GENERATOR	• SYSTEM OPERATOR • PLANNER • TRANSMISSION OWNER • GENERATOR	• TRANSMISSION OWNER • GENERATOR • SYSTEM OPERATOR • PLANNER

TRANSMISSION COMPANIES VERSUS INDEPENDENT SYSTEM OPERATORS (ISO)

Source: [Ocana, 2002]

Figure 2. 5 Transmission companies vs. independent system operator

the vertically integrated companies are privately owned, as in the US and Japan.

In terms of the transmission business, corresponding to forms of separation, three organizational approaches can be pointed out. Functional and accounting separation stays with vertical integration structure. Operational separation results in the utilization of the ISO approach. Lastly, the transmission company approach covers ownership separation (see Figure 2. 5).

2. 4 Open Network Access

Transmission and distribution lines provide the critical physical link that makes competition feasible. Thus, open access to the network and appropriate cost-reflective pricing is essential for the development of competition.

- **Transmission Pricing in Open Access Systems**

A key feature of the open transmission system, in an unbundled environment, is the need to charge all customers on a non-discriminatory basis for transmission services. An effective transmission pricing system must meet the following objectives:
- Allow sunk investment costs to be recovered (financial sufficiency);
- Provide adequate incentives for future long-term investment (long-term efficiency);
- Provide adequate signals for efficient network operation, or in other words, for the allocation of capacity to manage congestion efficiently (short-term efficiency);
- Avoid discrimination among users of transmission (competitive neutrality); and
- Promote simplicity and transparency.

The third objective, short-term efficiency, is particularly important. It is a key issue in the overall design of efficient electricity markets because, in addition to its direct role in the allocation of transmission capacity, it affects the dispatch of generation units. Transmission pricing may also have an impact on competition in generation, either facilitating it or distorting it.

Many pricing methods have been developed in order to meet these various objectives. Here, thanks to the document of [Ocana, 2002], we summarize two main competing approaches to the pricing of transmission services: non-transaction based and transaction based. The non-transaction based, or point tariffs, are independent of the commercial transactions that originate the transport of electricity. Point tariffs only depend on the energy injected or taken in each node. Point tariffs may be designed to reflect the costs of using the grid. In addition, point tariffs, such as nodal and zonal tariffs, that are sensitive to location, serve to manage congestion. Alternatively, the transaction-based approach, which consists of setting point-to-point tariffs, depends on the source and sink of each individual transaction. Contract path

Non-Transaction Based Pricing (Point tariffs)	Nodal
	Zonal
	Postage Stamp
Transaction Based Pricing (Point-to-point tariffs)	Contract Path
	Distance-related
	Etc.

Source: [Ocana, 2002]
Figure 2. 6 Pricing of network services

and distance related tariffs are two common examples of this approach. Transaction based tariffs are not, in general, cost reflective and do not serve to manage congestion. The following figure lists the main pricing techniques within each group, which are discussed and compared below.

Non-transaction Based Pricing
– Nodal pricing

The most complicated, but most accurate pricing method, is nodal pricing derived from the marginal cost theory. Nodal pricing can be based on short-run marginal (incremental) costs or long-run marginal (incremental) costs, depending on where the recovery of operating costs, capital costs, or expansion costs is desired. Short-run marginal pricing is discussed here. Similar procedures can be applied to other cases.

This method determines prices for power at each bus of the system, accounting for all costs and transmission constraints. The nodal prices are typically calculated as dual variables or Lagrange multipliers of an optimal power flow (OPF) calculation. Marginal transmission prices are derived directly from nodal prices:

$$Marginal\ transmission\ price = MP_i - MP_j$$

where MP_i and MP_j are the marginal nodal prices of electricity at buses i and j.

In a word, marginal transmission price is the measure of what it costs the grid to accept an additional unit of power at i and to deliver it at j. Short-run nodal prices change dynamically as a function of load distribution, system structure, and generation output patterns. Therefore, they must be recomputed on a periodic basis. This pricing scheme reflects only operating costs, therefore the recovery of capital and management costs must be added. A major advantage of this method is that the right operational pricing signals are revealed. The major disadvantage is the need for a separate revenue reconciliation exercise.

This pricing method has three further disadvantages such as: (i) intense real-time computational effort is required, (ii) the incentive problem: this method produces a perverse incentive for monopolizing transmission owners to cause or not

to relieve constraints, and (iii) the dramatic price volatility. Due to these disadvantages, no fully operative example of nodal pricing exists in the industry.
– Zonal pricing
Notwithstanding its compelling theoretical basis, the nodal pricing method has been deemed to be too complicated for applications, at least in the foreseeable future. Therefore, the zonal pricing scheme has been proposed and used as an alternative. This pricing method is a combination of the postage stamp method and the nodal pricing method and attempts to simplify the pricing process, while at the same time reflects the varying costs of supplying power to different areas.

Similar to the nodal pricing method, zonal pricing can utilize short-run or long-run marginal (incremental) costs, depending on the objectives, as mentioned before.

Generally, the nodal pricing method is first utilized to obtain the prices in all buses, and then a weighed average value of all nodal prices within a zone is set as the zonal price. This is done periodically to account for the change of system conditions. The nodal prices at all buses in a zone should be reasonably close to each other for the method to be meaningful. Otherwise, zonal boundaries should be adjusted. The Norwegian electricity market uses this approach for four or five zones. A long-run marginal-cost-based zonal pricing method is employed in the NGC in the UK for recovering capital costs.

Another important purpose of zonal pricing is congestion management. Congestion occurs when the dispatch of all pool contracted transactions in full would result in the violation of operational constraints. Under normal circumstances without congestion, all zones have the same postage stamp rate, and there are no inter-zonal charges if power is transported across zone boundaries. The ISO raises the price in the zone when congestion occurs, thereby providing a price signal to encourage users to sell or buy power in their own zone. Since the congested lines change with time, the zones are adjustable.

The determination of zonal prices is somewhat subjective and represents a major disadvantage, however the method is simple and easy to implement.
– Postage stamp pricing
A flat rate is set over pre-specified time periods, which gives the right to inject energy at any node of the grid and to take it from any other. This implies that non-price methods have to be applied to manage congestion whenever it arises. Nevertheless, postage stamp pricing has the advantage of simplicity and transparency. Prices are known in advance and are easily controlled by the regulator and the market because generation is separated from the market for transmission. This approach is widely applied in EU countries.

Postage stamp pricing, while institutionally simple and transparent, is generally inefficient. It ignores the impact of any particular transaction on actual system operations. As a result, it is likely to send the incorrect economic signals to users.

Transaction-based Approaches

Two common forms of point-to-point tariffs are: (i) Contract Path Pricing and (ii) Distance-Related pricing. Under Contract Path Pricing, prices are set for each transmission line in the grid. Each transaction is assigned a "contract path" over the grid joining the location of the buyer and the seller. The price charged for the transaction is the total price of the transmission lines crossed by the contract path. While this method may seem simple and intuitive, the contract path does not reflect the actual flow of electricity over the network or its cost. Thus, contract path pricing does not provide efficient management of congestion. This method is sometimes used in the United States.

Under Distance-Related Pricing, prices are set as a function of the distance between the buyer and the seller. The method is similar to contract path pricing and has the same pitfalls.

In general, transaction based approaches are unsatisfactory because they result in prices that neither reflect costs nor serve to manage congestion efficiently. In addition, they may also have anti-competitive effects. Their implementation requires the communication of information on commercial relationships that may be strategically sensitive. They may also favor discrimination against some competitors. For instance, distance related prices tend to impose larger costs on far away and foreign generators, which does not necessarily reflect the cost of transportation as a result.

- **Open Transmission System Operation**

Technically, the greatest challenge facing the deregulated and unbundled electricity supply system is the operation of the grid. This section will summarize[7] some issues that the system operator will face in routine system operation, and how these issues can be resolved in a market environment.

Dispatch

Under normal operating conditions, without any operating constraint violations, the major difference between a centrally dispatched utility and dispatch in the pool-type of electricity market is the replacement of costs by bids. When demand-side bidding is permitted, a slightly more complicated dispatch procedure is required, and the objective, theoretically, will change to the maximization of social welfare (in practice, just making an allowance for demand elasticity) rather than just bid cost minimization. However, if bilateral and/or multilateral transactions are permitted and/ or transmission congestion occurs, the dispatch methods become more complex.

---

7 Summarize from [Yoon and Ilic, 2000].

Congestion management is discussed in the next section.

Transmission Loss Compensation

If the ISO owns generators, transmission losses can be supplied at charges which will, essentially, be set by the regulator. If the ISO does not own any generators, compensation for losses must be procured from power suppliers through pool purchases, auctions, or contracts. In other words, the participants compensate for the losses. However, losses cannot be fully compensated ahead of time since the system load cannot be accurately forecast and loss distribution factors are only approximate. Ultimately, the residual losses must be compensated for through a real-time imbalance market.

System Control

The key responsibility of the ISO is ensuring system security while monitoring and controlling real-time system operation. During normal system operation, the ISO is responsible for maintaining system frequency and voltage within a tolerable range and ensuring that all operating constraints are respected. Most generators have the ability to change generation output automatically to respond to the supply/demand imbalance which is reflected in small system frequency deviations. However, in a commercial market, this service will not be voluntarily provided.

The ISO is also responsible for a second frequency control, automatic generation control (AGC). AGC smoothes over large frequency deviations, usually the result of large load variation and/or generation outage. The system control must ensure that there is a sufficient quantity of on-line unloaded capacity and/or quick-start generation capacity, and that this conforms to mandatory reliability criteria. These capacities must be provided or procured by the ISO.

Reactive power and voltage control are also of vital importance for secure operation. The ISO should first utilize the reactive power resources available in the transmission system, thereafter provision of extra reactive power by generators is necessary.

Ancillary Service Provision

Ancillary services are services provided in support of the basic service of generating real power and injecting it into the grid. These services also ensure that the supply of delivered power is reliable and of high quality. Ancillary services include, but are not limited to: reserve capacity, frequency control, AGC, reactive power and voltage control and black-start capacity.

Under a vertically integrated model, these services were simply and naturally bundled into the main activities of generation and transmission and the costs associated with energy generation and with ancillary services provision, were jointly recovered through the electricity tariff. In a competitive environment, this natural or organic link is fractured; giving rise to many debates about how these services should be procured. Two important considerations in procuring ancillary services are the costs of providing services and the value of the services to the system. Depending on the organizational structures in different electricity markets, ancillary services may be provided by the system controller or purchased.

Some ancillary services can be mandated, for example, all generators can be required to provide frequency control as a precondition of connecting to the network. However, mandatory provision of ancillary services is a little inconsistent with the objectives of unbundling. Competitive market-based procuring and charging for unbundled ancillary services has been practiced in several existing electricity markets, for example, the U.S. In the U.K., the pool operator either provides ancillary services or procures them from generators and charges users through the 'uplift'. Procuring ancillary services, separate from the energy market, is better suited for market structures when the ISO is separated from the PX, as in the case of California. In some electricity markets, both mandatory and market-based approaches are combined.

- **Congestion Management in Open-access Transmission Systems**

Congestion is not a new problem in power system operation and was a routine problem for the system operator in the traditional system. In electricity market environments, however, previously established practices for dealing with congestion can no longer be relied on since cooperation between market participants can not be guaranteed. In some electricity markets with bilateral and multilateral contract transactions this problem is more difficult to solve since these contract transactions introduce additional constraints on the system operator. Any control measures adopted by the system operator to eliminate congestion must not only be technically justifiable, but also be fair to users and commercially transparent.

Efficient methods initially utilized for congestion management are nodal or zonal pricing, which was mentioned before. Other methods currently used to deal with congestion include: market-splitting and counter-trading. Market-splitting and counter-trading are closely related to the operation of the physical day-ahead wholesale electricity market. Market-splitting entails the partition of nodes into different pre-defined congestion areas on either side of the constraint. New market prices are determined, one for each area, taking into account the initial bids on the day ahead physical market and the transfer capacity of the cross border line determined by the transmission system operators. The management of congestion by counter-trading is completely different. Indeed, counter-trading doesn't ask for direct modification of the equilibrium in the quantity and price of the day-ahead physical wholesale market. Counter-trading is a mechanism implemented through a separated voluntary market where generators bid for adjustments, upward or downward, in the day-ahead generation schedule. The principle of counter-trading is thus based on a buy-back principle which consists of replacing the generation of an "ill-placed" generator on the grid with the generation of a "better placed" producer. The sole buyer in this adjustment market is the transmission network operator that directly bears the costs of increasing and reducing generation. These costs, or a part of them, are then charged to national network users via the transmission tariff, depending on the regulatory model applied by the regulatory authority.

Whereas counter-trading and market-splitting have the same goal, relief of capacity constraint, they nevertheless differ in their trading process and in the resulting financial flows. We support the idea that the coordinated use of these two congestion management methods, with highly differentiated distributional consequences, gives an incentive to distort data concerning capacity limits in order to use a specific method rather than the other. Phrased differently, transmission system operators may benefit from the high level of flexibility in the assessment of the externalities that result from the uses of the grid and in the design of their security norms.

2.5 Competitive Electricity Market

The introduction of competition in the EI results in a sharp increase in transactions in electricity, as well as in the development of related financial contracts. This in turn leads to the development of organized electricity markets and related financial instruments. This section discusses about several kinds of trading arrangements, as well as the main components in such a trading chain.

- **Trading Arrangements**

It is possible to conceptualize and model the new trading structures into a few alternative categories as discussed in following section.

The Mandatory Pool

In the pool model shown in Figure 2.7, competition is initiated in the generation business by creating more than one Genco and is gradually brought to the distribution side where retailers could be separated from Discos and where consumers could be allowed to phase in a choice of retail supply. The transmission system is centrally controlled by a combination of an independent system operator and a power exchange (ISO+PX) which is disassociated from all market participants and ensures open access. The ISO+PX operate the electricity pool to perform price-based dispatch and provide a forum for setting system prices and handling electricity trades. Hedging contracts become a major option and are popularly used. The E&W system before NETA, Chile, Argentina, and East Australia fall into this category with some modifications to the basic structure.

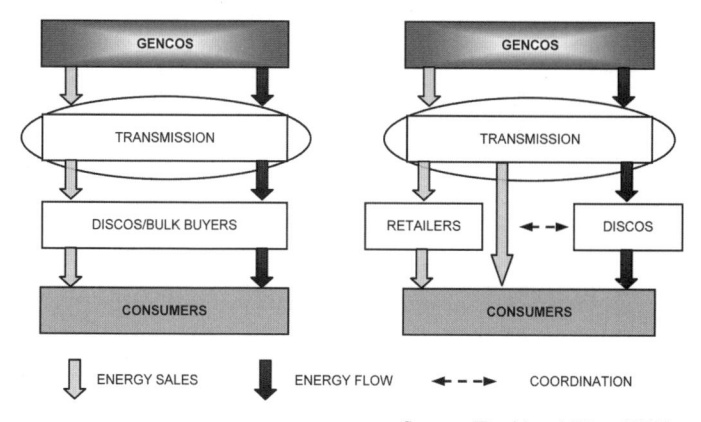

CHAPTER 2   49

Source: [David and Wen, 2001]
Figure 2. 7   Trading in the power pool

Pool and Bilateral Trades
The most likely system to emerge in the future is where a pool exists simultaneously with bilateral and multilateral transactions. The significant difference between this model and the pool model is that the transaction sector is unbundled into a 'market' sector and a 'security' sector (see Figure 2. 8). In the market sector, there are multiple separate energy markets, containing a pool market taken care by the PX and bilateral contracts established by the SCs. The ISO is responsible for system operation and guarantees system security, and in operational matters, holds a

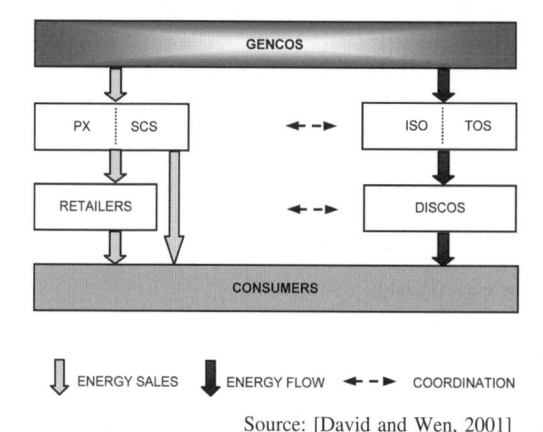

Source: [David and Wen, 2001]
Figure 2. 8   Trading with pool and bilateral contracts

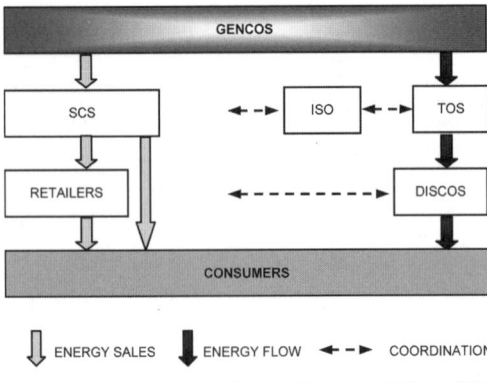

Source: [David and Wen, 2001]
Figure 2. 9 Trading with bilateral/multilateral contracts

superior position over the PX and SCs. The existence of a power pool is not mandatory in this model, however will invariably be the case. A market participant may not only bid into the pool, but can also make bilateral contracts with each others. Therefore, this model provides more options for transmission access. The California model is a representative of this category. The Nordic model and the New Zealand model also fall into this category with some modifications.

Multilateral Trades
Multilateral trades are a generalization of bilateral transactions where an SC or power broker puts together a group of energy producers and buyers to form a balanced transaction. In practice, multilateral and bilateral transactions will coexist with a power pool. Conceptually, the extreme case, where the concepts of pool and the PX disappear into this multi-market structure as illustrated in Figure 2. 9. Each market is managed by an SC or a broker under its individual rules. Differences in the rules of different markets can give rise to different strategies for participants. In this model, the ISO is restricted to operation and security. Only when system security is threatened does the ISO interfere in the management of contracted dispatches.

- "Spot" Electricity Markets

The physical nature of electricity does not allow for a true electricity spot market, that is, a market for immediate electricity delivery. Instead, transactions are scheduled some time in advance of physical delivery, for example one day, one hour, or five minutes in advance. Imbalances between scheduled and actual supply and demand that inevitably arise are handled following some predetermined

In an electricity market, prices are used both to co-ordinate the decisions of generators and electricity buyers, so that supply equals demand, and to ensure that these decisions are feasible given the physical constraints of the system.
Prices
Spot prices for electricity are set for each node of the grid. A number of cases are considered in increasing order of the complexity:
- Base case: In the simplest case, when there is sufficient generation and transmission capacity to cover demand and where transmission losses are ignored, there would be a single price for each time period:
  Price of energy (P 1) = Marginal cost of highest cost unit in operation
- At price P 1 there is generating capacity shortage: If setting the price to the marginal cost of the highest cost unit available results in a generating capacity shortage, the above rule cannot be applied. The price of energy then has to be increased in order to decrease demand. The price increase S needed to make demand equal to the available generation capacity is the difference between the marginal cost of generation and the marginal benefit of consumption. The price of energy is then:
  Price of energy (P 2) = P 1 + S
  S can be interpreted as the scarcity rent that covers the fixed generation costs
  Therefore, S provides incentives for investment in generation
- There are transmission losses: The price of energy (either P 1 or P 2, as it applies) would increase by a factor of (1 + Marginal Loss) at each node, reflecting that in order to supply 1 kWh, it is necessary to produce (1 + Marginal Loss)kWh. Thus
  Price of Energy (P 3) = P 2 × (1 + Marginal Loss)
- There are transmission constraints: The price at congested consumption nodes has to be increased to discourage use of the system, while the price at congested injection points has to be decreased to discourage use of the system. The magnitude of the adjustment, or "shadow price" of the constraint, is such that at no point of the grid, supply exceeds transmission capacity. The resulting prices are nodal electricity prices, discussed in detail in chapter 6. They include all the cases considered above as particular cases:
  Nodal Price = P 3 + Shadow price of the constraint at that node

Generation and Consumption
Because of the way wholesale prices are constructed, supply equals demand at each node and time period, and feasibility is ensured. Generators will produce energy only if their marginal cost does not exceed the price of energy at their injection point. Consumers will buy electricity up to the point where the price of electricity equals the marginal benefit of consumption.

Source: [Ocana, 2002]
Figure 2. 10 How an ideal spot electricity market would work

procedures, which may or may not be competitive. The Box below introduces the logical foundations of "spot" electricity markets.

- **Capacity Mechanisms**

While most other electricity markets are energy-only markets in which generators are only rewarded for actual energy supplied, some electricity markets use capacity mechanisms to procure generating reserves above market levels or to stabilize reserves over time, for example, capacity mechanisms were in use in the pre-NETA England and Wales market, Spain, Argentina, and the PJM.

Capacity mechanisms set a payment to generators in exchange for the generator being available to supply, if required. There are two broad kinds of capacity mechanisms depending on whether the regulator sets a price for capacity and lets the market determine the amount of capacity available or, conversely, the regulator sets the amount of capacity that has to be available and lets the market determine its price. These are known, respectively, as capacity payments and capacity requirements.

Capacity payments are used alongside a price cap that protects consumers against market power and compensates for insufficient demand side participation. Using this mechanism, it could reduce price volatility without altering the average price and reserve levels. However, price caps are controversial, as they provide incentives for generators to locate outside areas with low caps and expose generators to some quantity risk.

- **Derivatives Markets**

The introduction of competition in the EI results in a large increase in not only electricity transactions, but also in the development of related financial contracts, which leads to the development of related financial instruments.

Financial contracts play a key role in providing insurance for market players against price volatility in electricity markets. Prices in competitive electricity markets move rapidly in response to changing demand and supply conditions and, as a result, are volatile. Price movements have a healthy effect in efficiently accommodating supply and demand, but may have a negative effect on market players.

Electricity contract markets reallocate price and quantity risks. Electricity contracts are primarily financial instruments outside the scope of electricity regulation. In the past, the risks involved in the electricity supply were bundled with electricity itself. One of the benefits of reform is that risk is unbundled from the provision of the basic good and can be, to a certain extent, separately and flexibly managed and priced. Electricity contract markets are primarily financial markets and do not require specific electricity regulations. However, as in other commodity futures markets, financial electricity contracts may need to be backed by financial penalties and guarantees to ensure that the contracts are honored and may need to be subjected to various regulations to ensure transparency.

It has been suggested that contracts may help curb market power. The idea is that financial contracts reduce the generators' incentives to set high prices in the wholesale market because the price that a generator receives is set in the contract, not in the wholesale market.

Most financial contracts take the form of forward contracts, futures contracts, and option contracts. Forward contracts are bilateral agreements to deliver energy at a given price. They work in combination with a competitive wholesale market and are one of the simplest forms of derivative instruments that are used to transfer, or "hedge", price risk. The parties involved agree to a price for that day (the "strike" price) for "delivery" of a certain amount of energy later. Therefore, the settlement of a forward contract can be made without physical delivery. If the market price at "delivery" is higher than the strike price, the seller of the contract compensates the buyer for the difference. If the price is lower, the buyer of the contract compensates the seller. A forward contract settled in this way is called a contract for difference. Futures contracts are analogous to forward contracts with the only difference that they are standardized and traded in an organized market.

Option contracts give the right, but not the obligation, to buy or sell electricity at a certain price. The price has two components: an option fee equivalent to a kW charge payable when the contract is signed, and an exercise price to be paid for each kWh actually delivered. An option contract does not need the reference of a spot electricity price to be implemented. Compared with forward contracts, option contracts have the advantage (to the seller) of partially hedging quantity risk as fixed generating costs can be covered by the option fee.

Financial contracts are currently in use in electricity markets. For instance, forward and option contracts are extensively used among traders in the England and Wales pool, NordPool and Australian NEM. These contracts are settled for differences. Electricity futures are traded in NordPool, the US (NYMEX, PJM) and Australia (Sydney Futures Exchange). There is a plethora of far more complex and advanced derivatives that can be used to hedge all kinds of price risk.

CHAPTER 3
ELECTRICITY INDUSTRY PRIVATIZATION

Privatization is an instrument of economic policy, which could be applied to any sector of the economy. While the competition seems specified in the case of electricity industry, privatization in the industry is not truly distinct from other sectors, except the large size of the privatized deals in this sector. In this chapter several fundamental ideas on the economic analysis of privatization will be reflected upon from the definition of privatization, its strategies and implementation methods, to the political and economic consequences of privatization. Several crucial issues for privatization in the electricity industry are also discussed in this chapter.

3. 1   Definition of Privatization

- **Definition of Privatization**

Privatization is an instrument of economic policy. Like most prominent policy concepts, "privatization" has a cluster of overlapping meanings. The narrow definition refers to privatization at the level of the firm or units within it, while the broad definition refers to privatization at the sectoral level (e.g. telecommunication, electricity, social security, etc.).

In this book, the term is used in its widest sense, as defined in [Weizsacker et al., 2005], to refer to all initiatives designed to increase the role of private enterprises in using society's resources and producing goods and services by reducing or restricting the role that government or public authorities play in such matters. This is often carried out by a transfer of property or property rights, partial or total, from public to private ownership. However, it can also be done by arranging for governments to purchase goods and services from private suppliers or by turning over the use of financing of assets or delivery of services to private actors through licenses, permits, franchises, leases, or concession contracts, even when ownership remains legally in public hands. There are even cases such as "build-operate-transfer" contracts, where the private sector creates an asset, operate it for a certain length of time, and then transfers it into public ownership.

- **Forms of Privatization**

[Weizsacker et al., 2005] divided privatization into four broad categories, ranking

from "weak" to "strong":
- Putting state monopolies into competition with private or other public operators. In this category, the governments amend the rules to allow private enterprises to engage in specific activities. It results in competition between private and public operators.
- Outsourcing, in which governments pay private actors to provide public goods and services. The second category includes all cases in which governments contract with private enterprises to supply specific goods or services previously provided by public agencies.
- Private financing in exchange for delegated management arrangements, often with a view to transferring ownership to the state after a period of profitable use. The third category involves the financing of infrastructure by the private sector, which increasingly includes other projects such as schools, hospitals, and museums; these include build-operate-transfer (BOT) arrangements, concessions, and lease contracts.
- Transfers of publicly owned assets into private hands. Historically, governments have sold public land to private owners or granted more limited use rights. Recently, we have witnessed a raising tide in sales of all or partially public-owned enterprises, such as national airlines, telecommunications, and energy corporations.

Figure 3. 1 illustrates several service delivery models, from fully public to fully private, with the areas to be managed. Among the four categories mentioned above, the first and the second are normally considered as pre-privatization. This book aims to deal with privatization in the electricity industry, so it will mainly cover the two latter categories, private financing and transfers of public owned assets.

- **Privatization Techniques**

The privatization techniques considered here fall into the fourth category mentioned above, transfers of publicly owned assets into private hands. Even though there are some variations, the most commonly applied models are as follows:
- Public offerings

Privatization offerings have principally, but not exclusively, been a feature of privatization in developed markets where stable enterprises with established track records can be prepared relatively easily for access to liquid capital markets. An offering prospectus is normally prepared by the investment bank acting for the government. The bank may also, on its own or as part of a syndicate, underwrite the offering which may be on a fixed price or a demand-led basis (whereby the price of the shares is calculated to reflect the strength of demand from the public). The securities may be marketed either domestically or internationally.

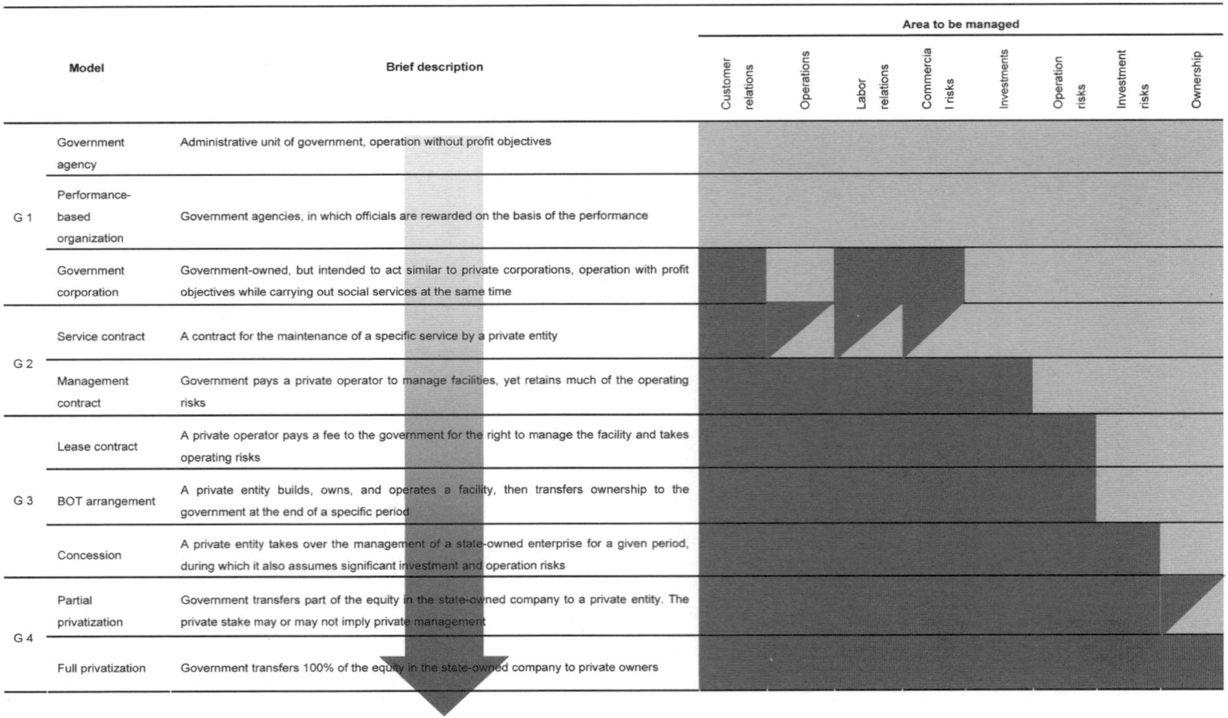

56   Part I

	Model	Brief description
G 1	Government agency	Administrative unit of government, operation without profit objectives
	Performance-based organization	Government agencies, in which officials are rewarded on the basis of the performance
	Government corporation	Government-owned, but intended to act similar to private corporations, operation with profit objectives while carrying out social services at the same time
G 2	Service contract	A contract for the maintenance of a specific service by a private entity
	Management contract	Government pays a private operator to manage facilities, yet retains much of the operating risks
G 3	Lease contract	A private operator pays a fee to the government for the right to manage the facility and takes operating risks
	BOT arrangement	A private entity builds, owns, and operates a facility, then transfers ownership to the government at the end of a specific period
	Concession	A private entity takes over the management of a state-owned enterprise for a given period, during which it also assumes significant investment and operation risks
G 4	Partial privatization	Government transfers part of the equity in the state-owned company to a private entity. The private stake may or may not imply private management
	Full privatization	Government transfers 100% of the equity in the state-owned company to private owners

Area to be managed: Customer relations, Operations, Labor relations, Commercial risks, Investments, Operation risks, Investment risks, Ownership

Figure 3.1 Range of service models from fully public to fully private

Source: Adapted from [Weizsacker et al., 2005]

If it is the first offering of securities by the company, it will often be referred to as an "initial public offering" or "IPO", as opposed to a "secondary public offering". Advantages of public offerings are perceived to include the attainment of widespread share ownership and the general transparency of the process making the privatization more easily acceptable politically. Disadvantages include associated expenses, delays, and the prospect of misjudging investment interest, which results in underinvestment in the shares of the company, causing the IPO to fail. The prospect of failure leads to the price of securities being marked down to ensure success.

– Private sale of shares

Through this popular technique, the government sells all or part of its share in a state enterprise to purchasers selected in one of a variety of ways, such as: (i) open or closed tenders, perhaps with pre-qualification of bidders, (ii) auctions, and (iii) directly negotiated sales. To meet potential concerns over the lack of transparency, many countries have introduced compulsory procedures to be observed for private sales, covering matters such as price setting, the selection process, and terms of payment, particularly relating to payment guarantees or acceptable forms of security, where ongoing financing is to be provided by the government.

Private sales are sometimes employed as a first step toward a privatization offering, being a useful method of establishing a core of table shareholders and improving the company's quality before widening share ownership to the general public.

– Sale of government/enterprise assets

An asset sale by a government can either take the form of the sale of individual state /state enterprise assets/rights or the sale of all the assets of an enterprise. Similar to a direct share sale, the sale of assets can be effected by tender, auction, or direct negotiation with a pre-identified party. However, unlike share sales, assets are normally sold without liabilities. It may be possible to create a contractual arrangement under which the buyer is obligated to re-hire a portion of the workforce and/or to assume certain liabilities with creditors, although this can be convoluted. Once again, uncertainty over valuation gives rise to considerable concern on the side of the state.

– Management and management/employee buy-outs

A management buy-out (MBO) is the acquisition of a majority of shares in a company by its management, while a management/employee buy-out (MEBO) additionally involves the employees as shareholders. In many cases, such transactions are highly leveraged, with the assets of the company being used as security for funding from financiers. Strong cash-flow projections will inevitably be a precondition for financiers making funds available for an MBO/MEBO.

The major attraction of MBOs and MEBOs is considered to be the prospective improvement in productivity. A necessary feature of MBOs and MEBOs is the presence of a competent and intelligent management and committed workforce. It may nevertheless be necessary to educate the employees as to the nature and benefits of owning shares.

3. 2   Rationales for Privatization

[Wardle and Towle, 1996] outline some of the theoretical arguments on the advantages of private versus public ownership.
- Economic efficiency. Privatization is a response to the failings of state ownership. Under state ownership, enterprises are often used by governments to pursue non-commercial objectives which are inconsistent with efficient and financially viable performance. Governments have many objectives other than profit maximization. Further, government objectives can change from one administration to the next. Thus, the inability of the government to commit itself to a policy can significantly reduce the efficiency of a firm's operation and governance. On the other hand, private ownership, it is argued, is equated with a higher level of managerial supervision resulting in more commercial and timelier financial decisions. This results directly from more single-minded profit maximization objectives of private ownership.
- Public finance rationalization. In theory, privatization reduces a government's current account deficit as responsibility for enterprise funding is transferred to the private sector. Moreover, privatization revenues and the prospect of increased revenues from taxation of more profitable enterprises can solve the pressure to meet short-term budget deficit targets of the government. Furthermore, with revenues from privatization, government has in its hands a tool that can yield tax cuts for the populace and/or an increase in government spending in other politically sensitive areas, such as health, education, and law enforcement. These prospects are very attractive to a government seeking re-election.
- Wider share ownership. Wider share ownership is generally regarded as a good thing. There is a belief that increased participation generating wealth for the nation in some way creates a more democratic society.
- Elimination of political interference. It is often argued that state ownership of industry inevitably subordinates commercial efficacy to political objectives.
- IMF/World Bank requirement. In many emerging markets, both IMF and the World Bank have tied their funding programs to privatization of the state sector. It was often known as "conditionality".

   Looking at the arguments for privatization, we should also investigate several its arguments against it as well.
- Privatization tends to weaken the state and its capacity to care for social equity.

By weakening the state, privatization also erodes the significance of democratic participation at national and sub-national levels.
- Privatization subordinates the greater public good, including long-term ecological and cultural values, to commercial imperatives.
- The need of private providers to make a commercial return, in the form of profits, dividends, rent, and/or interest, adds to the cost of providing public services.
- Commercially optimal decisions are often suboptimal for public goals; competition forces providers to ignore externalities.
- The private sector never really assumes risks in the provision of public service. Where costs exceed revenue, private operators respond by demanding subsides, raising charges, cutting necessary investment and maintenance, or walking away.

- **Lessons from Privatization**

Privatization is not an end in itself. Privatization should be treated as a means of increasing efficiency and not as a way of reducing or undermining the role of the state. The lessons learned from privatization are summarized as follows:
- Development of good governance, strong regulations, and regulatory institutions
- Do not privatize what the public sector can still do
- Never privatize for ideological reasons
- Secure democratic control over regulatory institutions and enable the state to reverse privatization in cases of severe failure.
- Conduct privatization under through a series of step-by-step policies.

Before deciding on whether and how to privatize, all parties involved should consider the context and specific local factors relevant to the case at hand.

3. 3 Managing Privatization

Privatization is a complex and demanding process. The strategy and institutional framework for its implementation have to be designed to meet the circumstances and the challenges of each country. There is no "right" way of implementing privatization. Nonetheless, there are key elements that will tip the balance between success and failure. Such key elements were discussed in detail in [Shehadi, 2002]. A summary of this discussion follows.

- **Political Commitment from the Top Political Leadership**

Privatization is an intensely political process and political momentum is needed to overcome three obstacles which impede a successful transition: opposition, bureaucratic inertia and lack of coordination. Governments can overcome the most important obstacles to privatization mainly by providing political commitment to the process. The more committed to privatization governments are, and appear to be, the easier it is to deter and overcome opposition and inertia, and to elicit coordination. This political commitment to privatization has to be maintained throughout the process, while avoiding the temptation to micro-manage.

International experience also suggests that governments can facilitate successful privatization by: (i) Demonstrating support for privatization by having the highest executive authority in the country oversee the privatization process, coordinate various government departments, and negotiate with opponents to privatization, (ii) Building a case for privatization and articulating it to the public and to all stakeholders; (iii) Adopting a labor strategy that addresses labor's legitimate concerns, (iv) Following a transparent and fair privatization process to eliminate the possibilities of corruption, and (v) Hiring world-class qualified professionals and technical experts working on legal, financial, regulatory, and technical aspects, with no conflicts of interest, to prepare privatization transactions and hold them accountable for their performance.

- **Transparency and Fairness of the Privatization Process**

In every transaction, it is extremely important for transparency to be maintained for the success of a privatization program. The public should know if the program is being carried out in a fair and honest manner. A lack of transparency can lead to the perception of unfair dealing, even where it does not exist, and lead to popular opposition that could threaten not only privatization, but also the government's credibility in general.

Transparency is required at four levels. First, the laws and regulations supporting the program should mandate publicity and openness in the implementation of the program. Second, the selection of advisors should be public, competitive, and according to pre-announced terms of reference and selection criteria. Third, individual transactions should be conducted using well-publicized competitive bidding procedures, clear and simple selection criteria for evaluating bids, and disclosure of purchase price and the buyer, in order to encourage participation from the widest possible range of domestic and foreign investors. Competitive bidding helps maximize the proceeds from sales while maintaining public confidence in the integrity of the process. Fourth, the individuals, or the unit in charge of privatization, should be identified and held accountable; they should follow precise, detailed, and publicly announced processes and procedures for carrying out privatization transactions. The lack of transparency is often associated with corruption in privatization. Corruption can take many forms, but there are remedies, such as adequate monitoring and supervision of the program, that governments can adopt to minimize it.

Transparency has a price, in terms of the speed with which transactions can be implemented. Transparency requires that transactions be well-prepared and carried out through competitive procedures. However, a transparent privatization process will also attract greater interest from local and international investors and will help increase revenue from sales proceeds. The time and effort put into ensuring transparency, if put to good use, result in investment in privatization.

- **A Favorable Legal Environment**

Privatization requires a favorable legal environment. Existing rules and regulations pertaining to private business can either improve or reduce the value investors are willing to pay for an SOE. Governments should ensure that the commercial, financial, capital markets, and labor legislation, provide the proper environment for privatization. The drafting of a privatization law is often an opportunity for the government to go through this legal exercise and pre-empt legal problems before they arise.

In cases where privatization concerns an infrastructural sector (telecommunications, electricity, etc.) that is already governed by existing legislation, that legislation needs to be revised to allow private sector participation and to put an effective regulatory framework in place. Furthermore, the elaboration of competition law and policy prior to privatization will help avoid many problems down the road.

- **A Clear Strategy for Privatization**

Governments that have succeeded in privatization have gone to great lengths to prepare for each transaction and find the best strategy for privatization. Privatization requires significant policy decisions that are best taken on the basis of formal proposals, following consultations with all the stakeholders.

The most important task is to define the general objectives of privatization and specific transactions for privatization in particular. Is the objective to improve the performance of the SOE or the sector? Is it to have a more competitive infrastructure? Is it to maximize the proceeds from privatization in the short run? Is it to attract foreign and local investment? In setting program and transaction objectives, governments should examine the trade-offs involved between the various objectives and make informed decisions. They should balance the objective of maximizing privatization proceeds with other more important priorities such as improving the competitiveness of enterprises and the economy as a whole, promoting competition, and developing capital markets.

In light of the review of the candidates for privatization and after identifying the preparations needed for each transaction to go proceed, governments should define their overall privatization strategy. The strategy needs addresses a number of issues: (i) the candidates for privatization, (ii) the timeframe for each privatization, and (iii) the priorities for the overall program.

In the process of finding a strategy, the government also needs to provide a vision and a roadmap on the future role of the state, the roles and responsibilities of the various participants, such as government departments as well as civil society, and the private sector, in the privatization process, policy for each level of infrastructural privatization (market structure, regulatory design, etc.), the use of proceeds, and social and labor concerns and how they will be addressed. Within this context, the government can then define the appropriate privatization method.

- **Choosing the Appropriate Privatization Method**

There are many methods for privatization (please refer to Section 3. 1). There is no panacea or "one size fits all" solution. The choice of method and strategy for

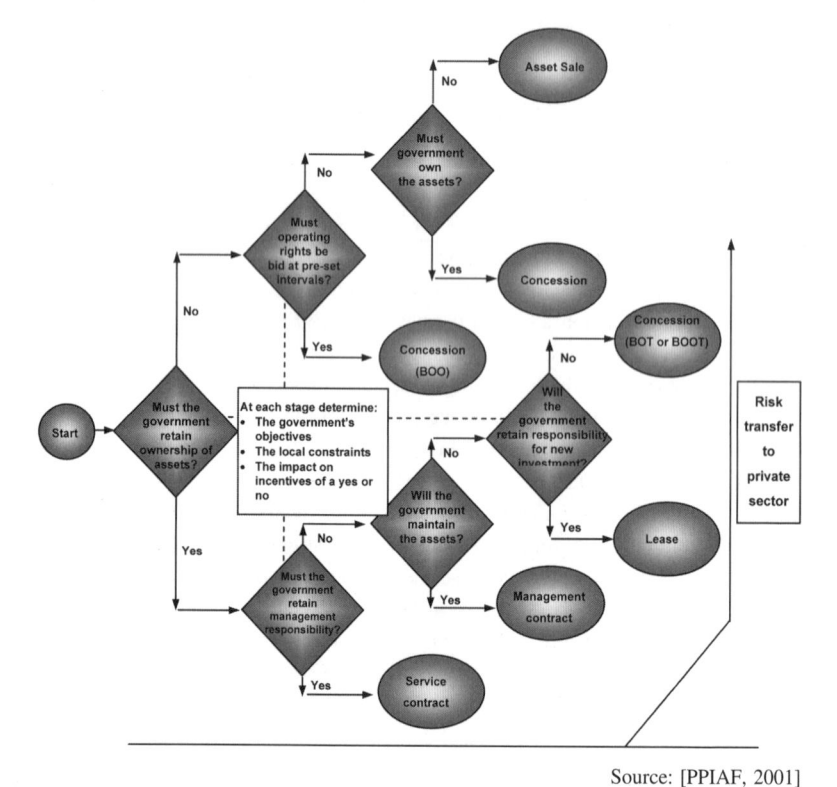

Source: [PPIAF, 2001]
Figure 3. 2 Choosing the appropriate privatization method

privatization is a very important determinant of the success of the transaction. The choice of the particular method of privatization has to be made in light of the sector or SOE characteristics, the government's policy objectives, and the state of the international market. The choice of method of privatization (see Figure 3. 2) is particularly important in the case of infrastructure.

- **A Professionally Managed Transaction**

Once the government has selected a method, privatization moves into its final stage: the transaction. The process will vary depending on the method. In essence, there are two different processes depending on whether the privatization method selected involves going to the financial market (Initial Public Offering, or IPO, for the initial listing on the equity markets) or to a more restricted market where bidders are pre-selected according to technical criteria (companies with vast experience in the sector, financial indicators, etc.). Each process requires careful preparation and extensive and costly advisory work.

Each process has its own requirements. IPOs have performed best where competent financial market regulatory authorities exist, when markets were liquid, when corporate governance was favorable (transparency in accounts, rights of minority shareholders well-defined and protected, etc.) and when the listed company had good management. Where any one of these requirements was not present, IPOs have underperformed, company performance did not improve, and, in some cases, insider trading was involved. In such cases, a trade sale of a block of shares of 25% or more to a "strategic" or core investor has worked much better, i.e., the investor was handed management control. Strategic investors were given both the incentives and the means to bring in good management and to turn the company around. Privatizations that have followed this process have seen share values increase within 18–24 months, corporate governance has improved, and the company was readied for an IPO.

To find the most appropriate strategic partners for an auction, whether for the sale of a block of shares or the award of a license, governments, with the help of their financial and legal advisors, have to define the pre-selection criteria, i.e., the criteria that will qualify interested investors for the auction. The criteria should be reasonable and announced prior to the pre-qualification. The auction process involves an exchange of information between the government and the market (investors and operators). The smoother and more complete this information exchange, the more likely that the auction process will deliver the required technical expertise and financial wherewithal to meet the expectations of the transaction.

- **Liberalization and Competition Before Privatization**

Economists concur that the benefits from privatization are maximized when an enterprise is privatized in a competitive environment. This is even truer of infrastructure privatization. For all the reasons mentioned above, PPI requires that governments address the issues of market structure to realize the benefits of privatization. It is best that "such structural reforms should be evaluated and decided before any final decisions are taken on the ownership question: it is easier to change the framework of competition and regulation before privatization."

The aspect of competition is presented in detail in Chapter 2 of this book, "Restructuring and Competition in Electricity Industry".

- **Establish Regulatory Framework**

The most difficult component of a successful privatization is to establish an effective regulatory framework. To be effective, a regulatory framework has to provide for stable and clear "rules of the game," credible enforcement of these rules, and reasonable rules, i.e., rules that balance the interests of the private firm and those of the citizen or user. To ensure a fair and level playing field, the regulatory authority should be at equal distance from all the operators in the sector, including state-owned operators. The regulator also needs to have a clear mandate to issue and enforce regulations. In addition to a well-designed regulatory framework, effective regulation also requires "regulatory capacity."

Regulation is one of three aspects of the reform. It is explained in detail in Chapter 4 – Electricity Regulation.

3. 4   Main Trends in Privatization

- **Privatization in the Entire Economy**

The economic and financial history of the post-war era has been characterized by privatization. A summarization of privatization milestones follows. One of the first privatizations in modern times was undertaken by the German government. In 1961, the government launched a policy of denationalization of the economy and sold its majority stake in Volkswagen through a public offering. The sale of Veba shares followed in 1965. Both offers were favorably received initially, but at the appeal of share ownership did not survive after the first cyclical stock market downturn in prices. The government had to bail out investors and reversed the state policy [Bortolotti and Siniscalco, 2004]. Other failed privatization attempts occurred in Chile and Ireland at the beginning of the 1970's.

Many people believe that the most important milestone in privatization history were those of Margaret Thatcher's conservative government in the United Kingdom in 1979. The label "privatization", which was originally coined by Peter Drucker, was adopted and replaced the name "denationalization" [Yergin and Stanislaw, 1998]. Privatization became established as a basic economic policy in the UK in 1984, after the successful initial public offering of British Telecom. Thanks to successful privatizations, the Conservatives obtained wide political support, which allowed them to successfully implement the process. At the end of the fourth Conservative legislature in 1997, all most SOEs were sold for more than 10% of the GDP 18 years earlier.

The perceived success of the British privatization program helped to persuade many other industrialized countries to begin divesting SOEs through public share offerings. During the 1990's, privatization policy became a fundamental element of the global economy, spreading not only in developed countries, but also emerging in less developed countries. First, the compound annual average growth rate was around 10% between 1990 and 1999, with global privatization revenues jumping from $25 billion in 1990 to $180 billion in 1999. Second, the number of countries that have implemented privatization policies has exceeded 110. Finally, privatization and private participation have touched almost every aspect of economic activity [Bortolotti and Siniscalco, 2004]. This study revealed the large cycle of privatization in the 1990's, which is now over. After the peak in 1999, global revenue fell sharply, at an average annual rate of 46%, and data from 2002 and 2003 seems to confirm with this negative trend.

In terms of regional trends, the international comparison of the number of deals shows that Western Europe has implemented the greatest number of transactions, followed by Central and Eastern Europe and the former Soviet Union, Asia, and Latin America. The breakdown of revenue by geographic areas shows that Western Europe accounted for 49% of global revenue, followed by Asia at 26%, and Latin America at 10%. The comparison between the number of deals and revenues offers interesting insights. It demonstrates that privatizations in Central and Eastern Europe and the former Soviet Union were numerous, but minor in scale. The opposite occurred in Asia, where relatively large scale privatization projects took place.

- **Privatization in Infrastructure Sector and Electricity Industry**

Infrastructure (telecoms, electricity/natural gas, transport, and water) accounted for half of all developing country proceeds between 1990 and 2003 (Figure 3.3). Infrastructure proceeds grew over time as more countries started to tackle these sectors in the late 1990s [Kikeri and Kolo, 2005]. They were overwhelmingly concentrated in telecommunications and power.

The most popular privatization methods utilized in the infrastructure sector included (i) Partial sale of a company to the public, (ii) Sale of a state-owned company to another company or consortiums, mostly to foreign investors, (iii)

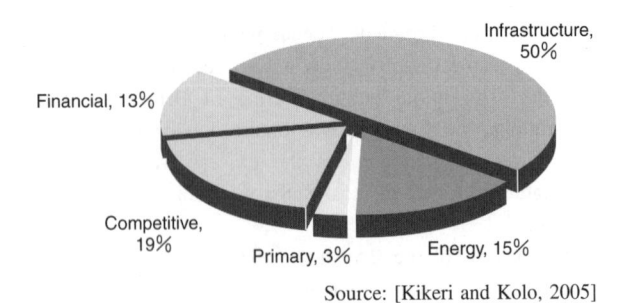

Source: [Kikeri and Kolo, 2005]
Figure 3. 3 Sectoral distribution of privatization proceeds: 1990–2003

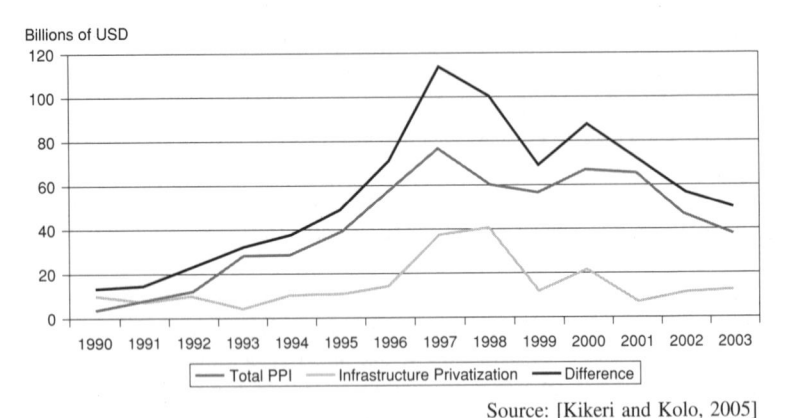

Source: [Kikeri and Kolo, 2005]
Figure 3. 4 The difference between new greenfield investments and privatization of existing infrastructure assets

Voucher schemes, and (iv) Investment vehicles in Greenfield projects such as: IPP, BOT, joint-venture, etc. One of the most important characteristics of the privatization process is the difference balance between privatization of existing infrastructure assets, which has raised $204 billion since 1990, and new green field investments, which were significantly higher at $350 billion (Figure 3. 4).

It should be note that, a large amount of proceeds from privatizing one sector does not mean that the sector is advanced in privatization. In the case of the infrastructure in general and the electricity in particular the large amount of proceeds at first is due to the fact that these sectors are capital-intensive industries.

Considering the number of privatization deals, it also presents that the electricity industry is one of the most important targets in privatization programs of

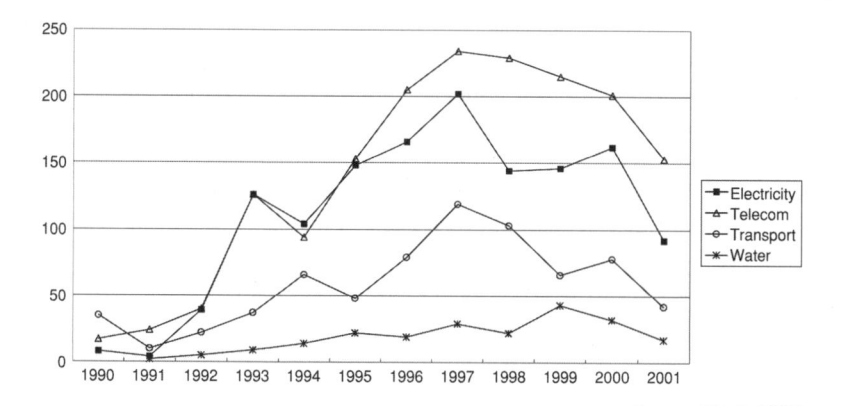

Figure 3. 5 Number of private participation transactions in developing countries

Source: [Iimi, 2003]

many countries worldwide. Figure 3. 5 shows the number of privatization transactions by primary sectors: electricity, telecommunications, transportation, and water and sewerage. It is in the telecommunications and electricity industries that private sector participation has attracted the most investment among public service industries.

However, in its privatization process, the electricity sector faces complex issues of pricing, access, and the quality of services. This calls for more of a case-by-case approach taking into account prevailing market structures and regulatory capacity.

3. 5 Specific Issues of the Privatization Process in the Electricity Industry

- **Main Characteristics of Privatization in the Electricity Industry**

Limited Acceleration of Privatization
Only a few countries have chosen to transfer ownership of industries or companies swiftly and completely. Argentina, the United Kingdom, Chile, and New Zealand have generally undertaken some of the most ambitious of privatization efforts by auctioning off companies directly to the public, thereby letting the market determine the value of these companies through the bidding process. British Energy is a famous example for this trend. However, for many reasons, it did not seem successful (refer to Chapter 9 for the British Energy case).

Privatization in the Electricity Sector is far from an End and State Ownership Continues to Exist Despite Increasing Emphasis on Private Sector Participation
Power utilities in nearly 85 developing countries are still owned and operated by the state. [Kikeri and Kolo, 2005] shows that there are 52 developing countries in total

with generation capacities between 29 megawatts (the Gambia) and 318 gigawatts (China). Almost 70 percent of them had not started or completed the process of including the private sector participation and a further 18 percent had just begun the process. In contrast, independent power providers had been established in 67 percent of the countries with another 21 percent planning to open electricity markets to IPP's. State ownership is most prevalent in low-income countries with weak institutional and regulatory capacities, such as countries in Africa, South Asia, East Asia, Europe, and Central Asia.

Privatization in the Electricity Sector has largely been Conducted through Partial Privatization

Except the UK, where electricity industry has been completely privatized, the process of EI privatization is slightly slow in most countries worldwide, even in countries reformed countries such as Australia and the countries in Scandinavia. Full privatization seems particularly difficult to achieve in civil law countries, has even been considered as unconstitutional. Partial privatization is the most favorable strategy.

In the electricity industry, partial privatization has been followed by such privatization methods as partial sales, voucher schemes, and investment vehicles in Greenfield projects. In which, investment in Greenfield project exist in most of countries in the world from developed countries to less developed countries. It especially seems to have been mobilized in Asian countries during the 1990s (refer Section 3. 8 for more a detailed discussion).

The formerly Communist countries in Central Europe seem to be favoring the voucher schemes. This method has been adopted whereby ownership of an industry is simply transferred to the general public with no cash exchanged. A lack of developed equity markets may have encouraged voucher schemes. After the initial distribution of vouchers, individuals have been allowed to buy or sell these vouchers, thereby encouraging the creation of stock exchanges. In some instances, the transfer of ownership has been implemented with labor and management being allotted favored shares.

In partial sales, the government gradually reduces its ownership in the shares of an electricity utility. Governments have often sold shares of a state-owned firm while still retaining a portion of the company (a "golden share"), thereby maintaining a limited degree of control over the company. Due to the important role of this method in the privatization process, the next section discusses the rationale behind partial sales, goes over to the famous "golden shares" phenomenon, and the state's control of in the industry.

3.6 Partial Sales – the Rationales

The divestiture of minority holdings appears to be a common feature of privatization worldwide. From 1977 to 1999, only 47% of the 2,459 deals reported in 121 countries involved the sale of the majority stock [Bortolotti and Siniscalco, 2004]. What is the rationale behind a government's decision to sell only minority shareholdings? Are there economic reasons for a partial sale or should they be preferred solely to the willingness to interfere in state-owned companies?

In a country where the fiscal conditions are good, a government has less incentive to privatize because it does not need revenue to finance the budget. Small stakes should then be associated with balanced budgets and lower debt.

However, more subtle explanations can clarify the use of partial sales. It includes institutional credibility, legal origin, and stock market development. First, governments may need to build credibility and investor confidence, subsequently it resorts to partial sales in order to signal its commitment. In countries under authoritarian rule by the military or one-party regimes, where political competition is absent or extremely limited, investment in privatized companies seems particularly risky. If the government holds a large block of shares, for example privatizes only a minority holding, private shareholders should be reassured that expropriation would also reduce the value of the investment for the public shareholder. Partial sales appear to be a strategy to signal government willingness to bear residual risk and not to interfere in the operating activity of the company in the context of high policy risk.

Second, investor protection can matter. We already know that where legal protection is poor, governments may be reluctant to sell majority stakes since they know that investors will discount the risk being expropriated by the managers of privatized firms. Moreover, full privatization seems particularly difficult to achieve in civil law countries due to the fact that electricity is considered a society product and needs to be controlled by state, such as the case of the Indonesian constitution.

Third, well developed financial markets can be functional for completing privatization. A large and liquid stock market can absorb the issuance of larger shares, so that governments can float a larger percentage of capital. A liquid market allows for the monitoring of managers through informative prices and through the threat of takeover, which in turn makes governments willing to sell large stakes, since the shareholders face less risk of expropriation. Clearly, this does not mean that underdeveloped financial markets make privatization unfeasible, but that governments are forced to use gradualism policies or to find other ways to implement privatization without depending on the development of the financial market.

Although partial sales do not imply a radical change in the control structure, they seem to have significant effect on the financial and operating performance of privatized firms. In a comprehensive analysis on privatization in India [Gupta, 2002], the author shows that the level and the growth rates of profitability and labor productivity improve with partial privatization, being negatively and significantly associated with a decrease in government ownership. Taken into account, the privatizations are truly partial sales, as a large majority the companies remained tightly controlled by the state after privatization. Given that privatization involved the listing of the privatized company shares, the public stock price produced valuable information for monitoring managers and alleviating agency problems, even if the company remained under government control.

The complete divestiture of ownership and relinquishment of control is a more complex and politically costly decision. Obviously, more money can be raised if the company is fully sold, but by doing so governments lose a powerful instrument for targeted redistribution, the right of having representatives on the boards in order to influence corporate decisions, and the power to safeguard public interests and national security.

3. 7 Golden Shares and Special Rights

In the above section, we discussed partial privatization. The state, as the largest shareholder in several essential industries, maintains control over management of privatized companies. In this section we outline the extent of the state's control in fully privatized companies via the use of "golden shares".

The sale of a majority holding is not itself sufficient to deter government interference in privatized companies. Governments can grant themselves wide discretionary powers over partially or over fully privatized companies. By exerting its right, the owner of "golden shares" can often influence the choice of management, and exert veto power over the acquisition of relevant stakes by private shareholders, even without owning the majority of stock in the company, or a single share of capital.

What is "golden share"? "Golden share" is defined as special powers granted to the state and the statutory constraints on privatized companies. Typically, special powers include: (i) the right to appoint members on the corporate board, (ii) the right to express consent or to veto the acquisition of relevant interests in privatized companies, (iii) other rights, such as consent on the transfer of subsidiaries, dissolution of the company, ordinary management, etc. The above mentioned rights may be temporary or permanent. Statutory constraints include ownership limits, voting caps, and national control provisions.

The content of golden shares, in fact, varies greatly across countries and companies. In some cases, governments' extraordinary powers are combined with substantial constraints so that golden shares become an extremely powerful instrument to control privatized companies. In other cases, the state's discretionary powers are more limited.

The main consequence of golden shares is the separation of property rights from control rights. When it occurs, controlling shareholders may abuse their power to expropriate outside investors. The financial incentives to reap private benefits of control are obviously moderated by equity of cash flow ownership, which should also enhance corporate valuation.

The separation of ownership and control induced by the golden share mechanism may be costly for governments which implement privatization policy because it discounts privatization prices. If so, why are golden shares so widely used? One possible reason can be found in the politicians' willingness to interfere in the operating activity of privatized firms [Shleifer and Vishny, 1994]. The opportunity cost in terms of lower revenue then has to be compared with the benefit of ongoing political control. In the first stage of the privatization process, golden shares may serve the purpose of defending national interests in strategic sectors (i.e. energy, utilities, national airline, etc.), by providing a shield against hostile takeovers by unwelcome investors, and giving newly privatized firms a limited time to adjust to the market.

A golden share works like a preference share, although it is more powerful. However, it seems to be less democratic. The golden share is normally held for a certain period, for example five years, but it is extendable for the good of public interest for another period when necessary.

3. 8   Investment Vehicles and the Role of IPPs

- **The Developing Role of IPPs**

IPPs have been a major source of new power generation capacity in many of the developed and developing market economies. Faced with serious capacity and energy shortages that cannot be remedied by public resources, many developing economies have also turned to private investors to expand electricity supply. In particular, this movement towards the use of IPPs has met a large and immediate financing gap by mobilizing direct foreign investment in the power sector. Properly structured IPP projects can also stimulate the development of local capital markets and bring the discipline of the capital market to support competition in the power sector.

IPPs are not the only answer to power sector development needs but, handled correctly they can undoubtedly make a major contribution and are an attractive option for financing future investments in power infrastructure. To date, a majority of IPPs have been developed in the absence of predictable and transparent regulatory mechanisms. To increase investment in the sector, host countries need to recognize that further development towards competitive markets will take place over time.

- **Critical Issues of IPPs in a Liberalized Market**
IPP programs have had the following desirable effects: (i) Where conducted in a fully transparent environment, IPPs have promoted open competition and thus encouraged lower cost development; (ii) Where projects have been soundly structured, most of the project risks have been the responsibility of the private sector; (iii) The private sector has provided wider access to capital markets, better management skills, access to latest technology and generally implemented projects faster than the public sector; (iv) Project financing is "off" the government's balance sheet, allowing governments to allocate scarce resources to other priority areas such as rural development, poverty alleviation, education and health; and (v) Governments have been able to benchmark existing public sector units against comparable private sector operations, for example in power generation. This results in both IPPs and existing public utilities improving their performance.

Thus, how is an IPP project soundly structured? [APEC, 1997] agreed to the following four themes as the best practices for establishing IPPs: (i) institutional and regulatory structures; (ii) tender/bid processes and evaluation criteria; (iii) power purchase agreements and associated tariff structures; and (iv) financing and its implications.

However, under the context of reform prevailing in the electricity industry, several main problem areas faced by IPPs should be recognized [Woolf and Halpern, 2001]: (i) A relatively long duration and fixed prices which are desirable to create a stable and certain revenue stream for the IPP, but at the same time freeze off-taker utilities. In other words they may create stranded costs reflecting above-market pricing; (ii) Because the IPPs are protected from market risk by long-term PPAs; they have little incentive to participate in the market. In some countries, their lack of participation would seriously affect the liquidity and indeed the success of the new market in capturing the benefits of market forces, particularly where the size of the IPP plant is a significant proportion of the total number of plants connected to the system; (iii) Forced contract renegotiation is extremely difficult, because of the legal sanctity and enforceability of the contract terms; (iv) In the early stages of any new wholesale market, prices and the extent of market risk will be difficult to predict, quantify, and mitigate. There will always be an initial period for the market to settle down and market rules will change to reflect operational experience. During this period, IPPs prefer the certainty of the power purchase agreements.

CHAPTER 4
ELECTRICITY REGULATION

Regulatory matters are not new; however, the practices in electricity regulation in a reform environment are quite recent. "Competent regulatory institutions are the linchpins of successful reform" [Kessides, 2004]. Regulation is not only one aspect of the reform process. Credible, stable regulation is required to achieve the benefit of privatizing and liberalizing the industry. Moreover, such regulation should be planned before privatization and the introduction of a market. In this chapter, first, a definition of regulation and its role in the new structure of the electricity industry will be discussed. After that, two dimensions of the regulatory process will be mentioned, including governance and instruments. Regulatory governance involves the creation of a transparent and predictable regulatory system. Regulatory instruments are mechanisms that pertain to pricing, subsidies and other operating policies. Both of these issues are considered under the new environment of competition and privatization.

4.1 Definition of Regulation

- **Definition of Regulation**

Literature on economic regulation has traditionally been based on the US experience of regulating natural monopolies. For more than a century, the presence of natural monopolies provided the main reason for imposing a set of regulatory mechanisms, such as entry barriers, obligations to serve, price, and quality controls. Under natural monopoly conditions the rationale for regulation is to remedy market failures and to avoid monopoly power. In the United States, regulation has always been seen as governmental intervention to prevent private decision-making and the monopolization of power against the public interest.

In 1887, the United States Congress established the first national regulatory agency, the Interstate Commerce Commission, to regulate railroad freight and passenger fares [Breyer, 1982].

In its broadest sense, regulation can be seen as all government instruments or interventions to protect public benefits, which include energy efficiency and environmental protection, promoting greater access to energy by the urban and rural poor, public -interest research and development activities, etc. The term regulation also refers to a set of rules and mechanisms established by a regulatory agency to limit monopoly power of infrastructure utilities.

- **Regulatory Functions**

There are several aspects that require regulation, including regulating competition (e. g. fairness for all parties involved in electricity industry), economic regulation (effective pricing), and technical regulation (ensuring standards) and regulation of public benefits. The functions and responsibilities of a commission include:
- General regulatory rulemaking;
- Rate setting, often called tariff setting;
- Utility system resource planning;
- Assuring system reliability;
- Environmental impact of resource utilization;
- Conservation and efficient use of utility and societal resources;
- Universal services;
- Consumer Protection; and,
- Nuclear safety

These functions and responsibilities are often at odds with one another. As a result, the commission is often faced with the task of balancing these competing objectives to develop a workable framework of regulation.

Traditionally, such regulatory functions are carried out by the government.

- **Regulation in the New Structure of Electricity Industry**

In the 1980s, deregulation was widely adopted, often as part of structural adjustment programs, with the aim of reducing the "regulatory burden" on the market economy and supporting the process of privatization and market liberalization. This situation raises a question: *Is there any contradiction between strengthening regulatory and market development? And what is the role of regulation in the new structure of the electricity industry?*

For two main reasons, the electricity industry requires regulation. First, to ensure fair treatment of customers who lack the protection that comes with competition. Second, to ensure that competitors have fair access to bottleneck network facilities controlled by incumbent service providers. If incumbents do not face regulatory constraints, they can use these facilities to control-or destroy-their rivals.

Opening energy markets to encourage competition and greater efficiency is a worthy goal, but it must complement other fundamental interests. It has become clear through recent experience that market forces do not always achieve public policy objectives of energy security and environmental protection. It was also recognized by the World Bank that 'less regulation is not necessarily better regulation' [World Bank, 2002].

This does not mean governments need to return to heavy-handed regulation. On the contrary, it is possible for governments to achieve many of the objectives in a manner that is consistent with competitive market reforms. The rehabilitation of regulation policy as a legitimate part of the development policy has focused attention on issues of regulatory policy design and implementation and the critical importance of regulatory institutional capacity in ensuring 'good' regulatory outcomes.

One of the most important tasks of regulators in a competitive environment is to establish an incentive regime for encouraging long-term investment. In energy industries, the economic characteristics of investments are long asset lives, the presence of sunk costs, and the economies of scale. The funding of long-term investment projects has no simple solutions. These depend on the restructuring process of the sector and on the return to private investment. Moreover, the mere entry of new actors in the power sector does not automatically assure an increase in the level of investment. As pointed out before, the restructuring process requires substantial changes in market structure, in regulatory mechanisms and in the management of energy companies, both the remaining state-owned and the private companies. The challenge of the new regulatory frameworks is to guarantee the achievement of four objectives: (i) increasing investments, (ii) improving the microeconomic sector performance, (iii) seeking the reduction of prices, and (iv) avoiding market power of the principal players.

However, those objectives can become contradictory due especially to incomplete contracts. For these reasons, as observed by [Sidak and Spulber, 1998], relationships between the regulator and the regulated remain in a permanent bargaining status.

Is it possible to combine regulatory and competition mechanisms in electricity industries?

As mentioned above, it is important to emphasize that the introduction of competition and the unbundling of electricity industries requires regulatory mechanisms allowing the access of competitors to the incumbent companies' networks. Even after the suppression of institutional barriers, it is difficult to guarantee competition with the presence of a company with strong market power. Under these circumstances, the presence of the regulator will be indispensable not only to deal with traditional regulatory aspects (price, quality, entry), but also with new key issues such as anti-competitive practices.

Nevertheless, as mentioned by [Mayer, 2001], there is a crucial conflict between control of monopoly abuse and providing incentives to invest and innovate, which existing regulatory frameworks have failed to resolve.

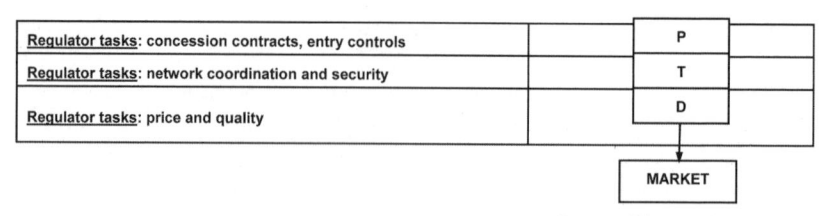

76   Part I

Regulator tasks: concession contracts, entry controls		P
Regulator tasks: network coordination and security		T
Regulator tasks: price and quality		D

MARKET

Source: [Pinto Junior, 2002]
Figure 4.1   Regulation in vertically integrated monopoly

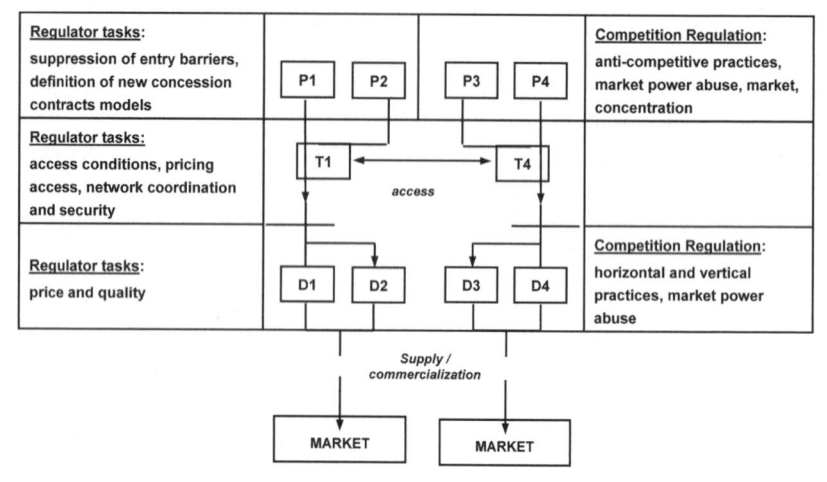

Regulator tasks: suppression of entry barriers, definition of new concession contracts models	P1 P2 P3 P4	Competition Regulation: anti-competitive practices, market power abuse, market, concentration
Regulator tasks: access conditions, pricing access, network coordination and security	T1 ← → T4 *access*	
Regulator tasks: price and quality	D1 D2 D3 D4	Competition Regulation: horizontal and vertical practices, market power abuse

Supply / commercialization

MARKET   MARKET

Source: [Pinto Junior, 2002]
Figure 4. 2   Regulation in network-opening model

Two main problems can be pointed to understand the regulatory challenges for creating a competitive environment: cross-subsidies and contracting problems. Both are related to investment characteristics, such as asset specificity, sunk costs, and the degree of vertical integration. Regulators have struggled to identify and then to eliminate these problems among the different activities of vertically integrated companies. Consequently, it is important to combine traditional regulatory mechanisms with anti-trust responses.

The Figures 4. 1 and 4. 2 present two types of models of industrial organization: the first is representative of a vertically integrated monopoly; the second illustrates an example of competitive market structures in the electricity industry.

As one can observe, the opening of networks requires the control of anticompetitive practices and new regulatory tasks.

CHAPTER 4   77

4.2 Regulatory Governance in Liberalized Environment

Regulatory governance involves the creation of a transparent and predictable regulatory system. An appropriate governance arrangement for reformed electricity markets would involve governments establishing policy objectives and allowing regulators, as independent institutions, to administer the rules and regulations to achieve those policy objectives, subject to appropriate accountability.

Credible, stable regulation is required to achieve the benefits for privatizing and liberalizing the electricity industry. [Kessides, 2004] lists the institutional prerequisites for effective regulation, including: (i) Separation of powers, especially between the executive and the judiciary, (ii) Well-functioning, credible political and economic institutions—and an independent judiciary, (iii) A legal system that safeguards private property from state or regulatory seizure without fair compensation and relies on judicial review to protect against regulatory abuse of basic principles of fairness, (iv) Norms and laws—supported by institutions—that delegate authority to a bureaucracy and enable it to act relatively independently, (v) Strong contract laws and mechanisms for resolving contract disputes, (vi) Sound administrative procedures that provide broad access to the regulatory process and make it transparent, and (vii) Sufficient professional staff trained in relevant economic, accounting, and legal principles.

The creation of regulatory bodies plays a crucial role in the process of building effective regulation. In the following sections, we will discuss about the institutional requirements of a regulatory agency, its structure as well as the role of this agency's commitment.

- **Key Characteristics of an Effective Regulatory Agency**

The structure, scope, and powers of a regulatory commission are key to successful restructuring of the industry. The key characteristics of a good regulatory commission include: coherence, independence, accountability, transparency, predictability, and capacity [Kessides, 2004]. These crucial features can not be developed overnight—achieving them will take time.

— Coherence

Regulations for each infrastructure sector should be complementary and mutually supportive. The laws guiding regulation must be in agreement, and regulations must be consistent over time. New rules should take into account previous ones, with amendments made to eliminate significant inconsistencies. Regulatory coherence requires that national regulators, ministries, and provincial and municipal regulators have clearly defined responsibilities, ensuring that the same agency always makes decisions involving specific aspects of regulation. Regulators should be required to publish statements explaining their goals and reasons for decisions on entry, pricing, and other industry behavior subject to oversight. Doing so forces the government to think through its long-term policy objectives and regulatory principles. It also enables firms and consumers to predict how they will be treated in the future, enhancing accountability.

– Independence

Effective regulation requires that regulators be largely free from political influence, especially on a day-to-day or decision-by decision basis. Agencies must be objective, apolitical enforcers of policies set forth in controlling statutes.

Still, complete independence for regulators is not possible or even desirable. The executive branch should be able to ensure that the regulators it appoints are sympathetic to its reforms and to administration policies. However, if regulators are not insulated from political intervention, the regulatory process may become politicized, decisions may be discredited, and policies may lack continuity. Compromise is needed to ensure that regulators are both independent and responsive to an elected administration's policy goals. Safeguards that can help achieve such compromise include: (i) Giving the regulator statutory authority, free of ministerial control, (ii) Setting clear professional criteria for appointing regulators, (iii) Requiring that both the executive and legislative branches participate in appointments, (iv) Appointing regulators for fixed periods and prohibiting their removal without clearly defined cause, (v) Staggering the terms of an agency's board members so that they can be replaced only gradually by successive administrations, (vi) Funding agency operations with user fees or levies on service providers, to insulate agencies from political interference through the budget process, (vii) Exempting agencies from civil service salary caps, to enable them to attract and retain well-qualified staff, (viii) Prohibiting the executive branch from overturning an agency's decisions except through new legislation or judicial appeals of existing laws.

– Accountability

A regulator's independence should be reconciled with its accountability. Allowing a regulator to set prices and quality standards gives it enormous power to redistribute rents. Without an accompanying obligation to respect previous decisions and the legal rights of all parties, a regulator has considerable leeway for opportunism. Thus, checks and balances are required to ensure that regulators do not become capricious, corrupt, or grossly inefficient. Citizens and firms should be able to find out who makes regulatory decisions and what guides them, and to voice their concerns. In addition, affected parties should be able to easily and quickly obtain redress if a regulator acts arbitrarily or incompetently. It is difficult to strike a proper balance between independence and accountability, but certain measures can help, for example, writing statutes that specify the rights and responsibilities of each regulatory agency, requiring regulators to produce annual reports on their activities which reviewed by independent auditors or legislative committees or allowing stakeholders to submit their views on requiring regulators, etc.

– Transparency

Citizens need transparent information to evaluate government performance. Thus, all regulatory rules and agreements, and the principles guiding them and future regulation, should be a matter of public record. This record must be accessible to all market participants, not just service providers, to inform the public of long-term business plans. Transparency helps induce investment by incumbents and new entrants and helps avoid costly, time-consuming regulatory disputes.

Transparency also protects against corrupt regulation. In addition, it makes citizens, especially those adversely affected by regulatory decisions, less likely to believe that decisions are corrupt. When regulatory decisions and principles are clearly written, the reasons for them are apparent. Moreover, corrupt decisions are easier to detect and harder to defend.

– Predictability

Behavior of regulatory agencies is predictable if they follow the rule of law, particularly with respect for precedent and the principle of stare decisis. Respect for precedent means that regulators reverse past decisions only if they have created significant problems. Stare decisis require that cases with the same underlying facts be decided the same way every time. Thus, regulatory decisions must be based on durable rules and procedures that will apply in future cases unless new information is obtained. Otherwise, market participants will lack confidence in regulation, undermining the size, scope, and quality of industry and related investments.

– Capacity

A regulatory agency's responsibilities should match its financial and human resources. Available financing reflects government willingness to support independent regulatory institutions. With the possible exception of very small and poor countries, lack of financial capacity is unlikely to be a genuine constraint, although failure to provide adequate financing for regulation is a more common problem.

Inadequate expertise is a much bigger challenge in many developing and transition economies. Well-developed economic, accounting, engineering, and legal skills are required for regulatory functions such as monitoring industry performance, analyzing cost data, dealing with information asymmetries, and analyzing the behavior of regulated firms. To overcome these deficiencies, regulatory agencies need to be given complete freedom to hire specialized staff. This may require exempting such agencies from civil service salaries and recruitment rules.

- **The Structure of Regulatory Institutions**

Several decisions must be made about the organization of regulatory governance. They include (i) centralized or decentralized and (ii) single or multi-overseeing sectors. Designing regulation involves trade-offs, each structure has advantages and disadvantages.

Centralized or Decentralized?

[Kessides, 2004] lists following strong points of decentralized design: allows local conditions and preferences to shape regulation, moves regulators to be closer to services, allows them to gather better information on users, and promotes competition among sub-national regulators to attract private investment. The most significant drawback of decentralized regulation is exposure to a greater risk of being captured by industry interests or local politicians.

Arguments for a centralized organization are: it makes the best use of scarce expertise and minimizes the fixed costs of regulation, reduces the risks of a regulatory race, and supports an efficient scale or scope of operation.

Analysis of the costs and benefits of either approach must reflect the country's institutional structure and the industry's technological features. However, several general conclusions can be drawn (i) small or poor countries may have only one effective tier of government, (ii) electricity networks are normally designed to operate nationally, thus in most countries transmission and several aspects of generation may be best regulated at the national level, while in large countries it may be more feasible to regulate distribution at the sub-national level.

Single or Multi-industry Agencies?

Establishing separate agencies has such advantages as (i) the unique economic and technological characteristics of each infrastructural industry can be recognized and regulators are able to develop deep, industry-specific expertise, (ii) mitigates the risk of institutional failure, and (iii) encourages innovative responses to regulatory challenges.

On the other hand, there are also benefits to using one regulator for several industries (i) it is possible to share fixed costs, scarce talent, and other resources, (ii) builds expertise in cross-cutting regulatory issues, (iii) reduces its dependence on any one industry and helps protect against capture. In addition, a multi-industry agency may be better able to resist political interference because its broader constituency gives it greater independence from sector ministers, (iv) enables issues emerging with these multi-utilities to be addressed, and (v) helps guard against distortions created by inconsistent regulation of directly competing utilities, such as electricity and gas, or for investment capital [Kessides, 2004].

So, deciding which structure for the regulatory agency involves numerous considerations — and no single approach will suit all circumstances. However, there are several rules. First, in economies with a small number of consumers and limited human and financial resources, the argument for merging regulatory responsibilities is strong. However, in large economies, the benefits of a multi-industry agency may be outweighed by concerns about insufficient industry focus and diseconomies of scale. Second, regulators in developing and transition economies multi-industry agencies with narrower responsibilities raise fewer concerns about inadequate industry focus or potential diseconomies of scale. Third, if market substitution occurs between the output of regulated industries, especially between electricity and gas-consolidation, regulatory responsibility may be stronger for certain industries. Fourth, scarcity of expertise and vulnerability to political and industry capture in developing and transitional economies also strengthens arguments for multi-industry regulation.

Commitment and Credibility

Regulation must be clear and credible, ensuring investors that regulators are committed to fair, consistent, and sustainable policies and procedures. According to [Kessides, 2004], commitment can be achieved through administrative procedures, concession contracts, and substantive economic restraints. However, a balance should be kept between commitment and flexibility. To encourage efficient performance, a regulatory system must be able to adapt its mandate and rules in response to new challenges, circumstances, information, and experiences. Such flexibility is especially imperative in sectors experiencing rapid technological and market changes.

The second dimension in regulation involves the regulatory functions and responsibilities. Not endeavoring to cover every function, the following section discusses several of the most important functions which are greatly influenced by the liberalized environment of the electricity industry.

4.3 Mechanisms to Regulate Prices

[Joskow, 1998 b] mentions five basic goals of price design: (i) Rent extraction or setting rates that strike a socially acceptable compromise between the interests of investors and consumers, (ii) Supply-side efficiency, providing signals and incentives for suppliers and investors to increase efficiency, (iii) Demand-side efficiency, providing signals and incentives for efficient consumption of regulated utility services, (iv) Revenue adequacy, allowing regulated firms to earn sufficient revenue to attract needed capital, and (v) Fairness, ensuring that prices are just and reasonable, and contribute to universal service goals without creating significant aversion to investment. Two alternative mechanisms for regulating prices are cost-plus and price caps. This section summarizes discussions of these approaches and the possible implications for pricing policies.

- Pure cost-plus. The regulated firm simply submits a bill for its operating expenses and capital cost (depreciation plus an after-tax return on its investment that equals or exceeds its cost of capital), and the regulator passes the costs in the prices charged on to the consumers. Prices are continuously tied to the accounting costs.
- Rate of return. The regulated firm's capital and operating costs are evaluated using a specific accounting system. Prices are set to cover these audited costs, plus a reasonable return on investment. Once the base prices are set, they are not adjusted automatically for changes in costs over time, they remain fixed until subsequent regulatory review.

Source: [Kessides, 2004]

Figure 4. 3 Cost-plus mechanism

- **Cost-Plus Regulation**

Until recently, cost-plus regulation (Figure 4. 3) dominated electricity utilities in many countries world-wide. Policymakers were attracted to this mode of utility behavior control because it: (i) seemed to be fair to both the regulated firm and its customers. It permits the firm to earn sufficient revenues, including a fair return on its investment, by passing its costs on to consumers through the prices charged. It is also designed to protect consumers from monopolistic pricing distortions; (ii) was likely to attract investment to the regulated sector because investors knew they would recover their operating and investment costs, perhaps with a return that exceeds the capital costs. These systems shift market-specific risks to consumers, satisfying the goal of revenue adequacy, (iii) kept utility services reasonably affordable by holding revenues close to costs.

However, cost-plus systems have drawbacks: (i) The firm has an incentive to engage in accounting contrivances and to pad its costs to convince the regulator to approve higher prices. Unless the regulator has a well-developed cost accounting system to audit the firm's costs, the firm can misrepresent them. If that happens, the regulator will set prices too high, frustrating its rent extraction goal, (ii) a pure cost-plus regime distorts the firm's incentives to minimize its because it has no incentive to do so (iii) the firm has incentives to expand its rate base by adopting excessively capital-intensive technology. All three drawbacks undermine supply-side efficiency. Regulators generally try to remedy these perverse incentives through regulatory lag, which imposes penalties for inefficiency and incorrect guesses, and rewards efficiency by permitting the firm to keep the profits it earns from cutting costs and improving performance, and a banded rate of return, allowing price adjustments only when returns fall outside that range, and efficiency audits/reviews.

- **Price Cap (Incentive) Regulation**

Given the weak incentives for productive efficiency under cost-plus regulation, many types of incentive-based regulation has been developed. Under these mechanisms, the regulator delegates certain performance related decisions to the firm, and the firm's profits depend on its performance as measured by the regulator.

Price caps are the main incentive mechanism [Baron, 1989], [Laffont and Tirole, 2000]. Their key purpose is to control prices, not the earnings, of the regulated firm (Figure 4. 4). Thus, this form of regulation does not make explicit use of accounting data. Under price caps the regulator: (i) defines a set of prices (or a weighted average of prices for different services) that the firm will be allowed to charge. The firm is free to price at or below these ceilings. (Price floors may also be set to prevent predatory behavior.), (ii) estimates the ability of firms in the regulated industry to limit cost increases and compares that with firms in other industries. The estimated differential in productivity between the regulated industry and the rest of the economy is called the X factor, and (iii) specifies a formula for adjusting prices (or the weighted average price) over time to reflect input inflation (easily observable changes in costs beyond the firm's control) and the expected rate of productivity improvement (X factor). Thus, price cap regulation severs the link between the firm's authorized prices and its realized costs.

In a typical price cap plan, related services and products are grouped into categories often referred to as baskets. Alternatively, all services may be bundled in a single basket. An overall price cap is set for each basket. Pure price cap regulation operates much like a fixed-price contract under which the firm is the residual

---

Price Constraints Imposed by Price Cap Plans

For a basket of regurated good or services, the typical price cap plan limits the weighted average (percentage) price increases to not exceed the difference between some measure of the general inflation rate and the specified productivity offset:

$$\sum_{i=1}^{n} w_i^t \left[\frac{p_i^t - p_i^{t-1}}{p_i^{t-1}} \right] \leq RPI^{t-1} - X$$

Where

$$w_i^t = \frac{p_i^{t-1} q_i^{t-1}}{\sum_{i=1}^{n} p_i^{t-1} q_i^{t-1}} \leq RPI^{t-1} - X$$

$n =$ the number of regulated goods or services
$t =$ year ($t = 0$ at the start of the price cap plan)
$p_i^{t-1} =$ the unit price of goods or services i in year t
$q_i^{t-1} =$ the number of units of goods or services i sold in year t
$RPI^{t-1} =$ the inflation rate in year $t - 1$ (most recent 12-month period)
$X =$ the specified productivity offset
$w_i^{t-1} =$ the proportion of the firm's total regulated revenue in period $t - 1$ derived from product i

---

Source: [Sappington and Weisman, 1996]

Figure 4. 4 Price cap mechanism

claimant for its entire cost savings [Laffont and Tirole, 1993]. The firm has strong incentives to pursue cost-reducing innovation, use the lowest-cost technology, operate with no waste, and report its costs truthfully [Weisman, 2002]. At the same time, consumers are protected because prices do not vary with the firm's reported costs.

Price cap regulation, however, also contains several problems, including: (i) Quality of service may be depredated due to the companies' cutting costs efforts. To avoid that, regulators must set standards for quality, monitor utility performance, and penalize poor quality, (ii) Concerns over excessive or low profits cause a convergence with ROR regulation in the time between rate cases increases. Taking this point into consideration, the regulator can increase the rate case frequency, making incentive regulation more closely resemble traditional (ROR) regulation, (iii) Shift of costs toward the most captive customers.

The disadvantages of both cost-plus and price cap mechanism and the methods to correct such weak points push the two mechanisms closer to each other, resulting in the use of a hybrid mechanism, as discussed in the following paragraphs.

- **Hybrid Mechanism**

Pure cost-plus and pure price cap mechanisms represent opposite regulatory extremes. Practical considerations and multiple regulatory objectives imply that neither is likely to be the most feasible or desirable regulatory scheme. Each trades incentives for rent extraction (with weight placed on consumer surplus) against those for supply-side efficiency (with weight placed on producer surplus), with cost-plus regulation focused on rent extraction and price caps focused on supply efficiency.

The optimal balance between these two goals depends on the cost of public funds. It is best to place the entire weight on supply efficiency only when the marginal social cost of taxation is zero, a condition that will never be met [Ergas and Small, 2001]. Most practical regulatory regimes are hybrid schemes that involve tradeoffs between supply-side efficiency, capital attraction, rent extraction, and demand-side efficiency (Figure 4. 5). These mechanisms aim to share cost benefits and burdens between the regulated firm and its customers.

Choosing between Rate of Return and Price Cap Regulation

From a public policy perspective, the choice between rate of return and price cap regulation is not nearly as clear-cut as once thought. Table 4. 1 illustrates the different features of rate of return and price cap regulation. However, there are several common agreements: (i) Cost-plus regulation is better for sectors with large investment requirements and countries with weak commitment capacity, (ii) Price cap regulation is better for industries with excess capacity supported by institutions with strong commitment powers, (iii) Price caps might be preferable in countries

Hybrid Regulatory Mechanisms are the most practical regulator systems which involve aspects of cost-plus and price cap mechanisms. Examples of hybrid mechanisms include:
- **Banded rate of return.** A range (band) of earning specified, and prices are set to generate earnings that fall within the range. Prices are not revised as long as earnings fall within the band.
- **Sliding scale profit-or-cost-sharing.** Prices are automatically adjusted if the firm's rate of return differs from a preset target. But without improving efficiency the adjustment is only partial. Thus, the firm and its customers share both risks and rewards. Alternatively, the rate of return can vary within a preset range without causing price adjustments. If the return falls outside the range, it can trigger profit (or cost) sharing.
- **Institutionalized regulatory lag.** Price reviews do not occur for a specified period, usually two to five years. During that time all investigations into the firm's earnings are suspended. Whereas the time between price reviews can vary significantly under traditional rates of return, under regulation it is known and fixed under institutionalized regulatory lag.

Source: [Sappington and Weisman, 1996]
Figure 4. 5 Hybrid regulatory mechanisms

Table 4. 1 Features of rate of return and price cap regulation

Feature	Rate of Return	Price Cap
Sensitivity of prices to cost	High	Low
Firm's flexibility to adjust prices	Low	High
Regulatory discretion	No	Yes
Regulatory lag	Short	Long

Source: [Armstrong and Sappington, 2005]

with poor accounting and auditing, scarce expertise, and high inflation.

Therefore, in practice, the optimal choice among regulatory mechanisms depends on a variety of factors: the quality of accounting and auditing systems, the availability of economic and technical expertise, the efficacy of the tax system, the sector's investment requirements, the government's commitment, institutional checks and balances, and overall macroeconomic stability. Some of these will change over time. Thus, different stages of national development have implications for the choice of regulatory regimes. During the first stage of regulation, with scarce expertise and poor auditing and monitoring, price caps with provisions for adjustment are likely the best choice. Initial prices might need to be high to attract capital and ensure firm viability, however increased investment and supply-side efficiency should compensate for them. This stage should be used to improve regulatory capacity and accounting and auditing systems.

Once these conditions have been met, the second stage can promote cost-plus mechanisms that facilitate large-scale, sustainable investment, especially if government credibility improves at a slower pace, and achieve the rent extraction goal in the face of continuing high costs of public funds. As development continues, with infrastructure system expansion, and nearly complete and enhanced commitment powers, the optimal solution is to move to hybrid regulation [Laffont, 1996]. Once infrastructure systems are developed, firms can do better by being less efficient (padding their costs), hence, it need for more powerful incentive schemes.

- **Grid Access and Transmission Pricing Regulation**
  (Refer Chapter 2, Section 2. 4 for a more detailed discussion.)

- **End-user Tariff Regulation**

Tariffs charged to end-users are traditionally regulated because utilities were traditionally monopolies with substantial protection from competitive entrants and there was no direct access by end-users to the electricity market (i.e., not retail supply competition). In a liberalized market, theoretically, tariff regulation is not required. However, experience shows that a time of non-regulated tariffs is still far away. Tariff regulation should protect customers that are unable to change their suppliers, such as households and small customers. This section summarizes several main points in determining tariffs for each customer class.

Ideally, economists would recommend setting tariffs for each customer at the marginal cost of providing electricity. If the marginal cost of providing an industrial customer is less than the marginal cost of delivering electricity to a household, then the household should pay a higher tariff. Unfortunately, different tariffs exist for different customers, i.e. price discrimination through economic efficiency can be viewed as inequitable. Also, charging different prices can involve cross-subsidization, i.e., charging one customer class a higher tariff so that another customer class can pay a lower tariff.

Price discrimination allows the separation of customers, negotiation of a maximum price for each customer class based on their willingness to pay, and thus captures all consumer surplus as profit. This is known as first-degree price discrimination. Second-degree price discrimination facilitates the incursion of different tariffs for different amounts of electricity, for example declining-block pricing or more popularly, increasing-block pricing applied for household customers.

Marginal Cost Pricing, Multipart Tariffs, and Peak-load Pricing
In practice, several pricing methods are utilized in the electricity industry. Average cost pricing, Ramsey pricing, two-part tariffs, peak-load pricing or time of use tariffs, and seasonal tariffs should be included.

Under the Ramsey pricing rule, customer classes are charged different tariffs based on demand elasticity. Customers with few purchasing electricity alternatives, such as residential customers, would be charged a higher tariff than customers who can purchase fuel and generate their own electricity. There might be an incentive for low-priced customers to sell electricity to high-priced customers. To avoid that situation, value-of-service pricing charges can be used.

Two-part tariffs allocate fixed costs among customers and use marginal-cost pricing to allocate variable costs. Although this approach has the benefit of signaling the correct price to the customer, determining how to allocate fixed costs is difficult. Two-part tariffs can be extended to a multi-part tariff in electricity pricing in order to reflect the different components of marginal cost in the electricity supply chain.

Peak-load pricing means applying different prices for on-and-off-peak electricity. This approach can solve the problem of underinvestment in base-load capacity and overinvestment in peak load capacity. However, implementing peak-load pricing based on time-of-day consumption usually requires new metering equipment. Two variations of this pricing are the time-of-day tariff and the time-of-year tariff, so-called seasonal pricing.

4.4 Regulation of Entry

A fundamental choice confronting all newly established regulatory commissions is whether to rely on licensing or generic rules as the primary instrument of regulatory control. A license-based system establishes most of the conditions of operation in the individual license documents. A rule-based system promulgates most conditions in rules of general applicability, supplemented by decisions in specific "cases" [USAID, 2000]. In theory, a license-based system has attributes of a contract between the government and the utility, with the terms set forth clearly at the outset. Meanwhile a rule-based system, offering the advantage of greater flexibility to meet changing conditions, depends on societal concepts of the due process of law. In fact, both flexibility and stability are essential attributes of effective utility regulation, so each system must find mechanisms to assure the apparent advantages of the other.

For licenses to be effective instruments of regulatory control, [USAID, 2000] lists the following conditions: (i) The license duration should be limited, especially in uncertain conditions, to a few years. Even under conditions of relative stability it should not exceed twenty years; (ii) The regulator should be able to terminate the license for noncompliance with license conditions following appropriate notice, an opportunity for correction and a public hearing, (iii) Transfer of the license without regulatory approval should be prohibited, (iv) The licensee should be required to supply a completely audited financial statement annually and the regulator should have complete access to the licensees books and records at any time, as well as the power to compel the prompt furnishing of all necessary information, (v) The property of the licensee should be subject to inspection by the regulator at any time, (vi) The regulator should have the power to resolve any disputes arising between the licensee and its customers, and perhaps also between the licensee and its suppliers of fuel and electricity, etc.

4.5 Regulation and the Mechanisms to Address Social and Environmental Issues

Governments must respond to numerous obligations. In many economies, especially developing ones, the electricity sector has been used as an instrument of social policy. This has taken various forms, ranging from subsidized tariffs for certain groups of consumers, to state-owned utilities serving as employment providers. This section analyses the changes that the restructuring process will likely bring about in this area as well as summarizes some of the regulatory mechanisms to deal with such issues.

In the traditional model, the state-owned vertically-integrated and monopolistic utility has functioned under equally diverse objectives, namely economic, financial, political, and social. Given this situation, it was natural for governments to incorporate social policy into energy policy and the operation of utilities. Traditional programs and practices to support public purposes include: (i) Universal service policies, including service to low-income customers and rural areas; (ii) Investments and other program support for energy efficiency in generation, delivery, and end-use services; (iii) Investments in and development of renewable, sustainable, and less-polluting generation resources; (iv) Support for research and development on electricity generation, delivery, use and impact; (v) Consumer protection and consumer education programs.

In this context of liberalization, electricity companies operate keeping the goals of economic efficiency in mind, but may reduce their attention to social benefits. This can cause a negative impact on both the emission of gases, energy efficiency, and energy research and development [APERC, 2000]. Electricity companies can continue to support public interest programs only in cases where it does not distort electric prices. Therefore, in a liberalized environment, a new set of mechanisms to address the social and environment issues should be developed.

Regulation plays a crucial role, providing the framework and necessary incentives for all actors involved in new industry structure. Effective regulation does not replace the function of the government; rather it provides the setting for it to pursue the government's policies. Across the globe, electric utilities, governments, and utility regulators have explored numerous mechanisms to deliver public interest programs in connection with electric service. Several mechanisms, which were developed in the context of vertically-integrated electric systems, such as applying Integrated Resource Planning, still have the greatest applicability in any type of single buyer industry structure. Others have been developed in connection with emerging retail competitive models. Following paragraphs discuss five methods for delivering public interest programs.

Energy Efficiency Programs
- Comprehensive energy efficiency and load management programs have been developed and widely implemented as part of utilities' Integrated Resource Plans;
- Jurisdiction-wide programs have been funded through system "uplift" charges, and administered through public efficiency agencies;
- Efficiency measures have been promoted through voluntary programs (e.g., the "Green Lights" and "Energy Star" programs) and mandatory building and appliance efficiency programs;
- Some jurisdictions have simple mandatory spending guidelines.

Renewable Electricity Generation
- Mandatory purchasing requirements at avoided cost;
- Support for renewable energy research and development through research consortia;
- Creation of a renewable energy fund to support new renewable energy production in response to a public bid offering;
- Establishment of a Renewable energy Portfolio Standard[8] applicable to all generators or retail electric sellers in a competitive electric market.

Research and Development
- Pooled funding, either voluntarily or through a mandate (e.g., a wires charge), to support public-purpose research and development;
- Tax credits for qualified R & D;
- Public expenditures through government agencies, universities, and grants to utilities and equipment manufacturers.

Emission Gases Mitigation
- Command and control (non-market interventions). Command and control systems are defined by sets of compulsory rules (quotas and standards) defining requirements on the level of emissions or directly governing energy use. Such systems are completed with a monitoring (control) component.
- Forced retirement of plant. This is a command and control approach to environmental policy, which has some associated costs and should not be

---

8 "Renewable portfolio standards," regulates a minimum fraction of electricity would be generated using non-hydropower renewable fuels. This method is popular in the US.

necessary in a free and competitive energy market.
- Taxes and fees. Environmental taxes are "priced-based" policy instruments. There are two kinds of tax instruments used for environmental objectives. Energy tax levies on total energy consumption. The less energy used the less emission of gases. That can be considered an indirect environmental tax. A direct tax includes a sulfur tax, nitrogen tax, or carbon tax. Taxes directly levied on the sulphur, nitrogen, or carbon content of fuel.
- Emission trading. Tradable permit or emissions trading is considered a "quantity-based" environmental policy instrument. An emissions trading scheme establishes a market price, and hence an opportunity cost for emissions because of the scarcity of a commodity. In this case the quantity of available "permits" is less than is necessary to accommodate business-as-usual emissions.
- Environment tax and emission trading also lead to international emissions allowances trading, "joint implementation", and the Clean Development Mechanism, under which "clean development" investments in developing countries "earn" allowances.

CHAPTER 5
WORLDWIDE ANALYSIS OF ELECTRICITY INDUSTRY REFORM
–THE DEVELOPMENT OVERVIEW AND THE NEW TREND OF REFORM

In this part, we will carry out an international comparison of electricity industry reform on a worldwide basis. At first, an overview on the development of reform will be discussed, including the progress of reform, successes and failures. Valuable lessons should be drawn from this review, as well as the new trend of electricity industry reform will be stated. Then, electricity industry reform will be reviewed for the viewpoint of several influential stakeholders who have had an important role in the reform process. The viewpoints are taken from various stakeholders in international multilateral institutions, multinational electricity companies, and labor unions.

5. 1   Electricity Industry Reform Overview

- **Progress on Reform and Private Participation**

The 1980's and 1990's are known as the "liberalization tide" period. EI reforms spread worldwide from developed countries, such as the United Kingdom, New Zealand, etc. to developing countries including those in Latin America and the Asia-Pacific region. Reform needs and pressures in developing countries are different from those in developed countries. In developed countries, there is a low operating efficiency is caused by the lack of competition due to monopolization of both the public and private power sectors. Meanwhile, developing countries must deal with power shortages and financial crises. To meet the growing demand of electricity for economic development, developing countries require huge investment capital, which is impossible to mobilize due to poorly developed financial mechanisms.

Although these programs have generally sought increased private participation, reform strategies and successes in attracting private investment have varied considerably across countries and regions. While electricity restructuring is spreading, many countries have taken few or no steps toward reform.

By 1998, 15 countries had substantially liberalized their electricity systems, and 55 had some liberalization under way or planned, however many of these reformers

Table 5. 1 Electricity reform by regions (1998) (% of countries where the reform has occurred)

Reform	East Asia and Pacific	Europe and Central Asia	Latin America and Caribbean	Middle East and North Africa	South Asia	Sub-Saharan Africa
State utility corporative	44	63	61	25	40	31
Enabling legislation passed	33	41	78	13	40	15
Independent regulator at work	11	41	83	0	40	8
Private investment	78	33	83	13	100	19
State utility restructured	44	52	72	38	40	8
Generation privatized	22	37	39	13	40	4
Distribution privatized	11	30	44	13	20	4
All reforms taken	41	45	71	17	50	15
Reform score (scale of 1–6)	2.44	2.70	4.28	1.00	3.00	0.88

Source: [Bacon and Besant-Jones, 2001]

were mature industrial countries. Of the 81 countries that had not taken any steps toward reform, many were developing or transitional economies [Bacon and Besant-Jones, 2001].

Even in Latin America, the leading region for private participation in electricity, reforms are far from complete. In 2001, the state still controlled significant portions of electricity activity in many Latin American countries [Millan, Lora and Micco, 2001].

Table 5. 1 presents a regional scorecard for electricity reform as of 1998, based on how many of the following steps had been taken by each country in a region: (i) The state-owned electric utility has been commercialized and corporatized (ii) Parliament has passed an energy law permitting partial or complete sector unbundling or privatization, (iii) A regulatory body, separate from the utility and the ministry, has started work, (iv) The private sector has invested in Greenfield sites that are being built or are in operation, (v) The state utility has been restructured or unbundled, and (vi) The state utility has been privatized, whether through outright sale, voucher privatization, or a joint venture.

Out of a maximum reform score of 6.00 (where all reform steps were taken), the average score was 4.28 for Latin America and the Caribbean, 3.00 for South Asia, 2.70 for Central and Eastern Europe and Central Asia, 2.44 for East Asia and the Pacific, 1.00 for the Middle East and North Africa, and 0.88 for Sub-Saharan Africa.

Privatization

The level of private sector interest has been extremely mixed across countries and regions, reflecting differences in reform efforts. Between 1990 and 2001 developing and transitional economies received approximately $207 billion in private investment in power projects. Over 43 percent went to Latin America and the Caribbean and about 33 percent to East Asia and the Pacific, while 2 percent went to the Middle East and North Africa and approximately 1.5 percent to Sub-Saharan Africa.

Private sector participation also varied considerably throughout the 1990's. Until 1997, electricity reforms and anticipated economic growth spurred enormous private investment in the sector. However, investment then plummeted, reflecting financial problems in many Asian, Eastern European, and Latin American countries. It is difficult to predict whether this reversal will persist [Jamasb, 2002].

Unbundling and Competition

Technological developments, particularly in electricity, have made it possible to unbundle infrastructure services and, in some cases, even to supply them competitively. The unbundling of electric utility generation, transmission and distribution systems allows electric and gas companies to form alliances to develop economical, integrated projects, a process which has led to an unprecedented number of mergers and acquisitions among natural gas and electric companies in recent years. Natural gas and electric utilities are merging to cut costs, to expand their service territories and to offer new multi-fuel services, as in the case of the planned merging of operations between Baltimore Gas and Electric (BG&E), which provides gas and electric service to the city of Baltimore and ten surrounding Maryland counties, and Potomac Electric Power Corporation (PEPCO), which provides electric service to Washington D.C., and two surrounding counties in Maryland. Within this unbundling process, electric generation is often divided among several companies, while high-voltage transmission companies are allowed to remain natural monopolies, whether public or private.

The earliest and most popular form to introduce competitive market into electricity industry was allowing new entrants in generation. This happened from 1990–1997 in most countries from Asia to Central America, the Middle East, and Africa. After 1995, the wholesale market, including power pool or bilateral trading, got the attention of many countries in Europe and North America and in addition to Australia and New Zealand. One to two years later, retail competition came on the scene and was characterized by phased expansion.

Regulation

[Kessides, 2004] emphasizes the importance of well-designed regulatory systems in the process of privatization, restructuring, and the new role of the State in a changing world. "Regulation can help protect consumers, workers and the environment. It can foster competition and innovation while constraining the use of monopoly power. Making the best use of the new options emerging for private provision of infrastructure and social services will also rely, often, on a good regulatory framework."

Privatization alters control, most obviously in the form of reduced state control and greater control by private investors. The privatization of natural monopolies requires governments to take special care to ensure that the monopoly power of the firm is restricted and cannot be abused by the private sector. This may require preparation of the sector for deregulation, and the establishment of regulatory policy and machinery to address issues such as opening the market to new entrants, interconnection charges to be paid for access to the fixed network or transmission system and economic regulation. Various forms of profit regulation have evolved to reduce the market power of private monopolies. The traditional approach to regulating utilities in the United States is through profit regulation, with the utility calculating and the regulator reviewing the expected operating costs for a normal year. An alternative form of regulation, via a "price cap", emerged during the privatization of utility industries in the United Kingdom, where the focus is on the future adjustment of prices relative to changes in the consumer price index. The flexibility and relatively greater ease of administration have made price caps a preferred form of regulation for both utilities and governments.

Autonomous regulatory commissions originated in the United Kingdom and the United States in the nineteenth century in the form of advisory railway commissions. By now, the institutional framework of regulation reveals a wide range of variations. In some cases, preference is given to ministerial regulation, as in Indonesia, Japan, Malaysia and several European countries, such as France and Germany, although the overall trend is now towards separation from the government. In other cases, there may be a single body covering all infrastructure sectors, which is well suited for countries with insufficient human, financial, and administrative resources, or specialized regulators for each sector, as in the United Kingdom and Argentina. Other countries have set up a single agency for closely linked sectors, such as gas and power, as in Hungary and the United States at the federal level, the Federal Energy Regulator Commission (FERC). In other cases, there is one multi-sector agency for all or most infrastructure sectors, as in Bolivia and the United States, which has public utility commissions at the state level. In some countries, there is no special regulatory body at all, as in New Zealand, where the Commerce Commission, the national competition agency, is responsible for the economic regulation of infrastructure on the basis of the country's competition rules. In Guinea and Côte d'Ivoire, the use of detailed and relatively rigid concession agreements, coupled with international arbitration procedures, has been preferred to the establishment of an autonomous regulatory body.

- **Regional Trends**

Reform strategies have also differed significantly across countries and regions.

Europe

Privatization follows the United Kingdom's model of rapid and extensive privatization, with the establishment of a specialized regulator, restructuring and unbundling their electricity systems and creating wholesale electricity markets. This approach is also being adopted in Eastern European countries such as, Bulgaria, Hungary, and Romania, and in the former Soviet countries of Armenia, Estonia, Georgia, Latvia, and Moldova. Other approaches found in Europe involve: (i) limiting reform to the creation of independent power producers as is Croatia and the Slovak Republic; (ii) providing third party access to a dominant utility as in the Czech Republic; and (iii) restructuring with plans for major divestitures as in Poland, the Russian Federation, and the Ukraine.

After nine years of negotiations, the Council of Ministers of the EU adopted on 19th, December 1996 a directive concerning common rules for the internal market in electricity. This directive provides for the gradual opening up of the electricity market over a six-year period. The first phase of this process, which had to be completed by the beginning of 1999, required each Member State to ensure that the market is open in proportion to its share in community consumption, as represented by final consumers using more than 40 GWh per year (per site) in 1997. This threshold will gradually be reduced to a level of 9 GWh of annual electricity consumption after six years. More precisely, under the electricity directive, EU Members had until February 1999 to open up a minimum of 25 percent of their electricity markets, accounting for the largest energy users, to competition. The minimum rises to 28 percent in February 2000 and to 33 percent in 2003. The European Commission will subsequently propose additional opening of the market which is due to come into effect in the ninth year of the directive. In fact, liberalization of the electricity market has proceeded much more rapidly than envisaged by the directive. According to the European Commission, the directive could have resulted in an opening up of more than 50 percent of the market by 1999.

The main thrust of the directive concerns the liberalization of the production of electricity and access to the market by third parties, which will allow certain electricity buyers to take advantage of competition through prices among producers. This gradual liberalization does not necessarily imply the privatization of operators, since European law is neutral concerning public or private ownership of enterprises. However, it does encourage privatization of production.

Latin America

Several Latin American countries restructured and unbundled their electricity systems and created wholesale electricity markets. The reform method used in the electricity sector is the so-called "Southern Cone" model, following legislation in Chile (1980) and Argentina (1992) which led to the splitting up of the industry into five functions: generation, dispatch, transmission, distribution networks, and supply, with deregulation at bulk and retail levels and a completely deregulated bulk sector, open to competition for generation. The process has been virtually completed in Argentina and Chile, and the model has been applied in Peru since 1993, and in Bolivia and Columbia since 1995. In Chile, 80 percent of generation capacity is private, as is all transmission and distribution. ENERSIS, a holding company, was established following the privatization of two state-owned electric utilities. ENERSIS has started to invest in other countries, such as Argentina, Brazil, and Peru. Similar reforms are under way in Brazil, Mexico, and other countries.

Central America

The countries of Central America and the Caribbean (Belize, Costa Rica, the Dominican Republic, Guatemala, Honduras, Jamaica, Mexico, Nicaragua, and Panama) have mostly adopted the model of independent power producers selling electricity to state-owned utilities.

North America

The electricity sector in the United States is made up of a mix of public and private ownership. Public ownership of electric utilities is still more prevalent than other forms of ownership. Some are quite large, although 75 percent of the market in terms of kilowatt-hour load provided is covered by investor-owned enterprises. In addition, many small municipal utilities are dependent on the supply from larger, private generation and transmission companies. As a whole, the utility industry has undergone a rapid process of restructuring and regulatory reform in recent years, driven by market liberalization. The Energy Policy Act of 1992 permitted competition in the energy sector, with California being the first state to open the generation of power to market forces through a wholesale pool system in which prices are likely to fluctuate every half hour. End-users are able to buy their electricity at a tariff governed by the pool price, but will also be able to use "repackagers", i.e. firms which bear the risk by buying in bulk from the pool. Power companies will also be allowed to pass on the entire cost of their "stranded" assets, which are often massive, in the form of price surcharges over a determined period of time. The Act has also led to a large number of mergers, including several involving both gas and electricity suppliers, as well as the pooling of other functions, such as billing, prompting fears that public interest would be compromised.

Asia

Many Asian countries have adopted variants of the single buyer model and invited private investment in generation through independent power producers, with negligible restructuring and reform. Countries such as Bangladesh, China, India, Indonesia, Malaysia, Nepal, Pakistan, Philippines, Republic of Korea, Sri Lanka, Thailand, and Vietnam employed the single buyer model. While almost 70 percent of private investment in electricity in Latin America and the Caribbean has been in divestiture projects, more than 83 percent in East Asia, the Pacific, and South Asia has been in Greenfield projects. However, given the poor state of the sector in many countries, IPPs demanded and received highly favorable "power purchase agreements", or PPAs, providing concessions designed to minimize their risks and guarantee returns on their investment. These included high electricity prices and rigid "take-or-pay" contracts, which required the utility to make a set payment for power, whether it was used or not, and the government guarantees payment for electricity. As a result, IPPs have been criticized as being built on the loss of socialization and the privatization of profit [Colley, 1997].

Africa

In Africa, countries have embarked on electricity sector reform as part of a larger program of structural adjustment with a focus on public sector reform. The privatization process is now well under way in at least ten countries in Africa, although so far it has not involved the transfer of shares or capital; it has been based on various alternative forms of privatization such as performance contracts, leasing contracts, concessions, etc. This approach was based on the small size of the sector in many African countries and the lack of strong regulatory framework [Covarrubias, 1996]. The model of IPPs has been introduced in Algeria, Côte d'Ivoire, Egypt, Ghana, Jordan, Kenya, Morocco, Senegal, and Tanzania.

5.2   The New Trend of Electricity Industry Reform

- **Successes and Failures**

Liberalization has become a worldwide trend. Many industrial, developing, and transitional economies implemented significant institutional reforms in their electricity sectors. Hopes for the results of the reforms are high. Although many reform efforts are still being developed and implemented, several success stories have emerged.

Evaluating reform process on a country level, electricity reform have been considered successful in several European countries such as Sweden, Finland, the Netherlands, and Spain, some parts of North America, for example, the PJM (Pennsylvania-New Jersey-Maryland) Interconnection and even in some threshold countries such as Argentina and Chile. Moreover, Finland is building the first nuclear power plant in a competitive market [Lammers, 2003].

Improve Operating Performance

At first, when properly designed and implemented, a combination of institutional

GWh Sales per employee
(measured against years from date of privatization)

- Chilectra
- Edenor
- Edesur
- UK RECs

Year from start of privatization

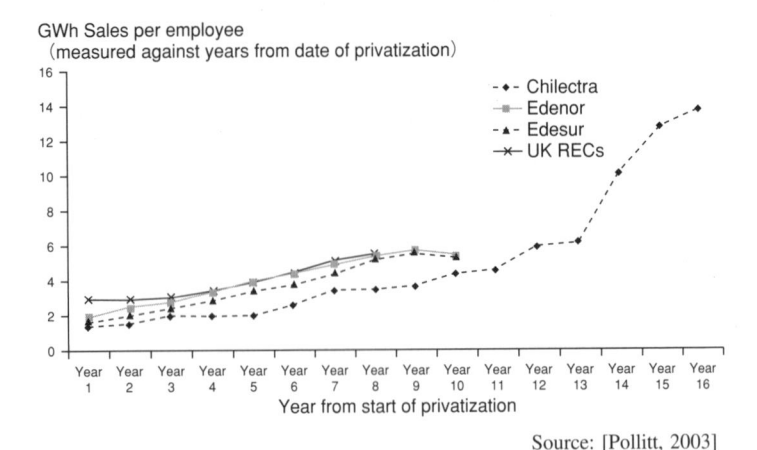

Source: [Pollitt, 2003]

Figure 5.1 The post-privatization labor productivity levels in electricity distribution in Argentina, Chile and the UK

reforms, such as vertical and horizontal restructuring, privatization, and effective regulation, can significantly improve operating performance. In several countries where electricity system reforms occurred, labor productivity increased in the area of generation and distribution in the first few years after liberalization. Labor productivity in the UK increased 60% by 1996. Figure 5.1 shows the rise in the post-privatization labor productivity levels in electricity distribution in several countries where reforms occurred.

In addition, the reforms have also had remarkable effects on the quality of supply. Technical and non-technical losses have been reduced and the quality of service has improved. Figure 5.2 shows the improvement in electricity losses in Argentinean electricity companies after privatization.

Increase Investment and Service Coverage
By relaxing the financial constraints facing state enterprises and establishing stable and fair regulation, electricity reforms have promoted investment and accelerated network expansion. In Argentina, installed capacity grew from 13,267 megawatts in 1992 to 22,831 megawatts in 2002, an increase of nearly 5 percent per year. The situation was similar for Chile's installed capacity. The impressive expansion of generating capacity in Argentina and Chile was achieved by private operators, while keeping prices low [Pollitt, 2003]. Before reforms, service coverage in Peru increased slowly from 44 percent in 1986 to just 48 percent in 1992. However, in the five years after following the introduction of reforms, service expansion accelerated considerably, and by 1997 the coverage was more than 68 percent

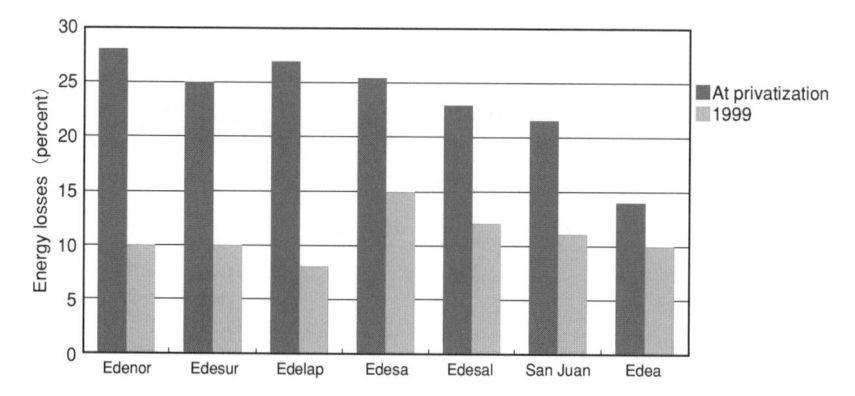

Figure 5. 2 Energy losses among Argentina's distribution companies, at privatization in 1999

Source: [Feler, 2001]

[Rudnick, 1998].
Moving Toward Efficient Pricing
Electricity reforms have better aligned prices with underlying costs to reflect resource scarcity, as efficiency requires. In many countries, this meant increasing prices that previously were too low [Joskow, 2003]. However, in several countries, such as those in Latin America, electricity prices have fallen as wholesale markets have developed and the entry of new generators has expanded supplies and increased competition. In Argentina, the average monthly price per megawatt-hour in the wholesale electricity market fell from about $45 (with peaks of more than $70) in 1992 to about $15 in 2001. Figure 5. 3 indicates a general decline in the average electricity prices for major customer categories in EU between 1997 and 2003. The reduction of average prices for households, small industrial customers, and large industrial users was 6%, 20%, and 9.5% respectively. Moreover, the figure shows that the order and magnitude of the prices for these customer groups seems to have come in line with the underlying costs of supply, which would suggest that residential prices should be higher than small industrial prices.

But . . . the Pendulum Swings Back
A series of financial crises, corporate scandals, stock market downturns, the California electricity crisis, and blackouts in North America and around the world occurred at the end of 1990's and especially at the beginning of 2000's. California experienced price spikes and 38 days of rolling blackouts between November 2000 and May 2001. PG&E and Calpex went bankrupt. Nearly half of the states in the US halted or reversed liberalization. Enron, once the 7th largest US corporation and

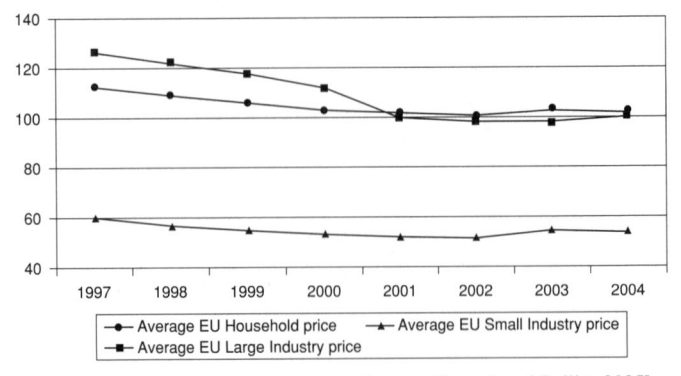

Source: [Jamasb and Pollitt, 2005]
Figure 5. 3 EU average real price (2004 euro/MWh)

world's largest energy trader, collapsed amidst a major accounting scandal. Moreover, American investors pulled out of Europe and British Energy is on the brink of bankruptcy, etc.

These events alarmed policymakers around the world, slowed EI reform and possibly impeded the development of competitive markets. Doubts over the effectiveness of liberalization and privatization policies grew. According to their level of liberalization, countries worldwide should be classified into five groups: (i) countries with the old structure largely intact and in a preparation period of reform, (ii) countries where reforms have been halted; (iii) countries where reforms are complete or seem irreversible, (iv) the European (EU) countries, and (v) the success stories: the UK and the Nordic countries. A trend towards the re-examination of reform strategy exists in each group. Even in countries where reforms have been successfully carried out, the advantages of a strong network and surplus generating capacity may run out due to a lack of new investment.

Reforms have many benefits, however, people also recognize that privatization and introducing competition raises a whole new set of issues. These include: (i) economic contradiction, (ii) sustainable development contradiction, and (iii) political contradiction.

Economic Contradictions
Experience with electricity reform so far revealed that creating a competitive market is an extremely difficult task. First, it is very costly. The majority of policies seem to be formed around the idea that competition is a "free lunch". For many products, the costs of competition may indeed be low compared to the benefits gained from actively participating in a competitive market, however, for the electricity industry the balance is very different. Costs of competition are various and often very high.

For example, the cost for IT systems and software in England's NETA, is US$2 billion. Secondly, electricity spot prices are highly volatile. Electricity cannot be economically stored and must be generated and delivered in real time to perfectly balance supply and demand that can fluctuate randomly. Since hourly market demand is highly price-insensitive, a supply shift along the demand curve can produce a large price swing. As well, a demand increase at times of capacity shortage can cause a large spike in prices. Besides the case of the California Energy crisis, where rate fluctuations drove prices up by 11 times, several similar situations exist. In Alberta, Canada in 1996, the average power pool price was under C$20/MWH. The price gradually rose to around C$40/MWH in 1999, then in 2000 it exploded to over C$130/MWH [King et al., 2006]. In 2001, the Alberta Government spent C$2 billion in monthly rebates at C$40 per retail customer to help reduce their bills. In Ontario, Canada, losses of C$730 million were seen due to the fluctuation of wholesale prices, while the retail rate remained frozen. In Singapore in 2003, the mean price of the spot market was S$92/MWH (S$1 = US$0.58 as of June 18, 2004), with most half-hourly prices lying between S$70/MWH to S$130/MWH. However, the price reached S$4500/MWH within a few trading periods.

Third, abuse of market power is common in deregulated generation markets. Market power is a seller's ability to raise the market price by economic withholding whereby the seller's offer prices that exceed marginal costs or physical withholding, whereby the seller withholds some capacity to raise the price of output from its remaining capacity. This was the primary cause for the California electricity crisis in May 2000 to April 2001 [King et al., 2006]. Last but not least, electricity deregulation causes inefficient generation and transmission investment. Deregulation decentralizes generation investment. While it shifts the investment risk from consumers to investors, it also creates uncertainties on the returns on and of investment. This increases the cost of capital and discourages investment in generation. Uncertainties can extend to transmission and distribution investment. Deregulation can harm reliability due to under-investment and the benefit from transmission congestion.

Sustainable Development Contradictions
Looking to the sustainable development impact, there is emerging evidence that reform has been designed to mainly address economic and particular financial concerns, with insufficient consideration for social and environmental issues. Social issues include the problem of unemployment and electrification or access to electricity for poor people. While reform results in higher labor productivity and the reduction of employees, which is beneficial to the entrepreneurs, it is clearly detrimental to employees and perhaps to the country's macro economy as well. Typically, it is very costly to develop the infrastructure needed to provide electric

service to rural areas. In a restructured electricity sector, distributors generally cannot commercially expand electricity into rural areas. This point poses a severe challenge to the reform process, especially in poor and less developed countries with relatively low rates electrification rates. Environmental issues connected with the reform process include long term public interest in sustainable alternatives, such as conservation and renewable energy, which can be neglected. Treating electricity as a commodity drives economic actors to focus on selling more kWhs rather than providing more services with fewer kWhs. For example, utility spending on demand-side-management programs in the US fell 45% between 1993 and 1998 [Byrne and Mun, 2003]. In the case of renewable energy investment, a dramatic reversal from a steady upward trend in development has been the clear result of restructuring in the US. Renewable energy generation in the US fell from 66 billion kWh in 1993 to 49 billion kWh in 1998, as utilities shut down or reduced output from renewable plants. In addition, since prices in the electricity market do not include environmental costs, older and highly polluting power plants have competitive advantages compared to the other modes of power generation. In the absence of strong environmental regulations, therefore, electricity reform appears likely to do more harm to the environment in the search for a cheaply priced electricity commodities. In fact, carbon dioxide emissions in the US electricity sector jumped 20% between 1995 and 2000 [Byrne and Mun, 2003].

Political Contradictions
While electricity reform initiatives aim to substitute self-regulating markets for political governance, the new model has turned out to be far from self-regulating. Regulatory measures needed for adequate supervision of market activities have proved to be more complex than those required under regulated monopoly regimes. The process of establishing regulatory agencies needed for electricity reform has tended to reinforce the authority of centralized and largely autonomous organization. This might diminish the range of local governments, especially in huge systems like those in the US, China, or India.

- **The New Trend of Reform**

Over the past twenty years of market-oriented development, the electricity industry has experienced many successes as well failures. Presently, the reconsideration of reform policy is a worldwide trend; however, re-regulation will never be totally accepted. In addition, competitive markets, if implemented correctly, offer benefits too big to ignore for governments, the electricity industry, and consumers. This period of slow-down may be considered as a spiraling of the development process. Electricity reforms are slowing down, but still moving forward.

But How?

The first valuable lesson gained form the reform process is that electricity is complex and competition needs to be «regulated into existence». The expertise, time, money, and effort it takes to create a functioning competitive market must not be underestimated. It is not enough to dismantle monopoly provisions and let the market get on with it, nothing will happen. A strong and independent regulator is vitally important. Another conclusion is that gradual reform is often better than shock therapy or sudden, drastic reforms.

It seems that agreement on step-by-step policies has been reached from various perspectives. A main argument is that restructuring of the electricity industry is a process, not an event, that should evolve in connection with local circumstances. These policies may create a smooth path that seamlessly connects vertically integrated utilities and fully unbundled market entities. It creates the so-called "the Third Way" of reform.

[Lammers, 2003] suggests: (i) Improve management. Utilities should be corporate (organized as companies under commercial law) and given discretion rights to act like commercial entities. National and local governments should refrain from interference in day-to-day operations. Accounting should be improved to give a clear picture of the companies' state. This reduces political/regulatory and business risk, (ii) Set realistic prices. In today's and tomorrow's harsher investment climate, cost-covering electricity tariffs are essential to attract investment. Where affordability is a major problem, a «lifeline» tariff can be used: a small, essential amount of electricity can be subsidized, but all consumption beyond that should be priced at marginal cost. The subsidy should not be a cross-subsidy; (iii) Create a viable company. The power company must be a certain size to be viable. In developed countries, the minimum size for an efficient electricity distributor is 3 million customers. Smaller firm incurs greater risk. Vertical and horizontal separation should not be carried out if the company and/or the market are too small. Competition can be encouraged through IPPs, (iv) Separate social activities. The commercial activities of the company should be kept separate from the social activities. «Lifeline» tariffs and rural electrification should not be financed through cross subsidies. There should be a separate organization for rural electrification. Rural electrification is necessary but expensive. Universal access to electricity is a key factor for the alleviation of poverty and for sustainable development. The public-private partnership should provide electricity for all.

WB methods for addressing new generation reforms include (i) Designing pricing policies that strike a balance between economic efficiency and social equity, (ii) Developing rules governing access to bottleneck infrastructure facilities, (iii) Adapting regulation to address emerging problems, changing circumstances, and new information in regulated infrastructure sectors, (iv) Finding new ways to increase poor people's access to services[Kessides, 2004].

Following the idea of the step-by-step policy of the third-way approach, this book in the following sections (please look for Part II and Part III) will endeavor to find an appropriate way for electricity industry reform in Asian countries and especially the case of Vietnam.

5.3 Reform from the Viewpoint of Several Main Stakeholders

- **Electricity Reform and the Policies of Multilateral International Institutions**

Multilateral international institutions are popular supporters for EI reform. Engaging EI reform as a loan's covenant, they often make a remarkable pressure to their borrowers, especially developing countries. This part of study covers development banks' policies, how such institutions change their policies, and the effects of these change on electricity reform.

World Bank and Asian Development Bank

The leading role that the World Bank (WB) and the Asian Development Bank (ADB) play in the privatization process of the electricity industry is widely known. The agenda of both banks is quite transparent and articulated in official documents related to their policy framework. Since the early 1990s, both the World Bank and the ADB have promoted the conventional neo-liberal model for electricity: the unbundling of vertically integrated public sector electricity utilities into separate generation, transmission, and distribution functions; the liberalization and privatization of generation, and privatization of distribution in some cases; and an autonomous regulator to take over the supervisory role from government.

Prior to 1993, World Bank funding for the power sector aimed at addressing the broad key objectives, including the provision of power service on the least cost basis and improving access to electricity. In 1993, there was a shift in policy, identifying the five guiding principles as conditionality for accessing loans from the Bank, which consists of transparent regulation, importation of services, commercialization, and corporatization, commitment lending, and investment guarantees [World Bank, 1993]. This policy agenda is shared by the ADB for its country operations in the energy sector of developing member countries (DMCs). The ADB's 1995 Energy Policy, as one primary focus, enabled private investment in the energy sector. In particular, it gave preferential allocation of ADB resources to DMCs willing to restructure their energy sector to improve efficiency and to attract private investments to meet incremental demand. It also supported BOT-type and joint venture projects between government utilities and private investors as key vehicles to private investment. In terms of policy sequencing, the ADB started with a thrust of commercializing the power sector beginning in 1981, then went on to push for

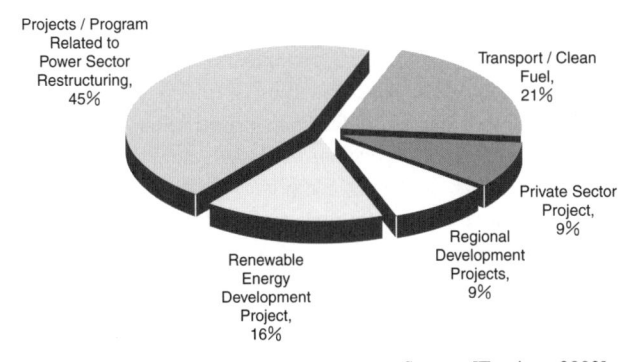

Figure 5. 4   Allocation of ADB loan for energy sector 2002

Source: [Tumiwa, 2003]

greater private sector participation through BOT and similar schemes in the early 1990s, and then finally on to greater privatization through vertical unbundling and changing the role of the government from a service provider to policymaker and facilitator. Loans are also provided on the principle of 'reform-linked assistance' [ADB, 2001].

The priority of reform can be seen in the ADB's lending pattern: by 2002 more than half the total energy loans supported power sector restructuring or privatization. Figure 5. 4 gives an example of the allocation of ADB loans in 2002. For the World Bank, the proportion was even higher, from 1993 between 75% and 93% of all power sector lending by the bank itself was to sustain restructuring or privatization [Thomas et al., 2004].

By 2003, the WB had acknowledged that 'one-size-fits-all' restructuring policies had failed to work as expected, partly due to political opposition and partly due to decisions by multinational companies to withdraw their investments in developing countries because they were perceived as too risky [Thomas et al., 2006]. A WB evaluation report in 2003 noted that by 1999 the WB's power lending portfolio had been recognized as one of the Bank's worst performers, with continuing problems of political and financial risk noted in South and East Asia. The report also observed that the bank's role in supporting privatization in the power sector as "less clear in the current global environment of sharply reduced private capital flows", referring to the withdrawal of multinational investors and reports of "risks of re-nationalisation". In 2003, both institutions returned their focus to infrastructure projects. However, both the ADB and the WB bank remain committed to the same principles of restructuring the power sector, including privatization wherever possible.

Japanese Bank of International Cooperation (JBIC)
While WB and ADB both strongly support privatization, JBIC, one of the most important donors to many developing countries, seems to be less aggressive about supplying loans to host countries to facilitate privatization. One pillar of JBIC's overseas economic cooperation operations is called the Private-Sector Investment Finance program, also called "support for private activity", is designed to supply private enterprises planning to undertake business in developing countries with funds provided as either equity investments or loans. However, in the OECD-DAC Development Partnerships Forum, 27–28 February 2002, Mr. Koji Fujimoto[9] said of JBIC's view on public-private partnership: "It seems conventional to consider that the private sector should cover the areas that the public sector cannot handle. Japan's ODA loans are provided for economic infrastructure development mainly in Asian countries as the foundations for promoting private sector investment. Recipient governments construct basic infrastructure facilities with Japan's ODA loans, followed by private investment. This relationship boosts economic growth".

Recently, JBIC, ADB, and WB have begun work on a Joint Flagship Study on Infrastructure development in East Asia towards the common goals of economic growth and poverty reduction in the region. The study, tentatively titled "Infrastructure Development in East Asia: The Way Forward," is the first such joint effort by the three major development institutions in this region in order to highlight the important role of infrastructure development in reducing poverty, and to review the role and impact of infrastructure provision in East Asia as well. It will explore various framework policies, such as new global public-private partnership, and ways of financing for more efficient infrastructure provision.

European Bank for Reconstruction and Development (EBRD)
The European Bank for Reconstruction and Development (EBRD) was established in 1991 to foster the transition towards market-oriented economies and to promote private and entrepreneurial initiative in the countries of Central and Eastern Europe and the former Soviet Union. Many of the projects approved have been in the energy sector (generation, distribution, and transmission), with privatization of state-owned enterprises being one of the primary beneficiaries of EBRD financing. The EBRD emphasis on the promotion of the private sector is complemented by its support for decentralized and efficient municipal services as a key ingredient of transitional process.

Asia-Pacific Economic Cooperation (APEC) forum
In its 1996 report, the APEC Business Advisory Council (ABAC) emphasized that

---

9   Mr. Koji Fujimoto is the Executive Director of Research Institute for Development and Finance, JBIC.

expansion of public-private sector cooperation is needed to meet the region's urgent need to expand, coordinate, and rationalize investment in infrastructure projects. In particular, it recommended that a series of infrastructure round tables be held in interested APEC economies to examine the infrastructure needs of the host country, identify and recommend corrections to remove impediments to private-sector participation in developing new infrastructure, disseminate the best regional practices and establish productive linkages between entities from both sectors that are able to meet particular infrastructure needs. Electricity demand in the APEC region is expected to increase by between 50 and 80 percent until 2010 and will require a least $1.6 trillion in investment capital. At the Second Meeting of APEC Energy Ministers, held in Edmonton, Canada, in August 1997, it was agreed that capital could not be provided by APEC governments and multilateral financial institutions alone and that business sector participation was essential. The ministers emphasized that this required the establishment of a predictable and transparent institutional and regulatory framework.

[Dubash, 2003] suggests that donors can play a critical role in nudging countries toward reform policies. At the same time, for donors to control the direction of reforms is inappropriate and undermines creation of a national consensus in support of reforms. To support electricity reform, donor agencies can: (i) continue to provide funds on a conditional basis the requires evidence of reform, but only if accompanied by domestic partnership over the type and content of those reforms, (ii) help governments conduct analyses of the scope for inclusion of public benefits concerning the reform process, recognizing that there is greater space for such analyses when governments are not facing crisis situations, (iii) provide support financial and capacity building for the reform process, or devise a mechanism of quality control and verification for the bidding processes and awarding of contacts. Such measures are necessary for countries with very small systems that lack the resources for developing institutional capacity and attracting foreign investors. Such countries prospects are low for private participation and the development of market-oriented solutions in the foreseeable future. They need support to facilitate gradual improvement or adopt measures, such management contracts.

- **Electricity Reform and the Viewpoint of Multinationals and Multi-utilities**

The increasing reform of the utility industry is being accompanied by another major phenomenon: the increasing role played by multinationals both in domestic economies and on an international level. Domestically, there is a trend of inter-industry competition that results in the establishment of many multi-utilities. In the United Kingdom, British Gas and Eastern Energy are now offering customers dual fuel service via electricity and gas contracts. Scottish Power offers gas, electricity,

Table 5.2 OECD multinational electricity companies active in Asia-Pacific

Company	Country origin	Activity	Assets	Countries Active
AES	USA	Generation	1666 MW	China, India, Pakistan, Sri Lanka
EDF	France	Generation	1684 MW	China, Laos, Vietnam
Tractebel	Belgium	Generation & supply	848 MW	China, Thailand, Laos
Enron	USA	Generation	204 MW	Philippines, Guam
Intergen	USA	Generation	1830 MW	China, Philippines, Singapore, Australia
Mirant	USA	Generation	2261 MW	Philippines
Transalta	Canada	Generation	280 MW	Australia
IP	The UK	Generation	3817 MW	Australia, Pakistan, Thailand, Malaysia
CDC	The UK	Generation	810 MW	Bangladesh

Source: [Thomas et al., 2004]

water, and telecommunication services to its customers.

Faced with greater competition in domestic markets, utilities are looking to diversify, as well as increase their international activities. A wide range of companies in France, Germany, Spain, the United Kingdom, and the United States own more than 60 percent of the electricity and supply companies in England and Wales. Table 5.2 lists OECD multinational electricity companies active in the Asia-Pacific in 2004.

However, as a recent Financial Times editorial on the decision by Entergy, a United States energy company, to cut back on its worldwide activities, including its sale of London Electricity, pointed out, being a global utility is harder than it looks. There are cross-border advantages in the utility industry, as in other sectors, but they do not apply to every segment or to every market.

[Thomas et al., 2004] also mentioned the remarkable withdrawal of the OECD multinationals from Asian electricity markets. Ten have withdrawn altogether from the region, and of the remaining nine, two are nationally owned, the EDF by the French government and the CDC by the UK government, and two, Enron and Mirant, are dealing with bankruptcy. This withdrawal is part of the worldwide trend of electricity multinationals withdrawing from ventures that they now consider too risky or unprofitable.

The replacement for multinational companies is Asian-based companies, which have started expanding or are seeking to expand internationally. This expansion implies that Asian companies are prepared to accept lower returns and/or higher risks than OECD multinationals. Such expansion may be supported by governments in the companies' country of origin as part of a policy to expand business activity in the region.

- **Electricity Reform and the Role of Employer Organization**

Reform of the electricity industry in most of countries has met strong opposition from employee unions. Utility employees may legitimately fear that they will lose their status and specific conditions of employment as a result of privatization or restructuring. They are often poorly informed about privatization and restructuring methods, and fear that they will be marginalized in collective bargaining and lose their job security and all or some of their social protections. The failures of several reform efforts in South Korea, Thailand as well as many other countries should be mentioned. Opposition from trade unions has delayed privatization plans of KEPCO and EGAT many times.

A lesson drawn from world-wide experience is that reforms need to be in balance with the government, the EI, and the public interest. "Public sector reforms are most likely to achieve their objectives of delivering efficient, effective and high-quality services when planned and implemented with the full participation of public sector workers and their unions and consumers of public services at all stages of the decision-making process" [Marcovitch, 1999]. Involvement of all stakeholders in privatization and the restructuring process is a prerequisite for successful reform. Active participation of employee representatives and the users of products and services undergoing changes are key to solving or easing many challenges of the reform process. Lingering disputes should be resolved through an open dialogue with interested parties, such as labor unions, etc. A joint labor-management-government research group should be established for decision-making in electricity industry reform. The reform process needs to be discussed within the enterprise, to explain the situation, identify solutions, avoid misunderstandings, and dispel fears. This opens the possibility for transforming the enterprise into an effective, results-oriented long-term coalition.

PART II.
ELECTRICITY INDUSTRY REFORM MODELING FOR ASIAN COUNTRIES

Part I has discussed electricity industry reform as a standard prescription and gave a worldwide overview of the twenty-year development of electricity industry reform with examples of achievements as well as failures. Moreover, it raised the urgent need to analyze the successes and failures associated with past reforms and to identify the instruments and policies that should guide ongoing and future efforts. Assessing the effect of reform efforts on performance requires systematic collecting cross-country data, defining and constructing basic economic measures for various aspects of reform and industry performance, and determining appropriate techniques for econometric estimation. Several studies have been conducted on the economic and social impacts of electricity industry reform. However, the recent empirical analyses mostly encompass OECD countries, Latin America, and Eastern Europe, where the outcome has been systematically monitored. Meanwhile, only the ADB and similar institutions have conducted a limited number of studies analyzing EI reform in Asian countries.

This part of the book endeavors to comprehensively analyze EI reform in Asian countries. Chapter 6 includes the facts and evaluation of EI reforms. The author tried to build a set of core indicators that characterize EI reform. Such indicators of course follow the main theme of this book and will be divided into three categories: regulation, competition, and privatization. The results are essential for the empirical analysis of industry performance in the next chapter. Chapter 6 is also extended to classify country groups based on the level of advances in electricity industry reform and also on the political and economic conditions their position in the three-dimensional matrix of restructuring, privatization, and regulation. Based on this classification, a representative of each country group will be selected for further analysis in Chapter 8. Chapter 7 will carry out an econometrics analysis, which allows policymakers to measure the impact of each reform policy options on EI performance. Based on that, this book intends to derive some predictions about the likelihood and the shape electricity industry reform change in several representative countries. Furthermore, the scope of this analysis covers nuclear power policy under the process of EI reform in Asian countries.

CHAPTER 6
THE EVALUATION OF ELECTRICITY INDUSTRY REFORM IN ASIAN COUNTRIES

Chapter 6 steps up to the second level of analysis, which emphasizes electricity industry reform in Asian countries. In this chapter, the author will try to build a set of core indicators for EI reform. Such indicators, of course, follow the main line of this book and will be divided into three categories: regulation, competition, and privatization. The results of this part will be very essential for the empirical analysis of industry performance in the next chapter. Chapter 6 is also extended to classify countries into groups based on status of electricity industry reform and on the political and economic conditions within the position of the three dimension matrix of restructuring, privatization, and regulation. Based on this classification, a representative of each country group will be selected for further analysis in Chapter 8.

6. 1   Overview

[Bacon and Besant-Jones, 2001] ranked Latin America and the Caribbean as the most developed region in terms of EI reform. European countries have made several achievements in reform, especially after the adoption of Directive 96/92, which dictates the creation of an "internal market for electricity". Reforms in Asian countries have progressed slowly. According to the scores from [Bacon and Besant-Jones, 2001], the pace of EI reform is only slightly quicker than the pace of EI reform in the African continent.

It should be noted that despite similarities in the fundamental direction of power sector reform, the environment in which power sector reform is taking place in Asia is different from that of the power liberalization reform in Europe and USA. First, the motivation for reforming the electric power sector is different. In USA and Europe, the main goal is to improve efficiency by means of deregulation and the introduction of competition, while in Asian countries, the objective is to reduce financial burden on the government. Accordingly, their focus is more on privatization and promotion of private investment than on competition. Second, while significant investments in electric power facilities were not required basically in Europe and USA, in Asian countries, substantial increase in demand for electricity, which is a product of economic development, requires a sizable increase in electric capacity. Third, Asian countries have such problems as tight electricity supply-and-demand, unstable power transmission systems, delayed development of power distribution networks, loss in the transmission and distribution of electricity and a financially fragile public utilities. Fourth, unprofitable investments for rural electrification, as well as for a large number of low-income people, will be required. Fifth, owing to the small market size there are a limited number of power generation and distribution companies, which is likely to cause problems in market power.

One of the reasons why developing Asian countries are hurrying to reform their electric power sector lies partly in the strong pressure of loan conditions from international financial institutions, such as the World Bank and ADB. Such loan covenants include (i) introduction of the market mechanism principle, (ii) pursuit of efficiency, and (iii) improvement in transparency, thus leading to the establishment of a system in which electric power of higher quality is supplied at lower price stably in the long run.

The pressure from these institutions, while necessary for the promotion of reforms and strong management, seems to be inappropriate for Asian countries when these institutions want to impose the same methods of reform used by countries in Europe and USA, without any consideration for the economic, financial, political, and cultural situation of each country. The probability that sector reform will fail is high, resulting in of the break up of publicly-owned electric power utilities and the intensification of public dissatisfaction. This is especially true of developing Asian countries where the investment environment, man-power, and institutional environment are not sufficiently developed.

While insight gained from reform experiences in more developed nations is necessary, it is more essential for Asian countries to know their status in terms of the long reform process. Next, a proper step-by-step schedule for the reform should be developed. The first issue will be systematically presented in the following sections. The second will be dealt with in Chapter 7 and 8.

6. 2   Build Criteria to Measure Regulatory Environment, Ownership and Competition Outcomes

The electricity industry is a network comprised of separate, yet connected and closely coordinated, potentially competitive and naturally monopolized activities. Although substantial variability exists in the development of individual reforms, they generally involve a combination of the following key elements, see e.g. [IADB, 2001], [Joskow, 1998 b], [Newbery, 2002]:
– Corporatization of state-owned utilities;
– Enactment of an electricity reform law;

- Regulatory reform, including adoption of incentive regulation for natural monopoly network activities;
- Establishment of an independent regulator;
- Unbundling of vertically-integrated utilities into generation, transmission, distribution, and supply activities, and where necessary, horizontal splitting;
- Provision of third party access to networks;
- Establishment of a competitive wholesale generation market;
- Liberalization of the retail supply market;
- Privatization of electricity assets;
- Definition of rules concerning consumer protection, allocation of energy subsidies, and stranded costs.

A World Bank survey on the state of energy reform in developing countries focused on six key steps: (i) corporatization or commercialization of the core utility; (ii) enactment of an 'Energy Law'; (iii) establishment of an independent regulatory authority; (iv) restructuring of the core utility; (v) private investment in Greenfield sites; and (vi) privatization [Bacon, 1999]. The survey suggests there is a logical sequence of development for reform steps, in which the first and most common step is corporatization and commercialization of a publicly-owned utility. The least logical or common step is privatization. It should be noted that not all the reform elements outlined above are appropriate for all countries.

Of course, these main steps, in turn, could be divided into many sub-criteria. [Jamasb et al., 2005 a] lists forty qualitative criteria required to construct a regulatory indicator, which is based on very detailed questions, most of which cannot be answered.

[Gutierrez, 2003] used eight factors, which are treated equally, to calculate the average regulatory framework index (RFI) for the telecommunication sector in Latin American countries.

In terms of market structure, [Jamasb et al., 2005 b] focused on eight areas with thirteen indicators.

Various analyses have different objectives for constructing their own list of criteria. However, in most of the cases, the un-weighted average indicator is used to evaluate success of the reform process. This averaging method helps researchers increase the aggregation level of indicator building. However, averaging also means it is unable to see the role of each sub-indicator to the total index. Another popular way to build indicators is to employ dummy variable. Using dummy variables, also impacts to the accuracy of the index. Recently, many researchers are trying to introduce numerous indicators, such as the share of IPP generation per total generation in electricity sector and the market share of three largest generators [Zhang et al., 2002] (Updated version). These indicators provide a clear picture of the reform process, however due to the difficulties in data collection, most research is limited to a single indicator by which the success of the reform process is measured. This can lead to a distortion of information regarding actual success of the reform process. As just mentioned above, [Zhang et al., 2002] (Updated version) uses IPP shares as a privatization indicator, however it covers only generation sector, not transmission or distribution. Moreover, the use of IPP shares as an indicator for privatization is only applied to Greenfield projects, not the privatization of state-owned companies. Other indicators are necessary.

- **Principals for Establishing Indicators**

In this book, the author proposes the following principals for establishing indicators:
- Serve for the analysis. The aim of the author is to evaluate and score the progress of reform in Asian countries and also make an empirical analysis to discover the effects of major reform polices on industry performance. Then, based on that, derive the appropriate policy implications for Asian countries. Meeting such objectives is useful for overcoming the two main weaknesses in the previous analysis.

  Therefore, indicators need to be well-defined, consistent, and comparable to be able to show the current state of reform, the trend over time, and more importantly, to answer policy and research questions.
- Be relevant to important reform issues. Three aspects of the reform process will be examined. One notable point of this research is that un-weighted average indicators will not be employed. Instead, a set of detailed sub-criteria will be utilized. Each dimension on the reform of regulation, competition, and privatization will be explained using two to five sub-criteria. Of course, applying such detailed criteria will make the analysis more complicated and increase the amount of work required for data processing. However, the main advantage is the effects of major reform polices on the industry performance can be discovered.
- Try to reduce the disadvantages of dummy variables. Instead of using one dummy variable for each of criteria, a set of two dummy variables can be employed. This method partially gives us the ability to examine the quality of reform step.
- Standardization among Asian countries. While in European or American countries, the statistical system are relatively developed and standardized, this is not the case in Asian countries. Therefore, many difficulties need to be overcome to collect and ensure the accuracy of comparative data between countries. An annex, presented at the end of this chapter, will provide a comprehensive illustration of the EI reform process in eighteen Asian countries. To ensure the

availability and comparability of data across countries, it should be accepted that there are several appropriate indicators of EI reform that cannot be applied. A detailed explanation is given in the following sections.

- **The Consideration of Essential Reform Steps**

As mentioned, the research approach of this book examines multiple dimension of EI reform. Therefore, this book considers reform issues from three aspects:

− Regulation. In this dimension, we consider two most common issues, legal and regulatory framework, which influences and supports EI reform in most countries. An electricity/energy act or law is generally recognized as a prerequisite for implementation of reform, while establishing independent regulatory agencies for overseeing functioning of the sector, such as pricing by producers, pricing for access to the transmission system, wholesale and retail pricing, controlling quality of supply, and protecting the right of consumers, is the linchpin of the reform. The independent characteristic of regulatory agencies can be seen in the transparency of the legal setting, the independence of financial resources (from regulatory fee), and the clarity of objectives and functions.

− Unbundling and competition. In this aspect, we examine five issues, including: (i) utility corporatization, (ii) structure unbundling, (iii) provision of third party access to networks, (iv) new entrants in generation, (v) establishment of a competitive wholesale generation market, (vi) liberalization of the retail supply market.

Corporatization of an electricity utility is recognized as the first step in the reform process. It has two parts: the removal of the utility from direct control that results from being a part of a ministry, and the creation of an independent corporation with the goal of behaving like a commercial enterprise (for example, maximizing profit). This step of the reform process has been used for a long time by electricity utilities in Western countries. However, it is still an important reform policy for developing countries in Asia.

The unbundling process follows the corporatization step. It results the separation of vertically-integrated utilities into generation, transmission, distribution, and supply activities, and where necessary, horizontal splitting. The unbundled structure is suitable for introducing competition through having enough participants and also makes it easier to privatize the sector.

Allowing new entrants in generation, opening transmission access, establishment of a competitive wholesale generation market, and liberalization of the retail supply market, respectively, are steps for introducing competition in the electricity industry.

− Privatization. The introduction and participation of the private sector in the electricity industry involves in two main components: (i) the entry of the private sector in new investment projects, and (ii) the privatization of existing companies. The participation of private investors in Greenfield projects has been seen as a crucial first step in the reform of the whole sector. In countries, where state-owned utilities dominate the EI, the share of IPPs reflects the structure of ownership. However, this is not the case in countries where EI is run by private companies, like Japan. Therefore to ensure the comparativeness across countries, we use the indicator of non-government proportion in electricity generation.

To implement privatization of existing companies, there are two components, including corporation offering its shares on the stock exchange and selling several part of its access, which can be a power plant and transmission or distribution companies.

- **Design Indicators**

An indicator will be built corresponding to each step of the reform process. To make a clear distinction between the level of reform in each country, and to measure the quality of the reform process, each indicator, in turn, should be presented at several levels. Some dummy variables can be utilized. For example, the first dummy variable identifies simply with the occurrence and the second explores the implementation process. However, using a large number of dummy variables may cause many difficulties in regression solving, especially the problem of complicated and instability. To arrive at a compromise, it needs to take into account the number of countries and the time period which will be included in the regression analysis.

In this analysis, for each aspect of reform, we only use two levels of measurement, for one reform policy, including (i) establishing a legal framework for electricity industry reform, (ii) corporatization of main electricity utilities, and (iii) sharing of non-government generation.

Regulatory Indicators
− Legal framework indicator:
  − Whether reform policy exists?
  − Whether the reform is provided by the Electricity/energy law?
− Regulatory body indicator. Whether it is fully independent?

Unbundling Indicators
− Corporatization core utility:
  − Whether corporatization starts?
  − Whether it completed?
− Unbundling of Generation, Transmission, and Distribution
  − Whether Generation, Transmission, and Distribution have been unbundled?
− Grid access
  − Whether TPA is allowed?

- New entrants in generation
  - Whether IPPs are allowed?
- The occurrence of whole sale market
  - Whether a power pool has been established?
- The occurrence of retail market
  - Whether a retail market exists?

Ownership Indicators
- Non-government generation. What is the share of non-government generation?
  - Whether the share is lower than 50% of total generation?
  - Whether the share is higher than 50% of total generation?
- Selling shares/bonds of core utility
  - Whether the core utility issues its shares/bonds?
- Separate and privatize a part of utility's assets
  - Whether existing utility launches the program of separating and privatizing a part of its assets (both in Generation, Transmission, or Distribution)?

Therefore, in this book the following eleven indicators to measure the reform are proposed:

Table 6.1 List of indicators for measuring reform outcome

Reform Elements	
1. Regulation	
– Legal framework for EI reform (R 1)	
+ Level 1	Y/N
+ Level 2	Y/N
– Independent regulatory agency (R 2)	Y/N
2. Unbundling & Competition	
– Corporatization (C 1)	
+ Level 1	Y/N
+ Level 2	Y/N
– Vertically unbundled (C 2)	Y/N
– Grid access (C 3)	Y/N
– New entrants in generation (C 4)	Y/N
– Wholesale market (C 5)	Y/N
– Retail market (C 6)	Y/N
3. Privatization	
– Non-government proportion in electricity generation (P 1)	
+ Level 1	Y/N
+ Level 2	Y/N
– Issuing corporation shares/bonds (P 2)	Y/N
– Separating and privatizing existing company (P 3)	Y/N

6.3 Comparison of EI Status in Asian Countries

As presented in the last section, the reform status of each Asian country is characterized by three dimensions, which are in turn divided into twelve indicators. In order to determine the existing status of EI reform, as well as make a comparison between countries, the most appropriate method should to be employed is the principal component analysis. However, since the data set consists of qualitative data, correspondence analysis should be used.

- **Multiple Correspondence Analysis and Software Selection**

Correspondence analysis has also been called correspondence mapping, perceptual mapping, social space analysis, correspondence factor analysis, principal components analysis of qualitative data, and dual scaling; these are largely synonymous terms, although techniques vary.

If more than two variables are involved, multiple correspondence analysis should be used. If the variables should ordinarily be scaled, use categorical principal components analysis.

Multiple correspondence analysis (MCA) is an extension of correspondence analysis (CA) to the case of more than two variables. The initial data for MCA are three-way or m-way contingency tables.

When utilizing three variables, a common approach to MCA is to combine the two least interesting variables. Values (categories) of the variable of maximal interest are rows; combinations of values of the remaining two variables are columns. Computations are the same as in CA, but different symbols are used for plotting, depending on the values of one of the two remaining variables.

Which software supports correspondence analysis?

Correspondence analysis is now supported by several programs, some of which are: SPAD, a French program preferred by many researchers who use correspondence analysis, SPSS (CORRESPONDENCE in the CATEGORIES module; when installed, look under Analysis, Data Reduction, Correspondence in the menus), SAS (PROC CORRESP), BMDP, SIMSTAT, EDA (Exploratory Data Analysis), whose ANACOR module supports correspondence analysis, ViSta from VisualStats, a free program, XL-Stat, an Excel spreadsheet add-in to do correspondence analysis.

In this book, we would like to use XLSTAT-Pro version 2006, the data analysis and statistical solution for MS Excel. XLSTAT offers a series of tools that can be used by students and experts alike. XLSTAT's user-friendly interface simplifies the use of complex data analysis techniques, such as Principle Component analysis (PCA), Correspondence analysis, linear, nonlinear, logistic and kernel regression, and

ANOVA (analysis of variance). Results are displayed in Excel sheets, making it easy to re-use the data for additional analysis or presentations.

XLSTAT includes a few tools that facilitate the visualization of data. The scatter plot allow for the plotting of two-dimensional data, while taking grouped information, or taking into account the superimposition of some points, and to easily label points.

- **Cross-Country Analysis and Group Classification**

Cross-country analysis will be done with the data of EI reform from 2004, the latest year. The data set contains 18 different countries which were analyzed on eleven variables: R 1, R 2, C 1, C 2, C 3, C 4, C 5, C 6, P 1, P 2, and P 3.

At first, we can try to see which variables contribute to the spread in the data. The amount of variation in the data that can be explained by different spread factors (Principal Coordinates) has to be investigated. The figure 6. 1 below shows how many PC's explain how much of the variation in the data.

Scree test. Since only the first few dimensions will have a clear interpretation, the researcher needs some criterion for determining how many dimensions to interpret. The first dimension will have the highest eigenvalue, explaining the largest percentage of variance. Later dimensions will explain successively less. If the eigenvalues are plotted, they form a curve heading toward almost 0% explained by the last dimension. The "scree test" is to stop interpreting dimensions when this curve makes an "elbow," often around the 3rd, 4th, or 5th dimension. The scree test is a bit subjective, but is widely used in correspondence analysis.

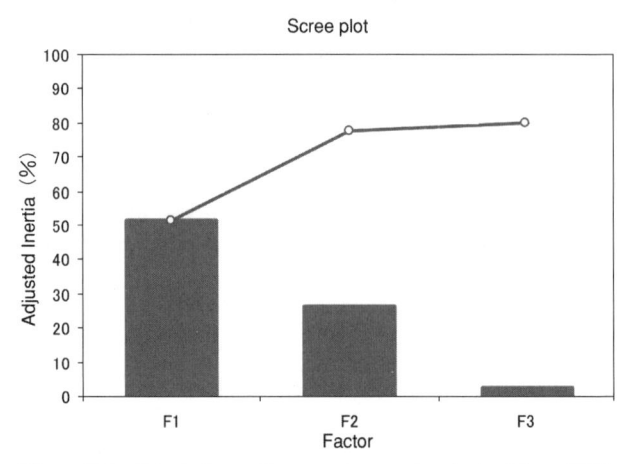

Figure 6. 1 Principal coordinates and scree plot obtained from MCA

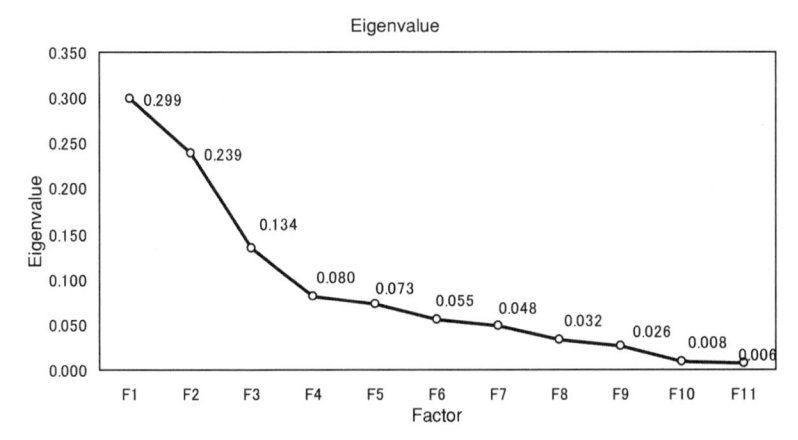

Figure 6. 2 Eigenvalue obtained from MCA

Other criteria are the Kaiser criterion (stop when the eigenvalue falls below 0.1) and the variance explained criterion (stop when 90%, others use 80%, is explained). From the Eigenvalue plot it is possible to say that with 3 principal coordinates; around 80% of the variation in the data can be explained. Therefore, instead of investigating all 11 variables in a constant way, we have now reduced the amount of variables to the main factors (principal coordinates).

The projection of variables on the factor plane circle is useful in interpreting the meaning of the axes. These trends will be helpful in interpreting the next map. To confirm that a variable is well linked with an axis, take a look at the squared cosines table: the greater the squared cosine, the greater the link with the corresponding axis. Figure 6.3 illustrates the projection of variables and the interpretation of corresponding axis.

In addition to Figure 6. 3, Table 6. 2 illustrates the assignment of indicators to correspondent coordinates: (i) The first factor, F 1, represents attempts to introduce unbundling and competition: F 1 from left to right shows an increase in unbundling and competition, (ii) The second factor, F 2, represents ownership and privatization: F 2 from bottom up shows an increase in ownership and privatization, (iii) The third factor talks about regulation framework for reform: F 3 from back to front (if plotted on a 3-dimensional chart) shows an increase in regulatory framework.

Now the question is how to visualize the differences between countries? To see if the status of EI reform in each country is different from each other, a case score plot is needed. This shows how different objects or cases, score according to the variables. Figure 6. 4 plots country "correspondent scores" in unbundling and competition against the "correspondent score" in privatization. The figure shows the

122   Part II

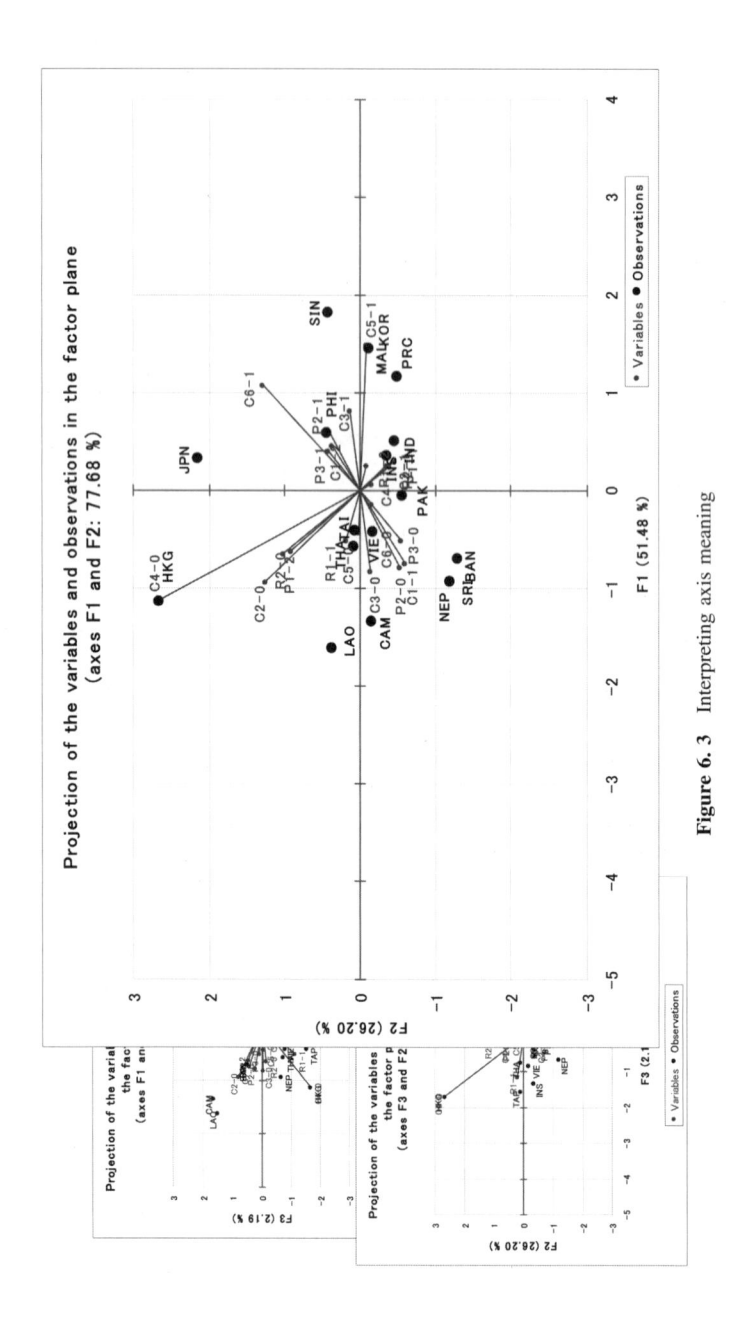

Figure 6.3 Interpreting axis meaning

Table 6.2 Reform process – correspondent factors

	Unbundling & Competition	Privatization	Regulation
R 1-2	0.129	0.017	**0.634**
R 1-1	0.129	0.017	**0.634**
R 2-1	0.157	0.399	**0.009**
R 2-0	0.157	0.399	**0.009**
C 1-2	**0.343**	0.229	0.107
C 1-1	**0.343**	0.229	0.107
C 2-1	**0.247**	0.454	0.185
C 2-0	**0.247**	0.454	0.185
C 3-1	**0.669**	0.021	0.004
C 3-0	**0.669**	0.021	0.004
C 4-1	**0.074**	0.417	0.167
C 4-0	**0.074**	0.417	0.167
C 5-1	**0.632**	0.002	0.005
C 5-0	**0.632**	0.002	0.005
C 6-0	**0.148**	0.207	0.213
C 6-1	**0.148**	0.207	0.213
P 1-1	0.187	**0.430**	0.100
P 1-2	0.187	**0.430**	0.100
P 2-1	0.489	**0.222**	0.044
P 2-0	0.489	**0.222**	0.044
P 3-1	0.209	**0.234**	0.004
P 3-0	0.209	**0.234**	0.004

four groups of countries distinguished by the level of two-dimensional reform. Using 3-D scattering, we can construct a better image of 3-dimensions reform for each Asian country. Figure 6.5 plots country "correspondent scores" in competition, privatization, and regulation. Similarly, regarding the 2-D plot, we see four different groups of countries, as follows:
- Group 1: Nepal, Sri Lanka, Laos, Cambodia, Bangladesh
  Countries where electricity industry is still under the old structure, even though they intend to promote reform.
- Group 2: Thailand, Indonesia, Pakistan, India, Vietnam, Taiwan
  These are countries in the first step of the reform process. Most electricity companies in such countries have been completely corporatized. They are mostly state-owned limited companies. The future plan for reform has already been issued and they may have strong movement in coming years.

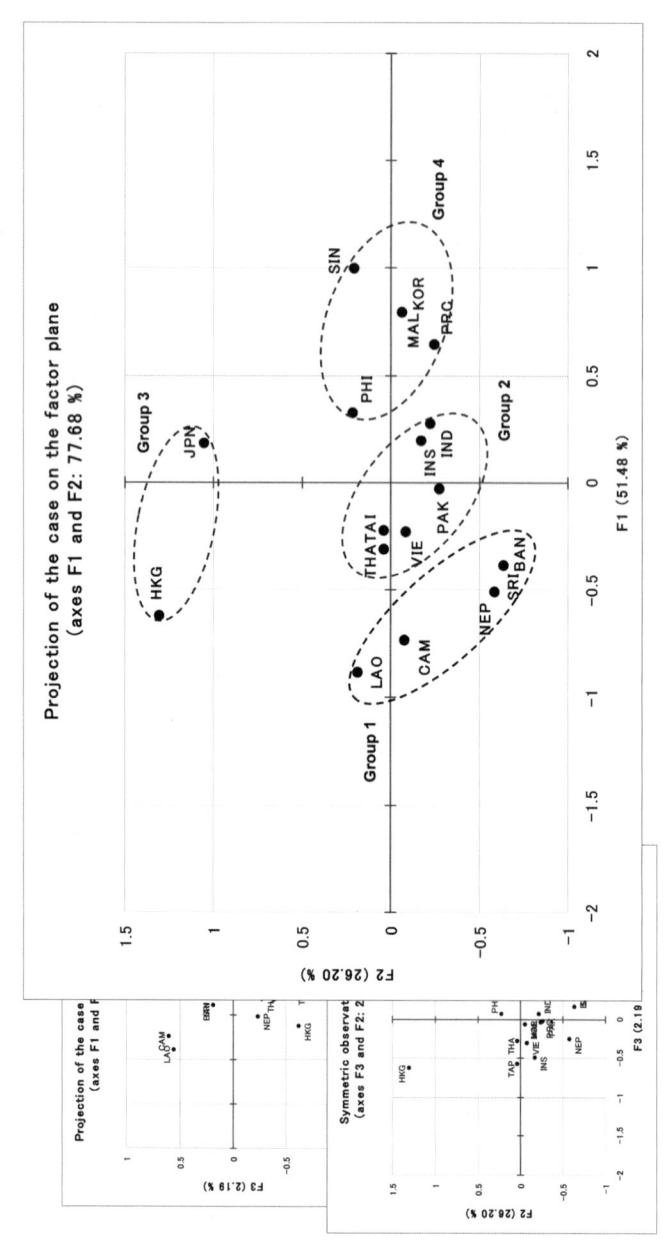

Figure 6.4 Two-dimensional visualization of country groups

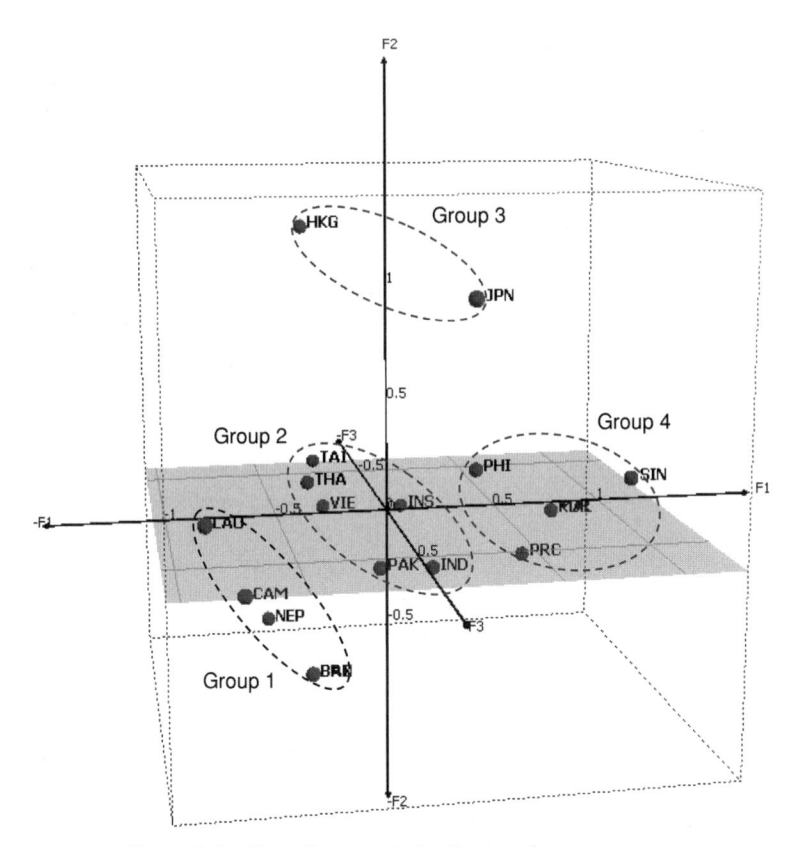

Figure 6. 5 Three-dimensional visualization of country groups

- Group 3: Japan, Hong Kong
  In these two countries, the electricity industry has long history of being run by the private sector. However, in terms of deregulation and competition, they are less developed.
- Group 4: Singapore, Malaysia, Korea, Philippines, China
  These countries have carried out the most far-reaching reforms in Asia, especially in terms of introducing competition and establishing regulation. However, this is not the case of privatization.

These groups reflect the different nature of electricity industry reform. The first group is characterized by a low level of competition, privatization, and regulation. The second group has initial reform experience, mostly related to the establishment of legal framework, corporatization, allowing new generation entrants, and planning

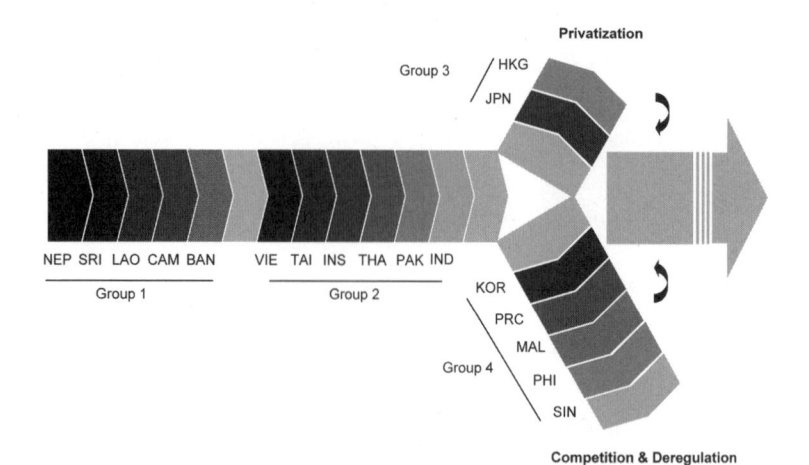

Figure 6. 6 Asian electricity industry reform in a group pattern

for new steps of reform via a regulatory body, introduction of a power pool and the launch of privatization. Group 3 is characterized by a very high level of privatization with a low level of competition and regulation. The final group is represented by a relatively advanced level of competition and establishment of a regulatory authority.

6. 4 Crucial Issues in Electricity Industry Reform of Asian Countries

Several common issues that are seen in Asian countries going through the reform process include:

- **Lack of an Effective Regulation**

Regarding the regulation indicator, most Asian countries have welcomed the introduction and passage an electricity law that covers industry reform. Indonesia, which passed the Indonesia Electricity Law 2002 (EL No. 20/2002), but was considered unconstitutional in 2004, is the only exception.

However, this is not the case with a regulatory agency. Competent regulatory institutions are the linchpins of successful reform. However, most Asian countries have paid little attention to creating such institutions. In some countries, the label "independent" is often applied too quickly to regulators, for example India. Although the legal framework is already set, CER and SERC have been established in most states, but the independence of such authorities has been questioned.

Another example is China; despite the establishment of an SERC supervise market competition in the power industry and issue licenses to qualified operators;

the Commission at this stage lacks enough authority to regulate the industry. Governing authority for approving electricity prices and construction of plants are still scattered among government departments, including the State Development and Reform Commission (SDRC) and the Ministry of Finance.

In another set of countries, there is no regulatory body. This is the case in Japan. The Electricity law has been amended many times, but it is still unclear on the position of regulatory agencies. Regulatory function are now scattered among many agencies, some of them are policy providers and some are neutral associations, which do not have enough legal authority to regulate power utilities as well as IPPs. The results were dismal.

- **Declining Trend of Investment Capital Mobilization from Private Sector**

According to the WB database, private investment in EI continuously increased until 1997 and then it decreased. It should be noted that recently the non-governmental electric generation still maintains a positive growth rate thanks to a number of contracts signed in the previous period and being put into operation at this time. The honeymoon of private investment seems to be over (see figure 6. 7) and few new projects are being negotiated. Why did private investment decrease?

Many Asian countries adopted variants of the single buyer model and invited private investment in generation through independent power producers, with negligible restructuring or reform. Here, several examples from the Philippines, India, and Pakistan will be examined. In these countries, independent power producers sell electricity to state-owned utilities through long-term contacts (PPAs) via governmental risk-sharing arrangements. This model proved successful for attracting a huge amount of private investment from multinational companies in the 1990's and solved shortages in electricity during this period. However, the Asian financial crisis in 1997–1998 dramatically devalued the region's exchange rates, and slowed GDP growth rates and electricity demand. Currency collapse doubled the cost of electricity under PPAs and made it unaffordable for the government. Many Asian governments tried to renegotiate their PPAs, or claimed that the deals were corrupt, causing foreign investors to lose their confidence. Such conditions caused OECD multinationals to withdraw from Asian electricity markets.

- **Coping with the Issue of Low Level of Rural Electrification**

In Asia, except for a few developed countries such as Japan, South Korea, or Singapore, most countries have a low electrification rate. For example, the rural electrification rate is 55% in Indonesia, 79% in the Philippines, 32% in Bangladesh (data in 2002–2003), 56% in Sri Lanka, and 79% in India (data from company websites in 2003).

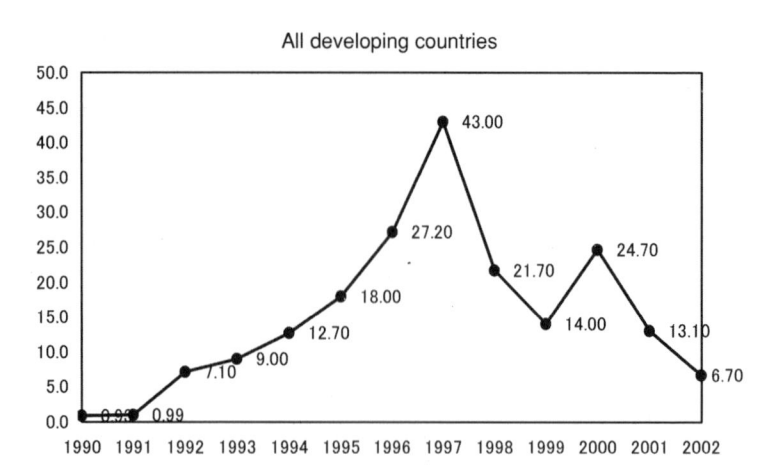

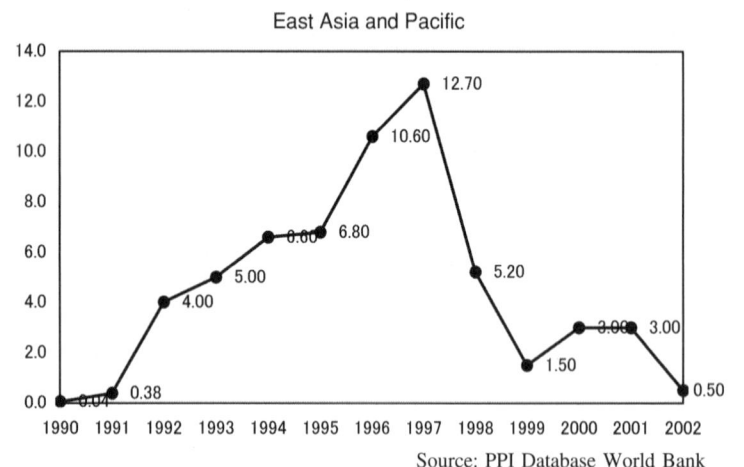

Figure 6. 7 Private investment trends in electricity in developing countries 1990–2002 (US$ billions)

Source: PPI Database World Bank

Typically, it is very costly to develop infrastructure to provide electric service to rural areas. In a restructured electricity sector, distributors generally cannot commercially expand electricity into rural areas. To achieve the government's objectives of rural electrification, special assistances or subsidies are required, such as a concessionary type of approach or a rural cooperative approach.

Most ASEAN countries going through the reform process endeavor to incorporate the electrification issue as one of the important objectives of reform, for example the Philippines, Vietnam, etc. However, adversely, there are several South

Asia countries ignored this issue, including Bangladesh, Sri Lanka, and India. For India, improving electricity access was not explicitly recognized as an objective of restructuring and regulatory legislation, which is perhaps a major shortcoming in the Indian reform policy.

Bangladesh and Sri Lanka are two countries that having very low rate of access to electricity. Meanwhile, unfortunately, having reviewed the existing legal and regulatory framework, gaps in the current legal and regulatory framework with respect to the regulation of rural electrification systems have been identified. Both the Electricity Reform Act and the Public Utilities Commission Act have not touched on the issue of rural electrification through renewable energy-based off-grid systems. Given that the GOSL policy is to expand access to electricity to 75 percent of the population by 2007, rural electrification is high investment priority. However, the lack of a clear legal and regulatory status already affected investments in both two countries. In particular, the proposed reforms require all generators and distributors of electricity to have a separate license to carry out these respective activities. The existing regulations prevent a single license holder from both generating and distributing electricity, which is the norm for off-grid renewable systems. There is a danger that the absence of a well-defined legal and regulatory structure for rural electrification projects, through renewable energy, will prevent new investments from taking place.

- **Issues Surrounding the Current Electricity Rates System**

In Asia, the subsidy and cross subsidy (coverage of charges for household use from charges for industrial use) to public power companies have distorted the whole electricity tariff system. In the Philippines, electric charges have been unbundled according to the costs, such as power generation costs and transmission costs etc. However Vietnam, Indonesia, Sri Lanka, and several other countries, have three common problems regarding the electric tariff systems; 1) level of electricity sales is lower than that of electric power supply cost, 2) electric tariffs are not unbundled and create bottlenecks in improvements in reliability, transparency, and efficiency, and 3) the national unification charge is adopted and the subsidy is not clarified (how much and where it is paid), compared with the charge in the cost base. The major issues at hand are what share of the electricity fee can be passed through to the end-users, when viewed in comparison with an affordability of household users and international competitiveness of industrial users.

- **Political Opposition and Growing Doubts over Reform Policies**

A number of Asian countries have now reversed or postponed their policies to liberalize and privatize the electricity industry. To elucidate this trend, in the

following part, we examine case studies from South Korea, Indonesia, Thailand, Pakistan and Sri Lanka.
South Korea
In 1999, the Korean government issued a three-phase restructuring plan that will lead to full retail competition after 2009. Phase 1 (~2002) – Competition for electric power generation: to divide the electric power generation unit into several subsidiaries and sell electricity based on cost competition through the Korea Power Exchange (KPX), gradually privatize the electric power generation subsidiaries from 2002. Phase 2 (~2008) – Wholesale competition: to divide the distribution and sales units into several subsidiaries for privatization. Phase 3 (after 2009) – Retail competition: transition to a fully competitive market where general consumers can directly choose their electric supplier.

The first phase of the plan has been implemented with the separation of five thermal power plant companies and one nuclear/hydro power company from the Korea Electric Power Corporation and the KPX has also been established. However, in terms of the privatization of electric power generation subsidiaries, the first sale of management rights of Korea South East Power Co., Ltd. failed. Moreover, strong opposition from the labor union exists. Under such pressure, a research team, under the control of the Tripartite Commission: the labor union, the industry, and the government, was set up to review information from fact-finding missions regarding electricity reforms in countries such as Brazil, USA, Canada, UK, and France. The team ruled against privatization and the splitting of distribution companies. The second phase of reform was postponed and it remains to be seen what steps the government will take in the future.

Indonesia
In 1998, a power sector restructuring policy was proposed. In September 2002, the new Electricity Law (EL No. 20/2002) replaced the former 1985 Law [EL No. 15/1985]. According to the new Electricity Law, the reform plan of 2002 includes the implementation of market structure and the encouragement of competition in electricity production and supply. First, the system would be a single buyer market and then in 2003, the market would be opened to multiple buyers and sellers. The timeframe to achieve full retail competition was set for 2003–2007.

In reality, however, three crises influenced reform of the Indonesian power sector: (i) the PLN financial crisis, (ii) the electricity crisis, and (iii) the legal reform crisis.

The PLN was burdened by a huge debt, which in 2003 consumed 40% of the annual government budget. That was a consequence of the Asian financial crisis of 1997–1998. However, the PLN failed because it could not afford its signed PPAs. The electricity supply crisis was mainly attributed to the PLN's poor performance and the unsolicited IPP projects. In 2005, many brown-outs occurred in Indonesia.

The governance crisis resulted from the absence of an independent regulatory body or strong regulations and law enforcement. More seriously, on December 1, 2004, the Constitutional Court decided that the Electricity Law [EL No. 20/2002] was contradictory to the 1945 Constitution and that the law has no binding legal force. Therefore, the Electricity Law of 1985 became the basis for regulation of the power sector. However, the law contains many points that cannot be applied to the current power situation in Indonesia. A new law based on Article 33 of the 1945 constitution is required. Until now the Indonesian government did not have any clear direction concerning the power sector.

Thailand

The cabinet approved a Master Plan for Privatization in September 1998, detailing the privatization of state enterprises related to communications, the water supply, transportation, and energy. The privatization in the electricity industry was set to occur in three stages: (i) 1999–2000: to corporatize EGAT, but keep it as a 100% state-owned enterprise, (ii) 2001–2003: to transfer EGAT into a holding company, with its business units established as operating subsidiaries, and launch the sales of shares in EGAT subsidiary generation companies, and (iii) beyond 2003: to establish a competitive power pool with retail competition in the electricity markets, and privatize all generation facilities.

The first stage of this plan was completed successfully. However, during the second stage, privatization plans faced extensive opposition from EGAT's labor unions, the danger of privatization in terms of higher prices, the risk of corrupt allocation of shares to cronies, and the risk of foreign control developing through buying of shares.

The government backed down and announced the cancellation of the EGAT privatization plans. The plans to liberalize the sector have been slowly abandoned, including the cancellation of the proposed power pool.

In 2005, the privatization plan was reviewed again. In May, the cabinet gave the green light to a long-delayed privatization plan for EGAT. It allows setting up EGAT PLC and then listing it on the Stock Exchange of Thailand. 75% of EGAT shares will be held by Ministry of Finance, foreign investors can hold no more than 5%, and the remaining 20% will be distributed to small-scale investors and the general public. To persuade labor unions, the cabinet agreed to increase EGAT staff salaries by 15%. However, on November 15, the Supreme Administrative Court accepted a petition by 11 civic groups to halt EGAT's IPO pending a review of the legality of two decrees authorizing the privatization of the country's largest power producer.

Pakistan
The Government has approved WAPDAs Strategic Plan for the Privatisation of the Pakistan Power Sector since 1992. Nevertheless, reforms in the power sector have been slow to materialize. Not much progressed from 1998 to 2004. WAPDA still exercises strong oversight and control over unbundled corporate entities. Heavy financial losses continue to burden WAPDA and KESC and drain the government budget. WAPDA and KESC have also failed to generate sufficient funds for investment in urgently needed transmission and distribution capacity. A regulatory agency for the power sector has been established, but it lacks predictability and transparency. Although reforming the power sector is part of the government's agenda, there seems to be a lack of political will to implement it aggressively.

Sri Lanka
Although the reform act has been enacted since 2002, until 2004 little progress seems to be have been made and the Sri Lankan President Chandrika Kumaratunga appeared to have ruled out privatization of the electricity industry. It is not clear whether the industry will be re-organized and how far private investors will be expected to meet the need for new power plants.

A lesson drawn from the reform experiences in Asia mentioned above is that reforms need to be in balance with the government, the EI, and the public interest. Understanding and participation by the people concerned with policy implementation is necessary. Lingering disputes shall be resolved through an open dialogue with the interested parties, such as labor unions, etc. The joint labor-management-government research group should be used for decision-making in electricity industry reform. Another important lesson is learning how to deal with oppositions to and doubts regarding electricity industry reform. The most appropriate solution for Asian countries may be a step-by-step policy. Further research in the next chapter will focus the effort to find the factors that influence electricity industry reform the most.

Appendix 6_1: A Brief Summary (one-page overview) of Electricity Industry Reform in Asian Countries

This section gives an overview of the development and present status of electricity reform in the following countries:
- Republic of Korea
- Thailand
- Indonesia
- Malaysia
- Singapore
- Japan
- Philippines
- China
- Taiwan, China
- Hong Kong, China
- India
- Nepal
- Bangladesh
- Sri Lanka
- Pakistan
- Vietnam
- Laos
- Cambodia

The summary is provided in a compact structure including general information, reform objectives, reform milestones, and reform elements.

Sources of information:
- Country maps: http://www.cia.gov/cia/publications/factbook/geos/xx.html#Geo
- All data by author compilation from WB, ADB, country reports, corporation websites, journal, etc.

REPUBLIC OF KOREA

- **General Information** (*Data in 2004*)
Population (million persons): 48.1
GDP/Capita (USD): 14,136
Installed Capacity of power plants (GW): 57
Electricity Generation (TWh): 342%
Nuclear generation: 38%

The electric power industry of Korea has been monopolized by Korea Electric Power Corporation (KEPCO), which is 51% government owned. KEPCO has the overall responsibility for generation (89% of capacity), transmission and distribution;

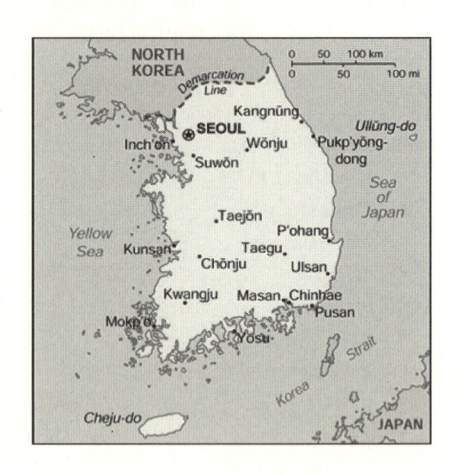

IPPs provide a small amount of capacity; co-generators sell surplus power to KEPCO. In 2001, the generation sector was separated from KEPCO and split into six companies and a power exchange, Korea Power Exchange (KPX) was introduced. The six generators remained in public ownership.

- **Objectives of EI Reform**
- To reduce the debt to capital ratio in the electricity industry.
- To gain financial and operational efficiency.
- To maximize the valuation of KEPCO's generation assets on privatization.
- To eliminate cross-subsidies resulting in distortions in pricing.
- To improve customer satisfaction with service.
- To create a distribution and supply structure which is successful in delivering inexpensive electricity.

- **Reform Process**

1969: – The exceptional case of the first IPP in South Korea, Korea independent energy corporation-Hanwha thermal power plant 2×162.4 MW.

Late 1970's: – Initial discussions on electricity sector restructuring. However, rapid demand growth did not allow any substantial discussion on this matter.

1989: – KEPCO stocks were listed on the Korea Stock Exchange. The government sold 21% of KEPCO shares to the public.

1993: – The Committee for management appraisal of government investment

companies decided to probe the management practices of KEPCO
- New independent generators were officially permitted. Hanwha extended its capacity in 1995.

1994: - KEPCO shares were listed on the New York Stock Exchange.

1994–1996: - Implementation of evaluation of KEPCO's performance efficiency in terms of operation and management. Privatization was recommended in a number of phases taking into account accelerating electricity demand and market power concentration after privatization. In addition, as a prerequisite, deregulation was acknowledged.

1997: - Establishment of the Committee for Electricity Industry Restructuring. The Committee consisted of experts from government, research institutions, and industry.

1998: - Announcement of the Privatization Plan for Major Public Enterprises in the wake of the financial crisis. As a part of the Plan, October 1998 was set as a deadline for finalizing the electricity industry restructuring plan.

1999: - The Base Plan was finalized and announced. It proposes the phased introduction of a competitive wholesale market leading to full retail competition after 2009.
- Under the Ministry of Commerce, Industry and Energy (MOCIE), the Electricity Industry Restructuring Bureau was created.
- A plan for initial grouping of KEPCO's power plants for divestiture was announced. The 42 thermal/hydro plants currently in operation or under construction will be divided into 5 subsidiaries of KEPCO and nuclear plants will remain under an additional subsidiary.
- Wholly amended to the Electricity Business Act to incorporate reform policy.

2001: - The Korea Electricity Commission was established to ensure a smooth transition to a competitive electricity market and a continuously well functioning market.
- The Korea Power Exchange was set up to operate the competitive electricity market (Cost-Based Pool).
- KEPCO non-nuclear generation was broken up into 5 wholly-owned generating.
- Subsidiaries, with the intention of privatizing one of them (at least partially) by 2002 and
- The rest (at least partially) beginning in 2002.

2002: - Failures happened in the efforts to divide distribution assets of KEPCO and privatize the generation company.

- **Reform Elements to Date**

Reform Elements	
1. Regulation	
– Legal framework for EI reform (R 1)	
+ Level 1	Y
+ Level 2	Y
– Independent regulatory agency (R 2)	Y
2. Unbundling & Competition	
– Corporatization (C 1)	
+ Level 1	Y
+ Level 2	Y
– Vertically unbundled (C 2)	Y
– Grid access (C 3)	Y
– New entrants in generation (C 4)	Y
– Wholesale market (C 5)	Y
– Retail market (C 6)	N
3. Privatization	
– Non-government proportion in electricity generation (P 1)	
+ Level 1	Y
+ Level 2	N
– Issuing corporation shares/bonds (P 2)	Y
– Separating and privatizing existing companies (P 3)	Y

THAILAND

- **General Information** (*Data in 2004*)
Population (million persons): 64.2
GDP/Capita (USD): 2,547
Installed Capacity of power plants (GW): 26.06
Electricity Generation (TWh): 127.5%
Nuclear generation: 0%

The current electricity industry structure is essentially a state-owned generation and transmission company monopoly, the Energy Generating Authority of Thailand (EGAT), selling wholesale to two state-owned distribution companies, the Metropolitan Electricity Authority (MEA) and the Provincial Electricity Authority (PEA). In 2004, EGAT produced around 50% of all electricity generated. In 1992, as the first efforts in privatization process, EGAT separated off some of its own capacity into EGCO, with 22% of installed capacity.

- **Objectives of EI Reform**
  - To ensure energy availability.
  - To encourage private investment.
  - To promote free market competition.
  - To implement energy conservation.
  - To reduce the environmental impact.

- **Reform Process**

1992: – The government established the National Energy Policy Council (NEPC).
 – The government also passed a master plan for state enterprise sector reform, which provided a framework for restructuring and privatization of certain sectors, including energy.
 – The EGAT Act was amended to end EGAT's monopoly on generation and to permit the private production and sale of electricity.
 – EGCO was created with EGAT as its parent and the off taker of power generated by EGCO's Rayong 1,232 MW Combined Cycle Power Plant

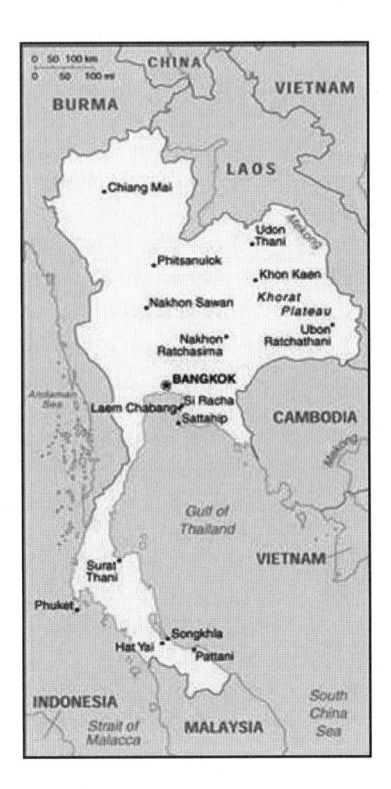

and 824 MW Khanom Power Plant.
1994: – EGCO was privatized and listed as a public company.
– The first round of IPP solicitation was issued. EGAT would buy up to 5,800 MW of capacity from IPPs for the period 1996–2003. Power Purchase Agreements (PPAs) has been signed with 7 IPPs with a total of 5,944 MW of electricity sold to EGAT.
1996: – Separation of generation, T&D businesses
1997: – 14.9% of EGAT's shares in EGCO were sold to CLP
1998: – Master Plan for State Enterprise Sector Reform
1999: – Guidelines for an Independent Regulator
2000: – EGAT established a subsidiary, RGCO Holding. 55% of its shares were sold through public offering. RGCO purchased the Ratchaburi Power Plant from EGAT.
– The liberalization plan, which was named the Electricity Supply Industry Reform and Thailand Power Pool' was approved by the cabinet. The main element of this plan is power pool establishment by 2003. However, it was put on hold when the new government came into being in 2001.
– Draft Energy Industry Act currently in Parliament
2001: – The plans to liberalize the sector have been slowly abandoned, including the cancellation of the proposed power pool.
2003: – With a change in government, Thailand reversed its intention to institute competitive power pooling into its electricity market structure. It is now concentrating its efforts on corporatizing and privatizing EGAT, which was scheduled to be listed on the Stock Exchange of Thailand in 2005.
2004: – A series of demonstrations and strikes were organized by the union. Union members feared privatization would cause prices to increase. They also feared the risk of corrupt allocation of shares to cronies and foreign control developing through buying of shares.
– The government backed down and announced the cancellation of the EGAT privatization plans.
2005: – The privatization plan was reviewed again, and in May, the cabinet gave the green light to a long-delayed EGAT privatization plan. It allows setting up EGAT PLC and then listing it on the Stock Exchange of Thailand. 75% of EGAT shares will be held by Ministry of Finance, foreign investors can hold no more than 5%, and the remaining 20% will be distributed to small-scale investors and the general public. To gain favor with the labor union, the cabinet agreed to increase the salaries of EGAT staff by 15%.

- On November 15, the Supreme Administrative Court accepted a petition by 11 civic groups to halt EGAT's IPO, pending a review of the legality of two decrees authorizing the privatization of the country's largest power producer.

- **Reform Elements to Date**

Reform Elements	
1. Regulation	
– Legal framework for EI reform (R 1)	
+ Level 1	Y
+ Level 2	N
– Independent regulatory agency (R 2)	Y
2. Unbundling & Competition	
– Corporatization (C 1)	
+ Level 1	Y
+ Level 2	Y
– Vertically unbundled (C 2)	Y
– Grid access (C 3)	N
– New entrants in generation (C 4)	Y
– Wholesale market (C 5)	N
– Retail market (C 6)	N
3. Privatization	
– Non-government proportion in electricity generation (P 1)	
+ Level 1	Y
+ Level 2	N
– Issuing corporation shares/bonds (P 2)	N
– Separating and privatizing existing companies (P 3)	Y

INDONESIA

- **General Information** (*Data in 2004*)

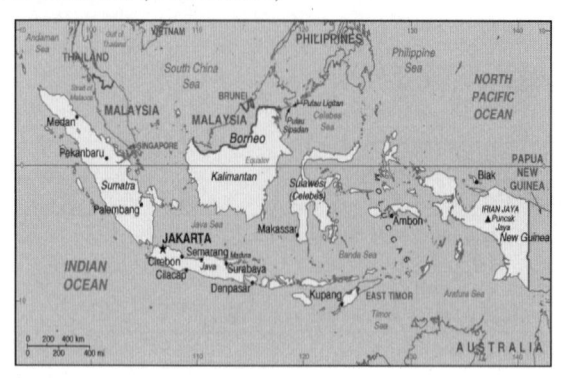

Population (million persons): 216.4
GDP/Capita (USD): 1,190.5
Installed Capacity of power plants (GW): 25.12
Electricity Generation (TWh): 120.2%
Nuclear generation: 0%

State utility, Perusahaan Listrik Negara-Djakarta (PLN) handles generation, transmission, and distribution. PLN faces a huge challenge in delivering affordable and reliable electricity to this dispersed country. It operates more than 600 unconnected transmission and distribution systems. The largest connected systems are on the islands of Java and Bali. Around 60% of the total population lives on Java. PLN was burdened by huge debt, which in 2003 consumed 40% of the annual government budget, virtually bringing the company to brink of collapse as a consequence. This was mainly a consequence of the Asian financial crisis of 1997–1998. However, the PLN failed because it could not afford its signed PPAs.

- **Objectives of EI Reform**
- To encourage private investment to finance future growth and adequate transmission infrastructure, and to increase the efficiency of production.
- To reduce reliance on oil, by developing domestic hydro and geothermal energy, and diversifying more into coal and gas.
- To improve customer service.
- To decentralize electricity utilities d on a regional basis and corporatize generation, transmission, and distribution activities in Java-Bali.
- To conserve and use energy more efficiently.

- **Reform Process**

1985:	–	Electricity Law permits private sector, cooperatives, and public to participate in power generation.
1989:	–	WB Power Sector Institutional Development Review recommends PLN pursue a strategy of "deregulation, decentralization, and competition" to move from "bureaucracy to enterprise".
1990:	–	First IPP (Paiton I) approved.
1992:	–	Implementation of regulation of the law passed in 1985. President Soeharto decreed that the private sector could participate in the generation, transmission, and distribution of electricity again.
	–	Issuance of PLN I through public offering. Funds collected are used to expand business.
1993:	–	Initial stage of power sector restructuring takes place.
1994:	–	PLN was converted from a general company (Perum) to a limited corporation (PT).
1995:	–	Establishment of PLN's subsidiary companies: PJB I, PJB II and P 3 B.
1994– 1997:	–	25 additional IPP contracts signed, a total of 27 contracts.
1997:	–	Economic crisis swept the country.
1998:	–	Power Sector Restructuring Policy (White Paper) released by Minister of Mines and Energy.
1999:	–	ADB's $400 million loan for Power Sector Restructuring Programme
2001:	–	Draft electricity bill was submitted to Parliament; PLN's corporate restructuring process is started.
2001– 2004:	–	Electricity tariff increasing every three months, totaling 24% increase per year.
2002:	–	Parliament passed the electricity bill and after it signed by the President become
	–	The Electricity Law No. 20/2002
2004:	–	The Court of Law for the Constitution passed a decree to cancel Government Legislation No. 20/2002 regarding electric power and reactivated Government Legislation No. 15/1985.
2005:	–	The Government issued Government Regulation No. 3 in 2005, regarding the changes made on Government Regulation No. 10 of 1989, regarding the supply and the use of electric power.

- **Reform Elements to Dates**

Reform Elements	
1. Regulation	
– Legal framework for EI reform (R 1)	
+ Level 1	Y
+ Level 2	Y
– Independent regulatory agency (R 2)	N
2. Unbundling & Competition	
– Corporatization (C 1)	
+ Level 1	Y
+ Level 2	Y
– Vertically unbundled (C 2)	Y
– Grid access (C 3)	Y
– New entrants in generation (C 4)	Y
– Wholesale market (C 5)	N
– Retail market (C 6)	N
3. Privatization	
– Non-government proportion in electricity generation (P 1)	
+ Level 1	Y
+ Level 2	N
– Issuing corporation shares/bonds (P 2)	Y
– Separating and privatizing existing companies (P 3)	N

MALAYSIA

- **General Information** (*data in 2004*)

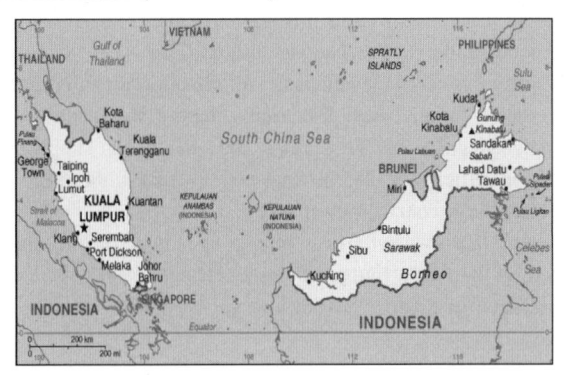

Population (million persons): 25.58
GDP/Capita (USD): 4,061
Installed Capacity of power plants (GW): 15.67[10]
Electricity Generation (TWh): 83 % Nuclear generation: 0

The Peninsular Malaysia electricity supply industry today is structured as follows. TNB and its wholly owned subsidiary, Tenaga Nasional Generation Sdn Bhd are involved in electricity generation, transmission, and distribution activities. There are 5 IPPs in the generation sector, and one licensed mini utility, the Northern Utility Resources, Sdn Bhd (NUR), involved in generating and distributing power to a franchised area in the north of Peninsular Malaysia. Sabah Electricity Sdn. Bhd. – SESB (Sabah) and Sarawak Electricity Supply Corporation- SESCo (Sarawak) are involved in generation, transmission, and distribution activities in East Malaysia. SESB was privatized and controlled by a consortium of companies at the end of 1998. Meanwhile, SESCo is still a statutory body, owned by Sarawak State Government. However, part of its equity is held by the private sector.

- **Objectives of EI Reform**
- To reduce government involvement in day-to-day commercial business activities.
- To lighten the burden on the treasury.
- To procure investment in the industry to meet the challenge of rapid demand growth.

---

10 Data in 2003.

- **Reform Process**

The electricity sector is undergoing substantial change, from a monopolistic, vertically-integrated industry managed by government utilities, to a sector comprising government-owned utilities, as well as private sector players.

1990:
- Corporatization of the national utility. The NEB is incorporated as a company; however the company still remains vertically-integrated.
- Electricity Supply Act, Electricity Supply Regulations, and Licensee Supply Regulation. To ensure the electricity supply industry will function efficiently and effectively, a framework was established to regulate the privatized power sector.

1992:
- 27% of NEB ownership was offered to public. At the same time, it was listed on the Kuala Lumpur Stock Exchange (KLSE).
- Generation opened to private players. IPPs were introduced on a direct negotiation basis.

1994:
- Electricity Regulations. To regulate operation and enhance the efficiency of the privatized power sector.

1997:
- Formation of TNB Generation. TNB no longer a player in generation. TNB Generation takes over all the major TNB power stations and sells power to TNB through PPAs, just like other IPPs. TNB is responsible on transmission and distribution sides.

1998:
- Privatization of Sabah Electricity Board

2000:
- Establishment of Energy Commission
- Establishment of IGSO
- Establishment of power pool

- **Reform Elements to Date**

Reform Elements	
1. Regulation	
– Legal framework for EI reform (R 1)	
+ Level 1	Y
+ Level 2	Y
– Independent regulatory agency (R 2)	Y
2. Unbundling & Competition	
– Corporatization (C 1)	
+ Level 1	Y
+ Level 2	Y
– Vertically unbundled (C 2)	Y
– Grid access (C 3)	Y
– New entrants in generation (C 4)	Y
– Wholesale market (C 5)	Y
– Retail market (C 6)	N
3. Privatization	
– Non-government proportion in electricity generation (P 1)	
+ Level 1	Y
+ Level 2	N
– Issuing corporation shares/bonds (P 2)	Y
– Separating and privatizing existing companies (P 3)	Y

SINGAPORE

- **General Information** (*data in 2004*)
Population (million persons): 4.24
GDP/Capita (USD): 25,192
Installed Capacity of power plants (GW): 8.92[11]
Electricity Generation (GWh): 36.81%
Nuclear generation: 0

State-owned companies continue to hold a monopoly over Singapore's electricity sector, although the restructuring and privatization process has begun. The three main generation companies are PowerSeraya, Senoko Power, and Tuas Power. All the companies are subsidiaries of Singapore Power and together generate 90% of Singapore's electricity.

- **Objectives of EI Reform**
- To reduce government involvement in day-to-day commercial business activities.
- To increase competition and efficiency in service sector.

- **Reform Process**

The electricity sector is undergoing substantial change in competition from a vertically-integrated industry managed by government utilities, to a wholesale market.

1995: – In Oct 95, the Government of Singapore corporatized the electricity and piped gas activities of PUB (Public Utilities Board). Singapore Power Ltd. includes three generation companies (Tuas Power Ltd, PowerSenoko Ltd, and Power Seraya Ltd), one transmission and distribution company (PowerGrid Ltd), and one Electricity retail company (SP Services Ltd, formerly known as Power Supply Ltd).

1998: – Formation of the Singapore Electricity Pool. The Singapore Electricity Pool (SEP) was implemented to introduce competition in the wholesale electricity market. PowerGrid Ltd, the Pool Administrator, operated the Pool to facilitate the trading of electricity between generation companies

---

11 Data in 2003.

and the retail company in a competitive environment.
- Tuas Power Ltd was floated in the Singapore stock exchange.

2000:
- Further deregulation of the Electricity Industry. The Government announced further deregulation of the electricity industry to obtain full benefits of competition. The key restructuring initiatives to be implemented included separation at the ownership level of the contestable and non-contestable parts of the electricity industry, the establishment of an independent system operator, and the liberalization of the retail market.

2001:
- Formation of an Energy Market Authority. The Energy Market Authority was formed to take over the responsibility of regulating the electricity and gas industries and district cooling services in designated areas from the Public Utilities Board.
- Formation of an Energy Market Company (EMC). Energy Market Company (EMC), a joint venture company of Energy Market Authority and the M-Co New Zealand, was formed to take over from PowerGrid Ltd as the Pool Administrator of the Singapore Electricity Pool.
- Divestment of Generation Companies. Singapore Power fully divested its generation companies, PowerSenoko Ltd and PowerSeraya Ltd to Temasek Holdings. The separation of the ownership of generation companies from the ownership of the transmission and distribution company PowerGrid Ltd was to enhance competition by ensuring a level playing field for all generation companies.

2003:
- New Electricity Wholesale Market. The new electricity wholesale market commenced operation. About 250 consumers, with the maximum power requirement of 2 MW and above, became contestable.

- **Reform Elements to Date**

Reform Elements	
1. Regulation	
– Legal framework for EI reform (R 1)	
+ Level 1	Y
+ Level 2	Y
– Independent regulatory agency (R 2)	Y
2. Unbundling & Competition	
– Corporatization (C 1)	
+ Level 1	Y
+ Level 2	Y
– Vertically unbundled (C 2)	Y
– Grid access (C 3)	Y
– New entrants in generation (C 4)	Y
– Wholesale market (C 5)	Y
– Retail market (C 6)	Y
3. Privatization	
– Non-government proportion in electricity generation (P 1)	
+ Level 1	Y
+ Level 2	N
– Issuing corporation shares/bonds (P 2)	Y
– Separating and privatizing existing companies (P 3)	Y

JAPAN

- **General Information** (*data in 2004*)
  Population (million persons): 127.69
  GDP/Capita (USD): 26,741[12]
  Installed Capacity of power plants (GW): 246.11
  Electricity Generation (GWh): 1,054.4%
  Nuclear generation: 27%

  The Japanese electricity industry is controlled by 10 vertically integrated regional companies. The two dominant companies are Tokyo Electric (TEPCO) and Kansai Power, with a third company Chubu Electric. These three companies own about 60% of Japan's 216 GW of generation plants. The remaining plants are owned by six other interconnected companies, Kyushu EPC, Tohoku EPC, Shikoku EPC, Hokuriku EPC, Hokkaido EPC, and Chugoku EPC. The other regional company, Okinawa EPC supplies Okinawa, but is not interconnected and owns less than 2 GW in generation capacity. The remainder of the plants are owned mainly by two companies, the Electric Power Development Corporation (EPDC, 16 GW) trading as JPOWER, and Japanese Atomic Power Company (JAPCO, 2.6 GW), which build plants using new or innovative technologies and sells their output to regional companies. The majority of EPDC was owned by government, with the regional companies holding the balance of shares. However, in October 2004, the government and the electricity companies sold their shares and the company is now an independent generator. 90% of JAPCO's shares are held by the regional companies and JPOWER.

- **Objectives of EI Reform**
  - To reduce the comparatively high electricity tariffs in Japan, and to correct the gap between domestic and foreign prices.
  - To improve the deteriorating load factor (due to a sharp increase in demand in the summer).
  - To adapt to the global energy sector reform trend.
  - Increase competition and efficiency in service sector.

---

12 Data in 2003.

- **Reform Process**
- *1995:* − First amendment of Electric Utilities Industry Law. Wholesale IPP entrance is allowed.
- *1996:* − The Program for Economic Structure Reform was adopted by the Cabinet.
  − First Round of IPP bid results were announced. 20 projects were accepted. Total capacity amounts to 3,046.9 MW.
- *1997:* − The Action Plan for Economic Structure Reform was adopted by the Cabinet.
  − Electric Utilities Industry Council (EUIC) established the Basic Policy Committee. EUIC is the advisory body to the Ministry of International Trade and Industry (MITI).
  − Second Round of IPP bid results announced. 16 projects were accepted. Total capacity amounts to 3,118.3 MW.
- *1998:* − The Committee released an interim report.
  − Third Round of IPP bidding. 2 projects were accepted. Total capacity accounts for 215 MW.
- *1999:* − The Committee completed a report recommending partial liberalization. It is recommended that partial liberalization should go along with ensuring universal service, reliability, energy security, and preservation of the natural environment.
  − Second amendment of the Electricity Utilities Industry Law
  − Fourth Round of IPP bidding. 4 projects were accepted. Total capacity accounts for 804.1 MW.
- *2000:* − Announcement of a wheeling tariff.
  − Amendment of the Electric Utilities Industry Law. Partial retail competition takes effect. Eligible customers accounted for 27.4% total demand.
- *2003:* − Amendment of the Electric Utilities Industry Law. Expansion of the liberalized market and the amount of eligible customers is increased to 14%, effective as of 2004.
  − The accounting unbundling in transmission sector was enforced.
  − Preparing for the establishment of a neutral system organization and power exchange.
- *2004:* − Privatization of JPOWER. The government and the electricity companies sold their shares and JPOWER is now an independent generator.
  − Electric Power System Council (ESCJ) is designated as a neutral transmission system organization that is a self-governing organization to secure fairness and transparency in the transmission and distribution

segment.

2005: – Japan Electric Power Exchange (JEPX), a non-mandatory type exchange, began operation.
– Liberalization was carried out for high or extra-voltage consumers.

- **Reform Elements to date**

Reform Elements	
1. Regulation	
– Legal framework for EI reform (R 1)	
+ Level 1	Y
+ Level 2	Y
– Independent regulatory agency (R 2)	N
2. Unbundling & Competition	
– Corporatization (C 1)	
+ Level 1	Y
+ Level 2	Y
– Vertically unbundled (C 2)	N
– Grid access (C 3)	Y
– New entrants in generation (C 4)	Y
– Wholesale market (C 5)	Y[13]
– Retail market (C 6)	Y
3. Privatization	
– Non-government proportion in electricity generation (P 1)	
+ Level 1	Y
+ Level 2	Y
– Issuing corporation shares/bonds (P 2)	Y
– Separating and privatizing existing companies (P 3)	Y

---

13 Wholesale market established in 2005.

THE PHILIPPINES

- **General Information** (*data in 2004*)
Population (million persons): 83.5
GDP/Capita (USD): 1,035
Installed Capacity of power plants (GW): 15.46
Electricity Generation (GWh): 50.55%
Nuclear generation: 0%

NPC is responsible for generation, transmission, and distribution. Distribution is performed by 27 private and municipally owned utilities, and also by 119 rural electricity cooperatives. The largest privately-owned distribution company is the Manila Electric Company, which distributes more than 75% of national sales. About 76% of villages are electrified and connected to the main grids, and the government aimed to extend electrification to all villages by 2004. NPC is under pressure from EPIRA to break up its vertically integrated assets into smaller sub-sectors, such as generation, transmission, distribution, and supply in order to prepare for eventual privatization.

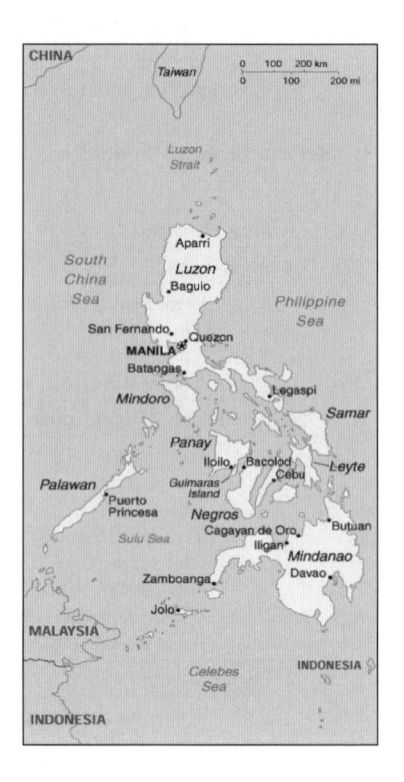

- **Objectives of EI Reform**
  - Increase the investment of private capital in the power industry, while minimizing the government's financial commitment.
  - Create an environment of competition and accountability.
  - Deliver competitive and affordable prices.
  - Improve operational and economic efficiency.
  - Make social subsidies transparent.
  - Share social and other costs among all users.

- **Reform Process**

1987: – Executive Order 215 amended the NPC's responsibilities to allow private entities to participate in power generation through co-generation, build-operate-transfer (BOT), and build-operate-own (BOO) schemes, though responsibility for strategic development remained with the NPC. Plants built under the NPC's power development program must sell their output to the NPC.

1990: – The Build-Operate-Transfer Law was approved to further encourage private sector financing of power infrastructure projects.

1992: – DoE is re-established, which included the Energy Regulatory Commission.

1993: – The Emergency Power Crises Act gave the President special emergency powers to contract with the private sector for additional necessary generation capacity to cope with power crisis.

1995: – IPPs are allowed to sell to distribution utilities directly.
– Congress started to develop a bill to define the future structure of the electrical power industry and the responsibilities of various agencies and entities.

2001: – After seven years of congressional debate and litigation, the Electric Power Industry Reform Act (EPIRA) came into force. The act has three main objectives: (i) to develop indigenous resources; (ii) to cut the high cost of electric power in the Philippines; and (iii) to encourage foreign investment. Passage of the Act set into motion the deregulation of the power industry and the breakup and eventual privatization of state-owned enterprises.
– The Electricity Regulatory Commission was established with significant autonomy and resources. This quasi-judicial body oversees the implementation of EPIRA, including authority over: (i) Regulating transmission and wheeling charges and retail tariffs for end-users, (ii) Granting and regulating certificates of compliance required of all industry participants, and (iii) Reviewing the unbundling of business activities.

2002: – Regulated transmission and distribution: unbundling of electricity rates for transparency.

2004: – Transfer transmission assets from NPC to TRANSCO.
– Transfer all other NPC assets and liabilities from NPC to PSALM.
– Sale of 5 power plants: Talomo, Agusan, Barit, Cawayan and Loboc under plan of privatizing 70% generation assets of NPC in 2006.

2005: – Sale of Masinloc (power plant).
– The DOE and the Philippine Electricity Market Corporation (PEMC) have become increasingly engaged in preparing for the commercial operation of WESM in Luzon. Prior to the commercial operation of the

WESM, a trial operation is ongoing to test the rules, systems, and procedures of the WESM, as well as ensure market participant readiness.

- **Reform Elements to Date**

Reform Elements	
1. Regulation	
− Legal framework for EI reform (R 1)	
+ Level 1	Y
+ Level 2	Y
− Independent regulatory agency (R 2)	Y
2. Unbundling & Competition	
− Corporatization (C 1)	
+ Level 1	Y
+ Level 2	Y
− Vertically unbundled (C 2)	Y
− Grid access (C 3)	Y
− New entrants in generation (C 4)	Y
− Wholesale market (C 5)	N
− Retail market (C 6)	N
3. Privatization	
− Non-government proportion in electricity generation (P 1)	
+ Level 1	Y
+ Level 2	Y
− Issuing corporation shares/bonds (P 2)	Y
− Separating and privatizing existing companies (P 3)	Y

CHINA

- **Overview** (*data in 2004*)

Population (million persons): 1,299.9
GDP/Capita (USD): 1,273
Installed Capacity of power plants (GW): 393
Electricity Generation (GWh): 2,085%
Nuclear generation: 2%

In 2002, the former monopoly, State Power Corporation of China (SPCC) was restructured through the separation of generation from the transmission. The generation assets of SPCC, which owned about 46% of the electricity generating assets was divested and most of its generating assets were transferred to five new portfolio Gencos: (i) China Huaneng Group, (ii) China Datang (Group) Corporation, (iii) China Huadian Corporation, (iv) China Guodian Group, and (v) China Power Investment Corporation. The transmission assets of SPCC were transferred to two grid companies, Southern Power Grid Company and State Power Grid Company. Five regional subsidiaries of State Power Grid Company were further established in North, Northeast, Northwest, East, and Central China.

- **Objectives of EI Reform**
- To improve the investment environment for private investment.
- To investigate and develop suitable models for generation markets.
- To reduce the risks to the utilities in contracting for current and future power.

- **Reform Process**

1984: – The first IPP was put into operation.
1985: – The "Provisional Regulations on Encouraging Fund Raising for Power

Construction and Introducing Multi-rate Power Tariff" were passed, removing the exclusive monopoly of the central government over investment in the power industry, allowing autonomous investment by sub-national governments, private enterprises and foreign companies.

1988: – The Ministry of Electric Power ("MOEP") was established.
– The government decided to institute a separation of governmental and business management functions through establishing regional and provincial power companies alongside respective power bureaus.

1996: – The government promulgated the Electric Power Law, which legalized the status of power enterprises as commercial entities and established the legal basis for private ownership.

1997: – The State Power Corporation of China (SPCC) was established to hold the state's ownership rights in the power sector and to support a commercial asset-holding relationship.

1999: – The State Council issued a directive to separate generation from transmission and distribution, and chose the single buyer model for restructuring China's power industry to create generation service market with power suppliers competing through open bidding.
– The SPCC initiated a limited experiment of wholesale market competition in six provinces. However, the experiment was soon halted because the quick return to a tighter power market in 2001 absorbed excess capacity and alleviated any immediate pressure for competition.

2001: – A proposal that suggested introducing competition into the power industry by breaking up the SPCC was made by the State Development Planning Commission (SDPC).

2002: – The vertically integrated SPCC was broken up into two government-owned grid companies and five state generation companies, created through the transfer of SPC assets.

2003: – An autonomous government regulatory commission was established. The State Electricity Regulatory Commission (SERC) is responsible for overseeing the industry and fiercely regulating prices[14].

2004: – SERC put a regional power exchange (PX) market into trial operation in China's northeastern and eastern provinces.
– SERC set up six regional power regulatory bureaus: North China,

---

14 Although the SERC was set up to supervise market competition in the power industry and issue licenses to qualified operators, the Commission at this stage lacks enough authority to regulate the industry. Governing authority for approving electricity prices and construction of plants is still scattered among government departments, including the State Development and Reform Commission (SDRC) and the Ministry of Finance.

Northeast China, Northwest China, East China, Central China, and South China.

- **Reform Elements to Date**

Reform Elements	
1. Regulation	
– Legal framework for EI reform (R 1)	
+ Level 1	Y
+ Level 2	Y
– Independent regulatory agency (R 2)	Y
2. Unbundling & Competition	
– Corporatization (C 1)	
+ Level 1	Y
+ Level 2	Y
– Vertically unbundled (C 2)	Y
– Grid access (C 3)	Y
– New entrants in generation (C 4)	Y
– Wholesale market (C 5)	Y
– Retail market (C 6)	N
3. Privatization	
– Non-government proportion in electricity generation (P 1)	
+ Level 1	Y
+ Level 2	Y
– Issuing corporation shares/bonds (P 2)	Y
– Separating and privatizing existing companies (P 3)	N

TAIWAN (CHINA)

- **Overview** (*data in 2004*)
Population (million persons): 22.7
GDP/Capita (USD): 13,529
Installed Capacity of power plants (GW): 34.6
Electricity Generation (TWh): 181.25%
Nuclear generation: 21%

The Taiwan electricity industry is dominated by the nationally owned Taiwan Power Company (Taipower) which is a fully integrated company. There are two major IPPs, the 2400 MW Mailiao plant owned by Taiwanese interests and the 1300 MW Ho Ping power plant jointly owned by Taiwanese interests and China Light & Power (Hong Kong).

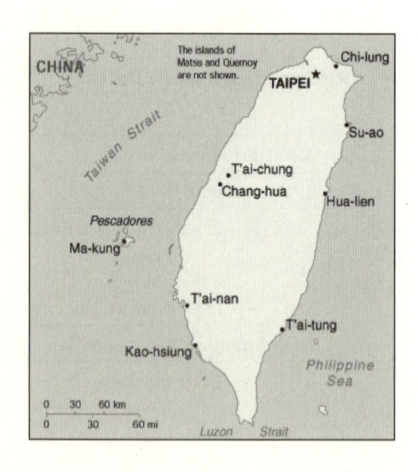

Plans to privatize and split up Taipower, in place since the 1990s, have been continually delayed, and privatization is not expected before 2006. The nuclear and hydro plants (about 5 GW of each) will not be privatized. Taipower has no major foreign investments.

- **Objectives of EI Reform**
  - To clarify procedures for project approval.
  - To achieve future investment with private funds.
  - To restructure the electricity industry into separate generation, transmission, and distribution companies.
  - To introduce open access and enable independent producers to sell direct to customers.
  - To privatize Taipower to raise funds for other public infrastructure projects.
  - To achieve a reserve capacity margin of 20% as soon as possible.

- **Reform Process**

1994: – The MOEA produced documents entitled Operational Guidelines for Unbundling the Power Generation Industry, and Main Points in Handling Independent Power Program. These were intended to promote and publicize operating standards for independent power production.

1995: – Application Guidelines for Establishing IPPs were published to assist

with procedures and in the review of applications. The MOEA drew up a six-year power purchase plan designed to double capacity by 2002, with IPPs producing 30% of generation and a reserve margin of 20%.
- To promote liberalization, the Executive Yuan submitted a draft amendment to the Electricity Act.

1996: – A development conference of all major political parties concluded that state-owned companies, such as Taipower should be privatized before 2001.

1998: – Taipower's original license expired, to be replaced by one amended to recognize that generation, transmission, and distribution rights can be granted separately. Generation was already opening up with the IPP program.

1999: – Another IPP program was announced to encourage further LNG capacity of about 2.8 GWe.

2002: – Foreign investors are permitted to own up to 100% of an IPP. The MOEA reviews applications on a case by case basis.
- Delays in the implementation of the Electricity Act and calls for an Energy Act indicate that the original target dates set for Taipower privatization around 2002 will not be achieved. The Ministry of Economic Affairs set a tentative target for the privatization before 2006.

- **Reform Elements to Date**

Reform Elements	
1. Regulation	
− Legal framework for EI reform (R 1)	
+ Level 1	Y
+ Level 2	Y
− Independent regulatory agency (R 2)	N
2. Unbundling & Competition	
− Corporatization (C 1)	
+ Level 1	Y
+ Level 2	Y
− Vertically unbundled (C 2)	Y
− Grid access (C 3)	N
− New entrants in generation (C 4)	Y
− Wholesale market (C 5)	N
− Retail market (C 6)	N
3. Privatization	
− Non-government proportion in electricity generation (P 1)	
+ Level 1	Y
+ Level 2	N
− Issuing corporation shares/bonds (P 2)	Y
− Separating and privatizing existing companies (P 3)	N

HONG KONG (CHINA)

- **Overview** (*data in 2004*)
Population (million persons): 6.88
GDP/Capita (USD): 23,916
Installed Capacity of power plants (GW): 11.89
Electricity Generation (TWh): 37.13%
Nuclear generation: 0%

Hong Kong is served by two vertically-integrated, privately-owned power companies: China Light & Power (CLP) that mainly serves Kowloon and the New Territories; and the Hong Kong Electric Company (HEC) that mainly serves Hong Kong Island. These two regional utilities are regulated by the government under a 15-year Scheme of Control Agreements (SCAs), with provision for interim review every five years.

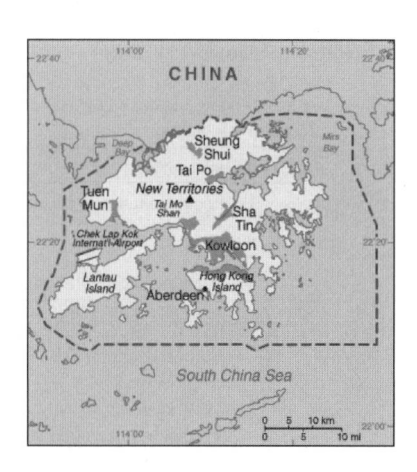

- **Objectives of EI Reform**
  - To ensure the reliability of supply, while monitoring the financial affairs of the utilities and regulating the returns on investment to acceptable levels.
  - To introduce some form of competitive elements after 2008 that demonstrate steps are being taken to ensure electricity is delivered at the lowest reasonable cost.

- **Reform Process**

1993: – Scheme of Control Agreements (SCAs) was established to regulate CLP and HEC. The agreements run from 1993 until 2008, with reviews in 1998 and 2003. Changes under the reviews have to be agreed upon by both parties. Under the SCAs, the obligations of these two investor-owned utilities are: (i) to provide adequate and reliable supplies at the lowest reasonable cost, and (ii) to permit the government to monitor their financial affairs and operational performance.

1998: – Review of the SCAs introduced: (i) some control on the construction of new generation capacity, and (ii) the power companies' environmental obligations.

– No progress on competitive elements is considered before 2008.

- **Reform Elements to Date**

Reform Elements	
1. Regulation	
− Legal framework for EI reform (R 1)	
+ Level 1	N
+ Level 2	Y
− Independent regulatory agency (R 2)	N
2. Unbundling & Competition	
− Corporatization (C 1)	
+ Level 1	Y
+ Level 2	Y
− Vertically unbundled (C 2)	Y
− Grid access (C 3)	N
− New entrants in generation (C 4)	N
− Wholesale market (C 5)	N
− Retail market (C 6)	N
3. Privatization	
− Non-government proportion in electricity generation (P 1)	
+ Level 1	Y
+ Level 2	Y
− Issuing corporation shares/bonds (P 2)	Y
− Separating and privatizing existing companies (P 3)	Y

INDIA

- **Overview** (*data in 2004*)
Population (million persons): 1,086
GDP/Capita (USD): 631.65
Installed Capacity of power plants (GW): 126.34
Electricity Generation (TWh): 556.8%
Nuclear generation: 2.9%

India's electricity industry consists of two levels: state and central.

At the central level, the role of central government has been as a regulator and also as the owner of the three power companies: (i) National Thermal Power Corporation (NTPC). NTPC generates about 25% of all electricity in India. It owns 22235 MW of operating plant and has 5610 MW under construction, (ii) National Hydroelectric Power Corporation (NHPC). NHPC owns 2475 MW of operating plant, 4322 MW under construction with 2420 MW of plant held in joint venture under construction, and (iii) Nuclear Power Corporation of India (NPCIL). NPCIL owns 2770 MW of operating plant and has 3060 MW of plant under construction. The Power Grid Corporation of India Ltd (POWERGRID) is the central transmission utility of India which transmits about 45% of the total generating capacity in India. POWERGRID also takes the role of a central transmission company to undertake all functions of transmission, planning, and supervision of the inter-state transmission system. At the state level, the vertically-integrated state electricity boards (SEBs) are under drastic reform.

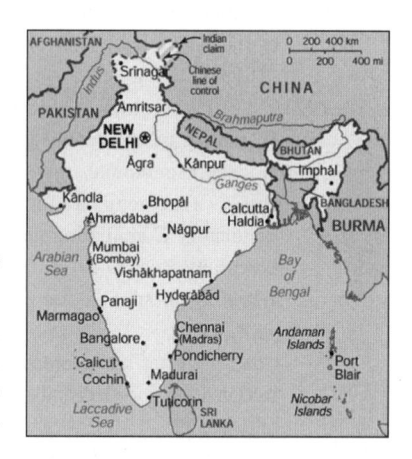

- **Objectives of EI Reform**
- To increase transparency, accountability, and viability of the industry,
- To facilitate private sector participation.
- To promote a competitive market.

(Unfortunately, improving electricity access was not recognized explicitly as an objective of restructuring of the electricity and regulatory legislation, which is perhaps a major shortcoming in the Indian reform policy.)

- **Reform Process**
1991: – Indian government issued the Policy on Private Participation in the Power Sector. This policy allows private sector participation in

generation with foreign investors allowed 100% ownership.
- Electricity Laws (Amendment) Act reinforced the integration of the grid in India by giving more authority to the regional load dispatch centers (RLDC).

1992–1997: – Eight projects given 'fast-track' approval status and sovereign guarantees by the central government in order to add generation capacity in a short timeframe.

1995: – Orissa Electricity Reform Act establishes the Orissa Electricity Regulatory Commission and provides for unbundling of the Orissa State Electricity Board.

1996: – World Bank support for Orissa Power Sector Restructuring project was approved.
- Chief minister's conference formulates a Common Minimum National Action Plan for electricity which indicates the intended direction of the reforms.

1998: – Chief Ministers' Power Reform Initiative reiterated the need for co-ordination between SEBs, as well as the importance of distribution reforms.
- Electricity Regulatory Commissions Act established a central independent regulatory commission, CERC (Central Electricity Regulatory Commission), and allows states to establish their own commissions, SERCs (State Electricity Regulatory Commissions).
- Electricity Law (Amendment) Act set up central and state transmission utilities. Transmission activity was given an independent status. POWERGRID was designated the central transmission utility (CTU) and the SEBs or their successors became state transmission utilities (STUs). The Act stipulates that the CTU and STUs are public companies. The CTU was given the responsibility for interstate transmission of electricity and for all planning and co-ordination of electricity dispatch.
- Andhra Pradesh, Karnataka, and Uttar Pradesh proceed with the preparation of the Electricity Reform Acts.

2001: – The 8th draft of the Electricity Bill was introduced in the Parliament. The main aim of the bill is to consolidate all relevant legal texts into one piece of legislation, and to accelerate the reforms and restructuring necessary to ensure a healthy power industry using market principles.
- Energy conservation bill is passed by the Parliament.

2002: – Independent regulatory commissions are established in 21 states.

2003: – The Electricity Act 2003 was passed in the Indian Parliament. This Act replaced the Electricity (Supply) Act of 1948 and the Electricity

Regulatory Commissions Act 1998, as well as other legislation related to the power sector.

2004: − The government sold 10.5% of the state electricity company, National Thermal Power Corporation (NTPC), to investors.

- **Reform Elements to Date**

Reform Elements	
1. Regulation	
− Legal framework for EI reform (R 1)	
+ Level 1	Y
+ Level 2	Y
− Independent regulatory agency (R 2)	Y
2. Unbundling & Competition	
− Corporatization (C 1)	
+ Level 1	Y
+ Level 2	N
− Vertically unbundled (C 2)	Y
− Grid access (C 3)	Y
− New entrants in generation (C 4)	Y
− Wholesale market (C 5)	N
− Retail market (C 6)	N
3. Privatization	
− Non-government proportion in electricity generation (P 1)	
+ Level 1	Y
+ Level 2	N
− Issuing corporation shares/bonds (P 2)	Y
− Separating and privatizing existing companies (P 3)	Y

NEPAL

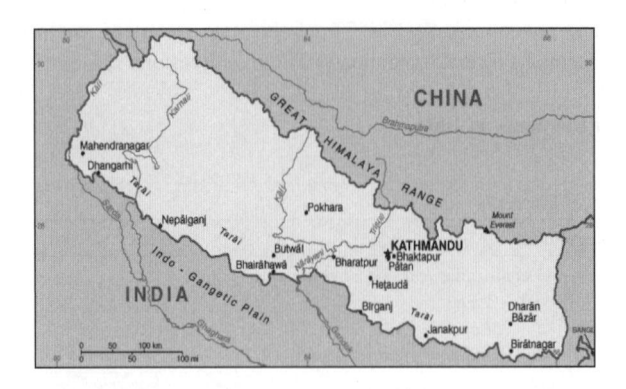

- **Overview** (*data in 2004*)
Population (million persons): 24.74
GDP/Capita (USD): 272
Installed Capacity of power plants (GW): 0.60
Electricity Generation (TWh): 2.64%
Nuclear generation: 0%

The power sector in Nepal is dominated by the Nepal Electricity Authority (NEA), a vertically integrated, Government-owned and controlled utility, with several independent power producers, some of whom also distribute power in areas adjacent to their plants. The story of power position in Nepal is that of highest potential, especially in hydro power and lowest consumption.

- **Objectives of EI Reform**
- To reduce the political intervention from government.
- To improve NEA's operational performance.
- To increase financial transparency and encourage all market players and potential investors. This is expected to help realize the country's abundant hydropower potential and enable export of power to neighboring countries.
- In the context of Nepal, there additional things must be done before the introduction of restructuring for competition is attempted, including to bring the supply up to the demand level so that power outages are ended, as well as providing reasonable access to electricity to people in rural areas.

- **Reform Process**
1992: – Hydropower Development Policy was enacted.

- Electricity Act was issued to allow the entry of Independent Power Producers (IPPs) in Nepal's Power Sector through non-recourse financing.
- The NEA act was also amended to "enable the NEA to function autonomously".

1993:   - A by-law decision noticed that the position of NEA has been replaced from a sole monopoly to one of the licensees with the responsibility to buy the privately generated power.

2001:   - Hydropower Development Policy 2058 made several amendments in the institutional arrangement of the power sector, which provides that Electricity Tariff Fixation Commission will act as a "Regulatory Body" with wider authority.

2005:   - Government confirmed that the legislative initiatives to enable the unbundling of NEA and the setting up of an independent regulator will be enacted.

- **Reform Elements to Date**

Reform Elements	
1. Regulation	
− Legal framework for EI reform (R 1)	
+ Level 1	Y
+ Level 2	N
− Independent regulatory agency (R 2)	Y
2. Unbundling & Competition	
− Corporatization (C 1)	
+ Level 1	Y
+ Level 2	N
− Vertically unbundled (C 2)	Y
− Grid access (C 3)	N
− New entrants in generation (C 4)	Y
− Wholesale market (C 5)	N
− Retail market (C 6)	N
3. Privatization	
− Non-government proportion in electricity generation (P 1)	
+ Level 1	Y
+ Level 2	N
− Issuing corporation shares/bonds (P 2)	N
− Separating and privatizing existing companies (P 3)	N

BANGLADESH

- **Overview** (*data in 2004*)
Population (million persons): 135.2
GDP/Capita (USD): 413.8
Installed Capacity of power plants (GW): 4.68
Electricity Generation (TWh): 20.06%
Nuclear generation: 0%

The power sector in Bangladesh is dominated by Bangladesh Power Development Board (BPDB), the state-owned utility which is responsible for generation, transmission, and distribution of electricity for most of the country. However, as a part of the reform process, several other players exist in both sectors: (i) in generation: BPDB+Independent Power Producers (IPPs)+Rural Power Company (RPC), (ii) in transmission: BPDB +Power Grid Company of Bangladesh Ltd. (PGCB), and (iii) in distribution: BPDB +Dhaka Electricity Supply Authority (DESA)+Dhaka Electric Supply Company Ltd. (DESCO)+Rural Electrification Board through Rural Electric Co-operatives called Palli Bidyut Samities (PBSs).

- **Objectives of EI Reform**
  - Increase private sector participation to mobilize finance.
  - Promote competition among the sector entities.
  - Increase the sector's efficiency.
  - Increase penetration and improve the reliability and quality of electricity supply.
  - Ensure a reasonable and affordable price for electricity.

- **Reform Process**

1991: – The Dhaka Electric Supply Authority (DESA) created to take over the electricity distribution system of the capital city Dhaka from BPDB as part of the unbundling process.

1992: – Industrial policy amended to open the power sector for private investment.

1993: – High power inter-ministerial committee constituted to recommend further reform measures to be undertaken. The Committee recommended unbundling of the sector according to functional lines and establishment of an independent regulatory commission.

1995: − Power cell, an institution under the Power Division, was created to design, facilitate, and drive reform measures.
− Preparation for the National Energy Policy.

1996: − The National Energy Policy was adopted. It recommended vertical separation of the sector and establishment of the Energy Regulatory Commission (ERC).
− Under the policy about 1200 MW IPP power was contracted, of which about 750 MW was put into operation.
− A mixed sector power company (72% share hold by 9 PBSs and 28% by REB) named the Rural Power Company (RPC) was established and produces about 140 MW and planned to add another 70 MW in the future.
− REB/PBSs started buying electricity from the private sector (3×11 MW).
− As part of the unbundling process: (i) Power Grid Company of Bangladesh (PGCB), created in 1996, took transmission business from BPDB. It had already taken over about 50% of the transmission assets and planned to take over all assets by the end of 2002, (ii) Dhaka Electric Supply Company (DESCO) created to take over a part of the Distribution Business of Dhaka City from DESA. It already took over the Mirpur area.

2002: − Corporatization process of the Ashuganj Power Station (2 nd largest power plant complex in Bangladesh) is started.
− Govt. approved Haripur (99 MW) Power Station's conversion into a Strategic Business Unit (SBU/Profit Center).
− Several distribution areas under the authority of the BPDB are converted into Strategic Business Units (SBU/Profit Center).

2003: − Ashuganj Power Company is established. It took over assets and liabilities of the existing operations of BPDB, Ashuganj Power Station.
− West Zone Power Distribution Company took over assets and liabilities of existing operations from BPDB.
− Parliament passed the new Energy Act 2003 as the basis for establishing an ERC.

- **Reform Elements to Date**

Reform Elements	
1. Regulation	
– Legal framework for EI reform (R 1)	
+ Level 1	Y
+ Level 2	Y
– Independent regulatory agency (R 2)	N
2. Unbundling & Competition	
– Corporatization (C 1)	
+ Level 1	Y
+ Level 2	N
– Vertically unbundled (C 2)	Y
– Grid access (C 3)	N
– New entrants in generation (C 4)	Y
– Wholesale market (C 5)	N
– Retail market (C 6)	N
3. Privatization	
– Non-government proportion in electricity generation (P 1)	
+ Level 1	Y
+ Level 2	N
– Issuing corporation shares/bonds (P 2)	N
– Separating and privatizing existing companies (P 3)	N

SRI LANKA

- **Overview** *(data in 2004)*
  Population (million persons): 19.46
  GDP/Capita (USD): 1,030.5
  Installed Capacity of power plants (GW): 2.59
  Electricity Generation (TWh): 8.04
  % Nuclear generation: 0%

  The Sri Lankan electricity industry is dominated by the government-owned Ceylon Electricity Board (CEB). This is a fully vertically-integrated company that generates, transmits, and distributes power. The CEB owns 85% of generation and distributes electricity to 2.4 million consumers, all except 350,000 in the Western and Southern coastal belt, which are supplied by another publicly owned company, the Lanka Electricity Company (Pvt.) Ltd. (LECO), working in the distribution sector. The CEB is wholly owned by the Government. The CEB, the Urban Development Authority, and the Treasury/Local Authorities are shareholders of LECO.

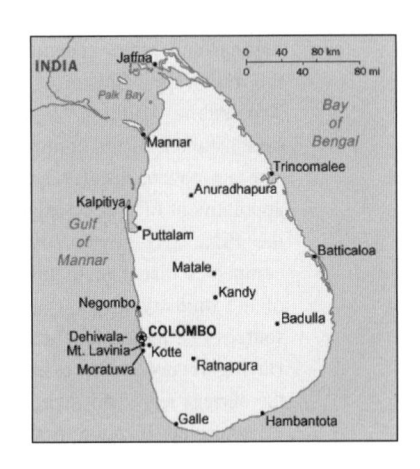

- **Objectives of EI Reform**
  - Increase private sector participation to mobilize finance.
  - Promote competition among sector entities.
  - Increase the sector's efficiency.
  - Increase penetration and improve the reliability and quality of the electricity supply.
  - Ensure a reasonable and affordable price for electricity.

- **Reform Process**

  1998: – The Government of Sri Lanka recognized the need for power sector reform and approved a restructuring model, which called for the unbundling of the power sector into separate companies for generation, transmission, and distribution.

  2002: – The Electricity Reform Act was passed. This is the overriding legislation for the sector, separating the functions of operation, regulation, and policymaking into separate entities by: (i) establishing the role of the Public Utilities Commission, (ii) unbundling the CEB and creating seven

new companies – five new DistCos (Lanka Electric Company will be incorporated within these), a new GenCo that will own CEB's existing thermal and hydro-generation assets, and a TransCo, and (iii) defining the role of the Ministry of Power and Energy.
- The Public Utilities Commission Act was passed and outlines the establishment of the commission to regulate public utilities in Sri Lanka. The act covers the: (i) appointment and removal of commissioners, (ii) appointment of a director general and staff of the PUC, (iii) objectives of the PUC, (iv) powers of the PUC to demand information from the companies that it regulates, (v) functions of the PUC, (vi) responsibility of the ministry to set general policy guidelines for the PUC, and (vii) sources of funding for the PUC.
- Having reviewed the existing legal and regulatory framework, gaps in the current legal and regulatory framework with respect to the regulation of rural electrification systems have been identified. Both the Electricity Reform Act and the Public Utilities Commission Act have not touched on the issue of rural electrification through renewable energy-based off-grid systems.

2003:
- The unbundling of the CEB is due to take place. New companies will be established and will begin operating as independent entities, with their own management structures and balance sheets. Assets will be allocated to the new companies, along with staff from the CEB.
- The MAC was established under the Ministry of Power and Energy. Its role is to monitor the commercial operations of the newly formed companies and to keep the relationship between the ministry and the companies at arm's length. The MAC will be responsible for appointing and dismissing board members for the companies and reviewing the corporate intent of the companies.
- The PUC has been established to regulate the electricity and petroleum industries and may later assume the regulatory functions for water supply and other utilities.
- However, late in 2003, the government announced that it would not sell the distribution companies immediately. However, adjustments have been made to the law to allow foreign ownership of utilities and the proposals will only work if a privately owned industry is anticipated.

2004:
- Little progress had been made and the Sri Lankan President, Chandrika Kumaratunga, appeared to have ruled out privatization of the electricity industry. It is not clear whether the industry will be re-organized and how far private investors will be expected to meet the need for new

power plants.

- **Reform Elements to Date**

Reform Elements	
1. Regulation	
− Legal framework for EI reform (R 1)	
+ Level 1	Y
+ Level 2	Y
− Independent regulatory agency (R 2)	Y
2. Unbundling & Competition	
− Corporatization (C 1)	
+ Level 1	Y
+ Level 2	N
− Vertically unbundled (C 2)	Y
− Grid access (C 3)	N
− New entrants in generation (C 4)	Y
− Wholesale market (C 5)	N
− Retail market (C 6)	N
3. Privatization	
− Non-government proportion in electricity generation (P 1)	
+ Level 1	Y
+ Level 2	N
− Issuing corporation shares/bonds (P 2)	N
− Separating and privatizing existing companies (P 3)	N

PAKISTAN

- **Overview** (*data in 2004*)
Population (million persons): 148.72
GDP/Capita (USD): 630
Installed Capacity of power plants (GW): 14.7[15]
Electricity Generation (TWh): 80.83%
Nuclear generation: 2.4%[16]

The electric power sector in Pakistan is still primarily state-owned. A privatization program is underway, but little progress appears to have been made to date. The main state-owned utilities are the WAPDA, and the Karachi Electricity Supply Corporation (KESC), which serves Karachi and the surrounding areas. Together, WAPDA and KESC transmit and distribute all power in Pakistan. In addition, there are a number of IPPs supplying electricity under PPAs negotiated in the 1990s. The government, under pressure from the World Bank, is now breaking up WAPDA, by creating 8 separate regional distribution companies, 3 generating companies, and a transmission company, with the prospect of future privatization of distribution, and encouragement of further private power developments, including privatized hydro-electric schemes. The government has sought the sale of KESC to private investors, but with the cost of rehabilitation, modernization, and expansion, investors are slow to take it on.

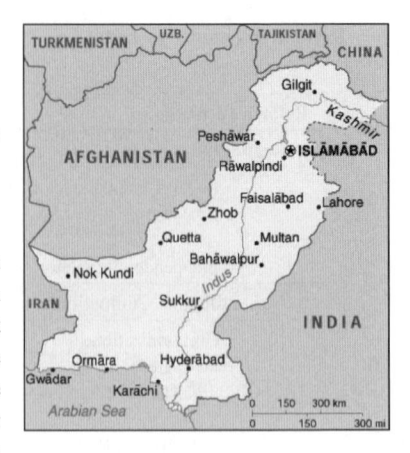

- **Objectives of EI Reform**
  - Enhance capital formation for the Pakistan Power Sector (PPS).
  - Improve the Efficiency of the PPS through competition, accountability, managerial autonomy, and profit incentives.
  - Rationalize prices and social subsidies, while maintaining certain socially desirable policies, such as rural electrification and low income "Lifeline" rates.
  - Gradual shift towards full competition by providing the greatest possible role for the private sector through privatization.

---

15 Data in 2003.
16 Data in 2003.

- **Reform Process**

1992: – The Government approved the WAPDA's Strategic Plan for the Privatization of the Pakistan Power Sector.

1994: – The Government of Pakistan formulated a power policy to permit the private sector to invest in the power sector and ensure sufficient generation capacity. The policy extended full flexibility to independent power producers (IPPs) to bring capacity on line as quickly as possible at predetermined power purchase prices. The government guaranteed implementation, fuel supply, and power purchase.

1997: – The government established the National Electric Power Regulatory Authority (NEPRA). The Authority is responsible for issuing licenses, franchising monopoly businesses, setting and enforcing performance standards and codes of practices, enforcing competitive policies, and setting charges for the monopoly parts of the industry.

1998: – Amendment of the WAPDA Act, which allowed the establishment of the Pakistan Electric Power Company (PEPCO) for unbundling of WAPDA's Power Wing into eight distribution companies (formed by existing area boards); three generating companies (comprising 11 of WAPDA's generating plants); and the National Transmission and Dispatch Company (NTDC).

– The second step of the government's plan is to (i) sell PEPCO's generating and distribution companies and (ii) privatize KESC.

– Nevertheless, reforms in the power sector have been slow to materialize. Not much progress has been made from 1998 to 2004. The WAPDA still exercises strong oversight and control over the unbundled corporate entities. Heavy financial losses continue to burden the WAPDA and KESC and drain government funds. WAPDA and KESC have also failed to generate sufficient funds for investment in urgently needed transmission and distribution capacity. A regulatory agency for the power sector has been established, but it lacks predictability and transparency. Although reforming the power sector is part of the government's agenda, there seems to be a lack of political will to implement it aggressively.

Another problem is the dispute between the Pakistani government with IPPs. In the early 1990's, 16 IPPs were set up in Pakistan by multinational companies in partnership with local investors, with the backing of the World Bank. In 1998, the Pakistan government charged several IPPs with alleged corruption, Hubco was the main target, the largest stock exchange quoted company in Pakistan, which was 26 percent owned by International Power, a UK energy multinational. The Pakistani government dropped the prosecution of Hubco and finally agreed to a revised price in 2000.

- **Reform Elements to Date**

Reform Elements	
1. Regulation	
− Legal framework for EI reform (R 1)	
+ Level 1	Y
+ Level 2	Y
− Independent regulatory agency (R 2)	Y
2. Unbundling & Competition	
− Corporatization (C 1)	
+ Level 1	Y
+ Level 2	N
− Vertically unbundled (C 2)	Y
− Grid access (C 3)	N
− New entrants in generation (C 4)	Y
− Wholesale market (C 5)	N
− Retail market (C 6)	N
3. Privatization	
− Non-government proportion in electricity generation (P 1)	
+ Level 1	Y
+ Level 2	N
− Issuing corporation shares/bonds (P 2)	N
− Separating and privatizing existing companies (P 3)	Y

VIETNAM

- **Overview** (*data in 2004*)
Population (million persons): 148.72
GDP/Capita (USD): 630
Installed Capacity of power plants (GW): 11.36
Electricity Generation (TWh): 46.2%
Nuclear generation: 0%

The electric power sector in Vietnam is still primarily state-owned. The main utility is the Electricity of Vietnam (EVN), a state-owned corporation for different power sector entities engaged in generation, transmission, distribution, and associated service functions. By 2003, the EVN consisted of seventeen power plants, four regional transmission companies, seven regional distribution companies, and several non-core businesses (46). In the past couple of years, the EVN has taken several initiatives to corporatize and commercialize, implementing organizational restructuring: transformation of the ownership structure for certain member units of the EVN into EVN-owned, single-member limited liability companies, joint-stock limited liability companies, or involve equalization of the company.

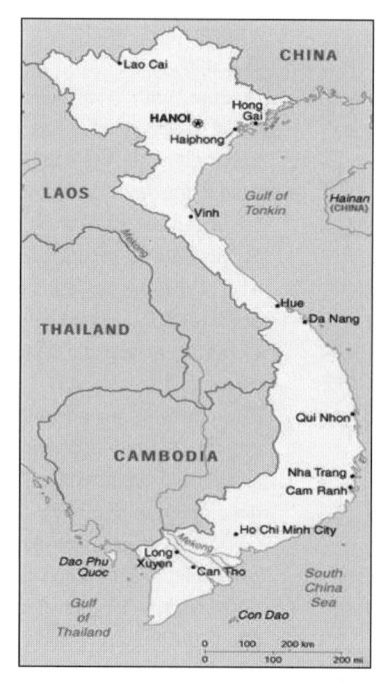

- **Objectives of EI Reform**
- Provide electricity access to the national economy and the entire population of Vietnam.
- Increase the operating/technical efficiency of the electricity sector to optimize the use of scarce investment resources.
- Ensure a reliable electricity supply of good quality.
- Resolve the mismatch between market-based production costs and state administered prices.
- Clearly delineate and separate state management functions and business management functions.
- Enable Vietnam to raise the necessary financing for power sector expansion to meet economic growth targets.

- **Reform Process**

1992: − The Government's power sector policy and reform objectives were first articulated in the Power Sector Policy Statement issued by the Ministry of Industry.

1995: − In line with the policy of renovating the State-owned enterprise management mechanism in the renovation period (Decision 91/TTg), the Electricity of Vietnam (EVN) was established in order to manage production and business of electricity in the entire country. This significant event was one of the initial efforts made by the Vietnamese government to separate the state from business management.

1997: − The Government updated its Power Sector Policy Statement. The updated policy statement outlined the objectives and strategy of the Government and the Ministry of Industry for development of the power sector in Vietnam.

− The power sector reform strategy included transforming the power sector enterprises into efficient commercially run entities that are financially strong and creditworthy, and have management autonomy in operations developing a regulatory framework (including promulgating an electricity law), adopting cost-based tariffs, and promoting private sector participation in the power supply.

1998: − EVN start to buying electricity from IPPs. The two largest BOT power plants are Phumy 3 and Phumy 2. 2.

2001: − The Government issued an Action Plan in 2001 to implement the Decision made at the third session of the Communist Party Central Committee, IX. The key points of the Government 2001 Action Plan include: (i) restructuring and improving the efficiency of State Corporations, (ii) building several strong economic groups by: adjusting and adding a model Charter or organizing and operating State Corporations (as issued in Decree No. 39/CP), promulgating a Decree on the transformation of the current structure of State Corporations into the parent-subsidiary model, and drawing up a plan for establishment of economic groups.

2003: − Upon approval from the Prime Minister of the "Overall Restructuring, Reform and Development Plan of State-Owned Enterprises of EVN for the Period 2003 to 2005" (Decision No. 219/2003/QD-TTg) transformation of the ownership structure of certain member units of EVN into EVN-owned single-member limited liability companies, joint-stock limited liability companies will be carried out, or involve equitization of the company such as listing company's stocks in the stock market after restructuring.

2004–2005: – Creating new single-share limited liability companies such as the Ninhbinh Power Ltd., Haiduong Power Ltd, Company, and the Cantho Generation Ltd. Company.

- The EVN has equitized sixteen subsidiaries, in which the two most important ones are the Vinhson-Songhinh hydro power plant and the Khanhhoa electricity provincial department. Shares of these two companies were offered in the Vietnam stock exchange.

2005: – Electricity Law enacted. The law contains several provisions for regulatory framework, market structure and tariff reforms for the power sector.

- Launching of an internal trial market for power plants that belong to EVN.
- Upon approval from the Prime Minister of the Additional Study on Rearrangement and Renovation of SOEs under EVN (Decision No. 12/2005/QD-TTg), the EVN transformed 7 of its power plants into independent accounting units for further equitization.
- Launching of IPOs via auctioning of the Phalai generation company Ltd.
- Establishment of an electricity and gas regulatory agency within the Ministry of Industry.

- **Reform Elements to Date**

Reform Elements	
1. Regulation	
− Legal framework for EI reform (R 1)	
+ Level 1	Y
+ Level 2	N
− Independent regulatory agency (R 2)	N
2. Unbundling & Competition	
− Corporatization (C 1)	
+ Level 1	Y
+ Level 2	N
− Vertically unbundled (C 2)	Y
− Grid access (C 3)	Y
− New entrants in generation (C 4)	Y
− Wholesale market (C 5)	Y[17]
− Retail market (C 6)	N
3. Privatization	
− Non-government proportion in electricity generation (P 1)	
+ Level 1	Y
+ Level 2	N
− Issuing corporation shares/bonds (P 2)	N
− Separating and privatizing existing companies (P 3)	Y

---

17 Trial internal market launched in 2005.

LAOS

- **Overview** (*data in 2004*)
Population (million persons): 5.84
GDP/Capita (USD): 445.13
Installed Capacity of power plants (GW): 0.64
Electricity Generation (TWh): 3.77%
Nuclear generation: 0%

EdL is a state-owned corporation under the Ministry for Industry and Handicrafts which owns and operates the country's main generation, transmission, and distribution assets in the Lao PDR, and manages electricity imports into its grids and exports from its stations. EdL also has a project development role and has been the implementing agency for government hydropower power projects and in the case of IPP projects is the Government's shareholder.

- **Objectives of EI Reform**
- Maintain and expand an affordable, reliable, and sustainable electricity supply in Lao PDR to promote economic and social development.
- Promote power generation for export to provide revenue to meet GoL development objectives.
- Develop and enhance legal and regulatory framework to effectively direct and facilitate power sector development.
- Reform institutions and institutional structures to clarify responsibilities, strengthen commercial functions, and streamline administration.

- **Reform Process**
1997: – The Electricity Law was enacted.
   – A licensing system has been established to award concessions, including terms and conditions.
   – Facilitation of IPP development, especially in hydro power projects.
1999: – The Environmental Protection Law carries provisions for hydropower development in terms of regulations.
2000: – A restructuring proposal was offered by the Ministry of Finance to solve the problems facing EdL (cash flow deficit) and allow it to meet loan

covenants.

- **Reform Elements to Date**

Reform Elements	
1. Regulation	
− Legal framework for EI reform (R 1)	
+ Level 1	Y
+ Level 2	Y
− Independent regulatory agency (R 2)	N
2. Unbundling & Competition	
− Corporatization (C 1)	
+ Level 1	Y
+ Level 2	N
− Vertically unbundled (C 2)	N
− Grid access (C 3)	N
− New entrants in generation (C 4)	Y
− Wholesale market (C 5)	N
− Retail market (C 6)	N
3. Privatization	
− Non-government proportion in electricity generation (P 1)	
+ Level 1	Y
+ Level 2	Y
− Issuing corporation shares/bonds (P 2)	N
− Separating and privatizing existing companies (P 3)	N

CAMBODIA

- **Overview** (*data in 2004*)
Population (million persons): 13.54
GDP/Capita (USD): 363.3
Installed Capacity of power plants (GW): 0.273
Electricity Generation (TWh): 1.04%
Nuclear generation: 0%

The electricity sector is very small in size. The Ministry of Industry, Mines and Energy (MIME) is the government agency responsible for the coordination of electricity sector policy, planning, and development. The electricity supply consists of 22 small isolated power systems. These systems are divided into two parts; (i) Phnom Penh with the 100,000 customers and six provincial towns served by Electricite du Cambodge (EDC) and (ii) the remainder is served by MIME or by private companies under contract with MIME. Electricity is available to only 13 percent of the total population.

EDC was established by a Royal Decree in March 1996, as a limited liability company fully owned by the Government.

- **Objectives of EI Reform**
- Create favorable conditions for capital investment in the commercial operation of the electricity sector.
- Alleviation of shortages in reliable power and reduction in electricity costs for EdC's grids.
- Commercialization of the EdC.
- Improvement of rural electricity access.

- **Reform Process**
1996: - The Royal Government of Cambodia has initiated an electricity sector reform process through the issuance of a Royal Decree that separated the Electricite du Cambodge (EDC) from the Ministry of Industry, Mines and Energy (MIME). The EDC was made a wholly, state-owned limited liability company for the Government to generate, transmit, and distribute electricity throughout Cambodia.
1999: - The Cambodia Power Sector Strategy was promulgated by the Ministry.

This strategy provides a comprehensive development framework for the development of the power sector within a twenty-year period. It covers all aspects of the sector's development: the Generation and Transmission Master Plan, Rural Electrification, Electricity Trade, the sector's Regulatory Framework and Commercialization/Private Participation.

2001:
– Electricity Law was promulgated.
– The Electricity Authority of Cambodia (EAC) was created as an independent regulatory framework. The purposes of establishing the Independent Regulatory Framework are to: (i) create favorable conditions for capital investment in the commercial operation of the electricity sector, (ii) protect the rights of consumers to supplies of reliable electricity at reasonable cost, and (iii) ensure that the provision of electric power services are governed by the principles of efficiency, quality, continuity, adaptability, nondiscrimination, transparency and wherever practical, competition.

- **Reform Elements to Date**

Reform Elements	
1. Regulation	
– Legal framework for EI reform (R 1)	
+ Level 1	Y
+ Level 2	Y
– Independent regulatory agency (R 2)	N
2. Unbundling & Competition	
– Corporatization (C 1)	
+ Level 1	Y
+ Level 2	N
– Vertically unbundled (C 2)	N
– Grid access (C 3)	N
– New entrants in generation (C 4)	Y
– Wholesale market (C 5)	N
– Retail market (C 6)	N
3. Privatization	
– Non-government proportion in electricity generation (P 1)	
+ Level 1	Y
+ Level 2	Y
– Issuing corporation shares/bonds (P 2)	N
– Separating and privatizing existing companies (P 3)	N

CHAPTER 7
ANALYSIS OF EFFECTS OF COMPETITION, OWNERSHIP AND REGULATION ON ELECTRICITY PERFORMANCE IN ASIAN COUNTRIES AND PREDICTIONS ABOUT THE SHAPE OF THE INDUSTRY

Electricity industry reform has been talked since the beginning of 80's. Over twenty years of development, EI reforms have spread worldwide from developed countries, such as the United Kingdom, New Zealand, etc. to developing countries including those in the Asia and Pacific regions. In contrast, at the end of 90's and especially the beginning of 2000's, the view regarding reform of the EI changed. After a series of financial crises, corporate scandals, the California electricity crisis, and a series of major blackouts in North America and other areas of the world, the need to find out how EI reform should be carried out arose. Several countries re-examined their reform strategies, while others even temporarily delayed the privatization process.

Within this context and in the light of accumulated experience from countries around the world since the 1990s, the empirical literature on reform can make an important contribution to the debate and, more importantly, serve as a basis for policy-making decisions.

In this chapter of the book, an econometrics analysis will be carried out. Based on that, policy makers are able to address the impacts of each reform policy option to the EI performance.

7. 1 Reviews on Empirical Studies of Electricity Sector Reform

The empirical literature on EI reforms and industry performance started being written in earnest around 1997–98. Economists felt they had enough data to be able to test formal hypotheses and make substantive conclusions about developments in electricity industry reform, rather than merely speculate about them or rely on anecdotal, country-specific evidence.

In this subsection, we discuss empirical literature that uses econometric techniques to study electricity liberalization assessment. Although the objective of the author is finding the impact of each reform policy option to EI performance, the literature reviewed broadly covers two categories: (i) the determinants of reform and the key steps taken, and (ii) the effect of various reform steps on performance indicators. Several papers that do not directly address the impact of reform, but give insight into how to model and estimate the impact, and determinants, of reform are also included[18].

- **The Effect of Various Reform Steps on Performance Indicators**
[Zhang, Parker and Kirkpatrick, 2002]
This paper explores the role of competition, privatization, and regulation in 51 countries in Africa, Latin America, the Caribbean, and Asia, and aims to identify the effects of particular aspects of reform in developing countries on: (i) Net electricity generation per capita, (ii) Installed generation per capita (i.e. the BHS proxy for investment), (iii) Net electricity generation per employee (i.e. labor productivity), (iv) Generation relative to average capacity (i.e. capacity utilization), and (v) Final prices to consumers.

The empirical work uses data for 1985–2000 and applies a fixed-effects panel data estimation technique as in similar studies that have been carried out for telecoms. The objective of the paper and its methodology are good practice, particularly regarding the specification of the hypotheses to be tested. Besides the institutional variables, the regressions include standard control variables, for example per capita GDP, share of industry output in GDP, and an urbanization indicator. The problem is that the data for the key institutional independent variables are, as the authors recognize, low powered and poor quality proxy variables for the underlying effects that they wish to test.

The precise definition of the variables is as follows: (i) Privatization. Countries are given a score of 1 if there is any private sector generation and zero otherwise. This is more a measure of private participation than of privatization. Based on this system, India would presumably count as a "privatized" country, just as Chile would, (ii) Competition. Countries are given a score of 1 if there is either a wholesale generation market and/or if large users negotiate contracts directly with generators. This is more directly relevant than (i) but again, as the authors recognize, it would classify a system with a heavily dominant incumbent and one or two IPPs selling to industrial firms (e.g. Jamaica) as a competitive market, (iii) Regulation. Countries are given a score of 1 if there is a separate regulatory agency "not directly under the control of a ministry." The results in the reported regressions are interesting and stimulating, also but very preliminary. The problem is that the data

---

18 This section has taken some reviews on econometric studies from [Jamasb et al., 2005 a].

used for the key variables provides poor proxies. This is the case for investment and, especially, for regulation and privatization variables. Hence, the author's interpretation of their results for policy purposes cannot be sustained with this information set. It needs much better data.

The paper was updated in 2005 with a modification of reform outcome definitions. The updated version provides an econometric assessment based on a panel data for 36 developing and transitional countries, over the period 1985 to 2003. To overcome the drawbacks of dummy variables in the 2002 analysis, a new set of reform measurements are employed. Regulation is calculated based on a four-component regulatory index, competition is measured by the square root of the difference between 100 and the market share of the three largest generators, and privatization is measured by the share of privately-owned capacities (%).

The study identifies the impact of these reforms on generating capacity, electricity generated, labor productivity in the generation sector, and capacity utilization. The main conclusion is that privatization and regulation do not lead to obvious gains in economic performance on their own, though there are some positive interaction effects. By contrast, introducing competition does seem to be effective in stimulating performance improvements.

[Zhang, Parker and Kirkpatrick, 2005]
This paper studies the effect of the sequencing of privatisation, competition and regulation reforms in electricity generation using data from 25 developing countries for the period 1985 to 2001. A fixed effects panel data model is used. Two sets of hypotheses are tested regarding (i) The order between privatisation and competition, and (ii) The order between privatisation and regulation. The primary performance indicators used in the study as dependent variables are net electricity generation per capita, installed generation capacity per capita, electricity generation to average capacity (the capacity utilisation rate) and net generation per employee (labour productivity).

A set of dummy variables are used to measure privatization, competition and regulation. The dummy takes the value of 1 beginning the year when some privatisation in generation occurs, even if only partial. Similarly, competition was measured by a dummy variable that equalled 1 either when a wholesale market for electricity was introduced or generators were first allowed to sign contracts with distributors or large users. Regulatory effectiveness is measured by a dummy variable to indicate whether a country claimed to have an electricity regulatory agency not directly under the control of a ministry[19]. The regulation dummy takes

---

19 This kind of regulatorry agency is described as 'independent regulation', although the degree of independence from government can be quite limited in practice.

the value of 1 beginning from the year the regulator was established.

Two sequencing variables are also taking the form of dummies. One dummy measures the order between competition and privatisation, which equalled 1 from privatisation if the country introduced competition before privatising its generating capacity. The sequencing between regulation and privatisation is measured by another dummy, which similarly took the value of 1 if an independent regulator was in place before privatisation and beginning from the year privatisation occurred. Macroeconomic and demographic variables were included in the estimations as environmental controls, including GDP per capita, the degree of urbanisation and the degree of industrialisation.

The study finds that establishing an independent regulatory authority and introducing competition before privatisation is correlated with higher electricity generation, higher generation capacity and, in the case of the sequence of competition before privatisation, improved capital utilisation also.

[Steiner, 2001]
Steiner tests whether the regulatory environment, the degree of vertical integration, and the degree of private ownership have an impact on efficiency and on prices. Efficiency is measured by the capacity utilization rate and reserve margin in electricity generation. Liberalizing regulation, restructuring, and private ownership are expected to lead to improved efficiency, and lower industrial electricity prices and industrial/residential price ratios. These hypotheses are tested for a panel dataset of 19 OECD countries for the period 1987–1996. There are controls for previous commitment to generation technology and the control for GDP serves as proxy for electricity market size. There are no controls for institutions or for macroeconomic policy.

[Steiner, 2001] finds that utilization rate is positively and significantly correlated with both private ownership and unbundling of generation and transmission. The coefficient on third-party-access, however, is not significant. Contrary to expectations, the coefficient on ownership is positive and significant, which suggests that private ownership is not necessarily correlated with increased competition. The establishment of a spot market was found to lead to lower prices.

[Hattori and Tsutsui, 2004]
Arguing that the precise definitions of the indicators are critical to this kind of empirical work, this paper reproduces Steiner's model for the same sample countries, but for the period 1987–1999, changing slightly the definitions of regulatory reform indicators. In contrast to [Steiner, 2001], [Hattori and Tsutsui, 2004] finds that the existence of a wholesale market is, statistically speaking, significantly positive for prices, and third party access is significantly negative statistically. Also, in contrast to Steiner's results, the sign of the private ownership coefficient is found to be

significantly negative for prices. The sensitivity of results to subtle changes in the definitions of variables recommends caution in the specification of reform and performance variables. Therefore, as well as specifying the appropriate model to test relevant hypotheses, one should carefully consider how to define variables, especially how to represent the various dimensions of reform.

- **The Determinants of Reform**

[Bacon and Besant-Jones, 2001]

This analysis tests hypotheses on to the determinants of reform. First, that country policy and institutions are positively correlated with reform where reform is understood as a number of key steps, such as corporatization of the core utility or enactment of legislation. Country-level policy and institutions are expected to be positively associated with reforms. The country policy and institutional assessment variable is based on 20 indicators. However, the indicators are not clearly identified. Economic management and sustainability of reform are among those mentioned.

Second, that country risk is negatively correlated with reform. Once again this variable is not clearly identified in the paper, though it is stated that it is made up of nine indices, of which political risk and economic performance account for 25% of the weighting. These hypotheses are tested for a sample of 115 developing countries for 1998. Although formulated to address reforms in the electricity sector, they could refer to liberalizing reforms in general, as both policy/institutions and risk variables refer to overall management of the economy and of the public sector. In addition, some regional effects are detected, suggesting that countries in Latin America and the Caribbean are more likely to initiate reform, while countries in the Middle East and Africa are more likely to have taken fewer reform steps.

[Rufin, 2003]

The analysis is an econometric study of the institutional determinants of competition, ownership, and the extent of reform (dependent variables) in electricity sector restructuring. The institutional determinants (explanatory variables) used are different measures of judicial independence, distributional conflict, and economic ideology. The study uses a cross-section OLS regression analysis of a set of models with observations of up to 75 developed and developing countries that undertook some steps toward electricity reform during the 1990s. The study finds that the relation between judicial independence on the one hand, and competition and ownership on the other, is ambiguous; i.e. the coefficients are often insignificant or, when significant, their sign shifts across models. The results also suggest that greater distributional conflict is significantly correlated with a higher degree of monopoly, while in ownership models, the coefficients are mostly not significant.

An important issue is the definition and construction of variables that comprise different components. The dependent variables in the study aggregate relevant aspects of the sector, and do not separate the effect of institutional factors on potentially competitive generation and regulated transmission and distribution networks. There is also a temporal problem: the dependent variables seem to be calculated for 1998, while the variables that are used for the computation of the independent and control variables are for different time periods. The study does not cover endogenous elements that are likely to exist in this context.

[Drillisch and Riechmann, 1998]
The analysis tests the correlation between energy dependency and the choice of liberalization model, as well as commitments concerning the environment and the choice of liberalization model. The paper also incorporates the date in which reforms started. For each year of delay, starting from 1990, half a point is deducted from the liberalization index. The study is limited to a cross-section analysis and the effect of time is not explicitly incorporated into the model. [Drillisch and Riechmann, 1998] do not see reform as a sequence of steps, but use two types of models: one for the wholesale market and one for the retail market. There is no control for institutional conditions, macroeconomic policy, or the level of development or market size. Another drawback of the study is the very restricted dimensions of reform included in the liberalization index. For instance, it does not take into account corporatization, privatization, restructuring, regulatory change, and the introduction of an independent regulator.

7.2 Rationales of the Study

Several conclusions can be drawn from this brief review of econometric studies. First, most of the previous studies examine the impact of reforms on OECD countries and extensively reformed countries, such as those in Europe. A few studies research developing countries, mainly Latin America or transitional countries. Only a few Asian countries are considered in such analysis, but only on an individual basis. No econometric work appears to have been undertaken to test the effects of reform in Asian countries. This is probably because of a lack of consistent data for power sector in Asia and partly because of the difficulty in accurately assessing the various reforms adopted. Asia, a relatively conservative continent in regard to EI reform, is looking for moderate reform policies. This research will focus on analyzing the situation of Asian countries.

Secondly, in most of studies mentioned above, the aggregation method is utilized widely to obtain overall scores for the aspects of reform. While scoring may favor observing the position of each country in the reform process, reducing the number of variables and facilitating interpretation of test results obviously leads to the loss of information concerning differences across countries in the specific choices made for each variable.

Additionally, applying the aggregation method implies that the discrete variables are turned into continuous ones, i.e. all the qualitative data are considered as quantitative data, which is unacceptable from the strict viewpoint of econometrics.

To cope with the limitations of previous studies, this analysis is conducted based on a set of detailed data which is exempted from any level of aggregation, either between different variables or among levels of each variable.

Therefore, based on the review of previous studies, we have recognized the rationales for the study. Several other important conclusions drawn from that research are related to the methodology to carry out this analysis. At the current stage of EI reform, it should work better with panel datasets than with simple cross-section models, preferably containing pre- and post-reform data. However, estimating a model containing time-series cross-section data usually implies a complicated error structure. Problems such as serial and/or contemporaneous correlation and heteroskedasticity should be addressed in the estimation procedure. Ignoring the presence of serial correlation results in consistent, but inefficient estimators. Similarly, assuming homoskedastic disturbances, when heteroskedasticity is present, will still result in consistent estimators, but correction for heteroskedasticity makes the estimation far more efficient [Baltagi, 1995]. Dealing with contemporaneous correlation across panels is a more complicated matter, considering the availability of data.

The second notable issue is, when dealing with a varied sample of developed and developing countries and a larger time series, country and time effects should be included in the model in the form of a random effects specifications and a fixed effects specifications. For each of these specifications, country-specific effects are assumed to occur for constant term, so that slope coefficients are identical across countries. The random or fixed effects account for country-specific effects not otherwise included in the regressions. The random effects specification assumes that the relevant units, in this case countries, are drawn at random from a larger population, while the fixed effects specification is more appropriate if the focus is restricted to a specific set of units. The relative merits of the two specifications for panel data are discussed in many papers and it could be used Hausman tests to choose the effects specification. However, the review of the literature also shows the weakness of such tests and mostly the decision to use fixed or random effects should be based on the characteristics of country's data set. The fixed effects specification makes sense in the context of this study, if we consider that it consists of a sample of specific Asian countries, so that country-specific constant terms are fixedly distributed across cross-sectional units.

7.3 Data and Modeling

- **Data**

In this analysis, several hypotheses (please refer Section 7.4) were tested using panel data for 18 countries in Asia, including developed countries, like Japan or Singapore, to poor developing countries, such as Bangladesh. The analysis period covers 20 years from 1985 to 2004. The starting date for the study, 1985, was dictated by data availability, though little reform of the electricity sector began before this date. The final date, 2004, represented the last year for which data were available at the time the research was conducted. The choice of the sample countries was based on access to data and especially information on privatization, competition, and regulation in each country. Even so, not all data exist for all years for in all 18 countries.

Dependent Variables

This analysis is carried out to evaluate the impact of different reform policy options on industry performance. Therefore, performance indicators should be utilized as dependent variables for the purpose of economic modeling. Efficiency is difficult to measure in the electric supply industry. The choice of indicators used was driven partially by data constraints and measurement difficulties. The following indicators were used for this research.

- Net electricity generation per capita of the population (*gen*)

  Net electricity generation per capita measures the extent of electricity available to the economy. Most likely that the higher level of this indicator, the more matured industry is. It should be noted that this indicator has a strong relationship with the economic indicator. In fact the GDP elasticity is one of the most essential bases for forecasting electricity demand. Therefore, it requires incorporating economic indicators as a control variable in the model. It will be presented in following paragraph.

- Installed generation capacity per capita of the population (*cap*)

  This indicator, to some extent, shows the investment level in electric generation. Similar with to generation indicator, this indicator is also greatly influenced by economic indicators.

- Capacity utilization rate (*uti*)

  The capacity utilization rate is calculated as energy production divided by total average capacity and multiplied by 8760 hour. This is a measurement of capital productivity, which is a very important indicator in case of a capital-intensive industry like electricity generation. Greater usage of capacity implies greater (productive) efficiency, though this clause is not always true. Because a higher utilization rate also means a lower reserve margin for the system. Therefore, discussion of empirical results should not depend much on the

results from this indicator.
- Electricity loss ratio (*los*)
  Electricity loss ratio is one of the famous indicators used to measure the operating performance of the industry. However, one point that should be taken into consideration is the period of expansion for the electricity system, especially into rural areas, the electricity loss increase rather than decrease. This situation happens in many developing countries, for example India, Bangladesh, etc. It can make the interpretation of empirical results more difficult.
- Share of nuclear power generation (*nuc*)
  The share of nuclear power generation may not be directly considered as a performance indicator. However, it could show the level of technological development, the level of proper fuel mix, and especially the level of energy independence which is, in energy-imported countries like Japan or Korea, highly dependent on nuclear power.

Another candidate proxy for efficiency measurement would be labor productivity (output per unit input); however, the appropriate data regarding employment in the electricity supply industry is not available, and a measure of labor productivity gives little indication about efficiency in a capital-intensive sector like electricity. One more potentially useful measure of performance, quality of service, could not be estimated because of a lack of data. Similarly, we would like to have investigated the impact of reforms on the prices charged for electricity generated, but there is a lack of sufficient comparable data across our sample of countries to carry out such an analysis. Therefore, the empirical examination of such measurements is left for future work.

The indicators of generation, capacity, and capital efficiency were calculated based on data from Energy Information Administration, International Energy Annual 2003 database, and Asian Development Bank (ADB) – Key Indicators (www.adb.org/statistics). In some years, modifications are made based on updated data for country or industry relevant websites.

Independent Variables

The same set of reform data mentioned in Chapter 6 is again employed here to show the different reform policy options. Only one change happens occurs in the privatization indicator, the share of non-government generation, which changes from dummy tonumeric data in order to better evaluate the reform process.

Regulatory Indicators
- Legal framework indicator:
  - Whether reform policy exists? (R 1_1)
  - Whether the reform is provided for by Electricity/energy laws? (R 1_2)

- Regulatory body indicator. Whether it is fully independent?

Unbundling Indicators
- Corporatization core utility:
  - Whether corporatization starts? (C 1_1)
  - Whether it completed? (C 1_2)
- Generation, Transmission, and Distribution unbundling
  - Whether unbundling Generation, Transmission and Distribution happens? (C 2)
- Grid access
  - Whether TPA is allowed? (C 3)
- New entrants in generation
  - Whether IPPs is allowed? (C 4)
- The occurrence of whole sale market
  - Whether power pool establish? (C 5)
- The occurrence of retail market
  - Whether retail market exists? (C 6)

Ownership Indicators
- Non-government generation. How much is the share of non-government generation? (P1)
- Selling shares/bonds of the core utility
  - Whether the core utility issues its shares/bonds? (P2)
- Separate and privatize a part of the utility's assets
  - Whether the existing utility launches the program of separating and privatizing a part of its assets (both in Generation, Transmission or Distribution)? (P3)

Control Variables

As mentioned above, economic indicators have strongly influenced on the dependent variable, especially capacity expansion and electricity generation. Given to the availability of data, in this analysis there considers three economic indicators: GDP per capita, industry contribution to total GDP and the export contribution.

- **Modeling**

The empirical approach uses cross-country and time-series variation to examine the impact of each reform policy option, both in terms of regulation, restructuring and competition, and privatization on the industry's performance. The analysis also attempts to control for country-specific and economic variables. Section 7. 2 has discussed rationales of the analysis and selected a fix-effect model to capture the empirical analysis. Thus, taking into account the number of variables described above, it could be defined by the following model to explain each performance measure:

$$y_{it} = \alpha_i + \beta_{1\_1}(R\ 1\_1_{it}) + \beta_{1\_2}(R\ 1\_1_{it} \times R\ 1\_2_{it}) + \beta_2(R\ 2_{it}) + \beta_{3\_1}(C\ 1\_1_{it})$$
$$+ \beta_{3\_2}(C\ 1\_1_{it} \times C\ 1\_2_{it}) + \beta_4(C\ 2_{it}) + \beta_5(C\ 3_{it}) + \beta_6(C\ 4_{it}) + \beta_7(C\ 5_{it})$$
$$+ \beta_8(C\ 6_{it}) + \beta_9(P\ 1_{it}) + \beta_{10}(P\ 2_{it}) + \beta_{11}(P\ 3_{it}) + \delta(x_{it}) + \varepsilon_{it}$$

in which:
- y_{it} is the performance indicator of generation per capita, installed capacity per capita, capacity utilization rate, electricity loss or share of nuclear power in country i and year t
- $R\ 1_{it}$, $R\ 2_{it}$ is respectively the regulation options in country i and year t
- $C\ 1_{it} - C\ 6_{it}$ is respectively the restructuring and competition options in country i and year t
- $P\ 1_{it} - P\ 3_{it}$ is respectively the privatization options in country i and year t
- Two variables, R 1 and C 1, are shown by two levels R 1_1 and R 1_2 and C 1_1 and C 1_2, which will be treated as interactants.

To increase the variance of variables, logarithmic form is applied for all of non-percentage variables. The model is estimated using panel data techniques across countries and years.

Before conducting the regression analysis, the quality of the data set must be checked. To meet the requirement of the panel data techniques, the examination of the three most bothersome problems is required. They are multicollinearity, heteroskedasticity, and serial correlation.

Multicollinearity

Strictly speaking, multicollinearity is the violation of the assumption that no independent variable is a perfect linear function of one or more other independent variables. Perfect miconllinearity is rare, but severe imperfect multiconllinearity (where two or more independent variables are highly correlated in a particular data set being studied), while not violating classical assumption IV, still causes substantial problems. The major consequence of severe multicollinearity is to increase the variances of the estimated regression coefficients and therefore decrease the calculated t-score of the coefficients [Studenmund and Cassidy, 1987]. Hence, it is important to check if high multicollinearity occurs in the model, particularly in the panel data estimation. Several useful ways to detect multicollinearity have been agreed upon. It can be checked by the correlation matrix among explanation variables or using the variance inflation factor (VIF = 1/(1-Rk 2)) [Maddala, 2002]. A commonly given rule of thumb is that a VIF of 10 or higher may be one of the reasons for concern.

Heteroskedasticity

Heteroskedasticity is the violation of Classical Assumption V, which states that the error terms are drawn from a distribution that has a constant variance. As discussed by [Baltagi, 1995], assuming homoskedatic disturbances when heteroskedasticity is presented will still result in consistent estimates of the regression coefficients, but these estimates are no longer efficient. Also, the standards of these estimates are biased unless one computes robust standard errors correcting for the possible presence of heteroskedasticity. Hence, it is very important to assess whether the heteroskedasticity problem in the model is severe or not. In order to detect heteroskedasticity problem, it proposes to use White's procedure.

Serial Correlation

Serial correlation, or autocorrelation, is the violation of the classical assumption that the error terms are independent of each other. Pure serial correlation is serial correlation that is a function of the error term of the correctly specified regression equation. Impure serial correlation is caused by specification errors such as an omitted variable or an incorrect functional form. Because serial correlation in linear panel-data models biases the standard errors and causes the results to be less efficient, researchers need to identify serial correlation in the idiosyncratic error term in a panel-data model. The commonly used method of detecting first-order serial correlation is the Durbin-Watson d test, which uses the residuals of an estimated regression to test the possibility of serial correlation in the error term. Recently, a new test for serial correlation in random- or fixed-effects one-way models derived by [Wooldridge, 2002] is attractive because it can be applied under general conditions and is easy to implement. With the advantage that it could be connected to the STATA program, which will be selected for regression analysis and is presented below, it proposes use of the Wooldridge test to check the serial correlation problem in this analysis.

Results of such tests show that there are not any serious problems with multicollinearity, heteroskedasticity, and serial correlation. All of criteria are significant at the level of 95%. Detail results are presented in Table 7.1 below. However, considering the severe impact of heteroskedasticity in panel data, as suggested by [Gujarati, 2003], it is better to account for heteroskedasticity, even if it is not so serious. Therefore, White's heteroskedasticity-corrected estimations (robust standard error estimation) are employed in all regression equations.

- **Software Selection**

There is a great number of statistical software which can be employed for regression. With the characteristics of panel data analysis, STATA is proposed. This analysis is indebted to Prof. Shoichi Ito of Kwansei Gakuin University, for his kindness in allowing me to use his computer and the STATA program to conduct the empirical analysis.

7.4 The Research Hypotheses

Hypotheses on Privatization

Based on the theoretical and literature review, the hypothesis that each privatization policy option may lead to higher capital utilization, more capacity, and hence higher output, and result in lower electricity losses and have a negative impact on the share of nuclear power generation is proposed.

Hypotheses on Competition

It is proposed that more competition policy options may have a positive impact on capacity expansion, electric generation, and capacity utilization, lead to lower electricity loss and harm the share of nuclear power generation, except for the commercialization and corporatilization option, which may have positive impact on company capability and increase the share of nuclear power.

Hypotheses on Regulation

It is proposed that the maturity of regulatory framework will improve capacity utilization, raise investment in the industry leading to more capacity and more electricity generated, and reduce electricity loss. It may cause a reduction in the share of nuclear power generation.

7.5 Results and Discussion

- **Empirical Difficulties**

The results of the empirical analysis should be considered with caution. First, as discussed above, the performance measures are imperfectly represented by net generation per capita, installed capacity, capacity utilization rate, loss ratio and nuclear power share. Moreover, there are issues of cross-country comparability of the data. The IEA performance data are based on submissions from national administrations. Different countries have different classifications and reporting conventions, so that observations in a given performance data series may not have the same meaning across all countries.

Another possible source of bias is that the model does not control for technical matters of the electricity supply industry. For example, according to statistics of transmission and distribution losses (T&D losses) it fluctuates all the time. For developed countries like Japan or Korea, the electricity system has matured as the trend of T&D loss has declined. However, for some developing countries, such as India, transmission losses do not seem to decrease much with the ongoing effort to form an integrated network nationwide. The distribution losses may even increase with the expansion of network to rural areas. For those reasons, the result of the regression may not show the real impact of reform policy on the industry.

These errors in performance measurement are less serious than the problem of measurement of regulation and competition indicators. By using a dummy variable to measure regulation and competition policy options, this analysis attempts to avoid the problem of subjectivity and judgment when constructing the quantitative indicators from qualitative information. However, the most severe problem related to the dummy variable, as recognized by the author, is the capability to measure actual development, since there are only two levels, 0 or 1, except two interactive pairs, R 1 and C 1. Moreover, several indicators serve only as crude proxies of the policy options they are meant to measure. For example, the TPA indicator is based on the formal approach to network access, rather than actual use of Third Party Access; there may be no formal barriers to TPA, however simultaneously, a monopolist may be the only producer in a market, making TPA rules moot. The legal TPA may not result in actual entry, and if, the incumbent retains practical control of the market.

Another problem related to the model employed here is the lag of the variable. Several of the variables used in the model measure a policy, which should impact performance indicators for more than one year. To deal with that situation, employment of the dynamic panel data technique may be necessary and should be used for further analysis.

- **Empirical Results and Discussion**

Despite limitations related to the availability of data, the regression may offer several interesting results. The model was estimated for each of the dependent variables, namely, electricity generation per capita, installed generation per capita, capacity utilization, loss ratio, and nuclear power share. The fixed-effect with robust standard error was used to overcome the problem of heteroskedasticity, even though it is not serious. Table 7.1 presents the regression results.

Electricity Generation per Capita

The first two columns in Table 7.1 show the estimated coefficients and t value of each independent variable on net electricity generation per capita under the log form. Most of coefficients are signed positively, which is same with the hypothesis, except the variable of corporatization and retail competition. Figure 7.1 below extracts the coefficients, which are statistically significant at 10% (at least).

From the figure above, it can be observed that the policy of increasing non-governmental generation in the electricity sector (as a measure of privatization) gains the most positive impact to the dependent variable of electricity generation per capita. Second is the occurrence of legal framework for industry reform (both level R 1_1 and R 1_1 xR 1_2). All three privatization options show positive impact on net electricity generation per capita. For restructuring and competition, C 1 has a negative impact, and C 3 and C 6 are statistically insignificant. In terms of regulation options, while R 1 seems to have a good impact on the generation indicator.

Figure 7.1

Variable	Value
p1	0.45502
r1_1	0.12151
c4	0.11015
c5	0.08104
r1_1xr1_2	0.07608
p2	0.07547
p3	0.06039
c2	0.05563
c1_1xc1_2	0.16051 –

Electricity generation per capita

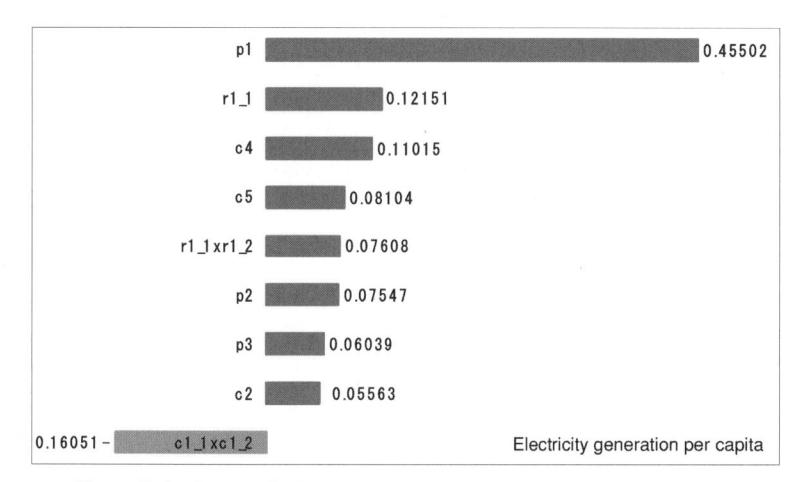

Figure 7.1 Impact of reform options on net electricity generation per capita

Variable R 2 is statistically insignificant.

Figure 7.1 also shows two notations: the negative coefficients of C 1 – commercialization and corporatization and the statistical insignificant of R 2 – establishing independent regulatory body. Commercialization and corporatization should strengthen power companies and therefore, improve their capability to meet demand. Regression results show a negative impact for both levels of corporatization process. Several explanations for this may be: (i) many countries in the analysis data set completed the corporatization process before the starting point of analysis period of 1985, such as Japan, Hong Kong, etc., so the impact of the variable could not be addressed, (ii) for developing countries, such as Bangladesh, Nepal, etc., corporatization has mostly been carried out on the first level, however the impact of corporatization may not be shown immediately in the generation of the first year. Even, in these poor countries due, to the lack of capacity building both in human resources and the institutional framework, most likely at the point of transferring from government dependents to independently operation, power companies may scope with many difficulties. Therefore the positive impact of this policy option could only be understood in longer term[20].

To explain the statistical insignificance of R 2, there also seems to be a lag in the establishment of regulatory bodies. [Jamasb et al., 2005 a] listed forty questions to be employed for measuring the efficiency of regulatory bodies. This means that establishment of a regulatory body is a complicated and time-consuming process. A

---

[20] This relates to the lag problem mentioned in the part of empirical difficulties.

dummy variable is a poor representative to measure this policy option. It may cause this variable to be statistically insignificant.

As expected, GDP per capita is positively correlated with electric generation and the larger the degree of industrialization in a country, the higher the average amount of electricity generation available to each citizen. However, the degree of openness of the economy, as reflected in the exports variable, was not statistically significant.

Installed Generation Capacity per Capita

The regression results on installed capacity per capita in Table 7. 1 (columns 3 and 4) are similar to those for electricity generation per capita. Again, most coefficients are signed positively, which is the same with the hypothesis, except the variable of corporatization and retail competition. Figure 7. 2 below extracts the coefficients which are statistically significant at 10% (at least).

Similar to the case of net generation per capita, the most influenced policy to installed capacity per capita, as well as the second most, is the share of non-government electricity generation and the occurrence of legal framework for industry reform (both level R 1_1 and R 1_1 xR 1_2). However, P 1 is only one of the three privatization options causing a positive impact on installed capacity per capita. For restructuring and competition, C 1 has a negative impact, C 4, C 5, and C 6 are statistically insignificant, while C 2 and C 3 show a strongly positive impact on the indicator. In terms of regulation options, while R 1 has a positive impact on the generation indicator, R 2 is statistically insignificant. It is also seen that from the Figure 7. 2, that completion of the corporatization process also has a negative

variable	value
p1	0.45911
c3	0.11758
c2	0.1079499
r1_1xr1_2	0.09592
r1_1	0.06125
c1_1xc1_2	0.0745137-

Installed capacity per capita

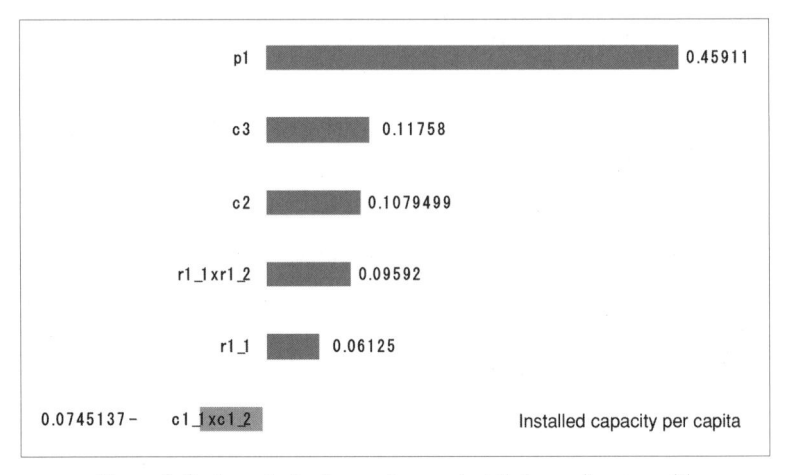

Figure 7. 2 Impact of reform options on installed capacity per capita

Table 7.1 The results from regression analysis

	Electricity generation per capita		Installed capacity per capita		Capacity utilization rate		Electricity loss rate		Share of nuclear power generation	
	Coef.	t	Coef.	t	Coef.	t	Coef.	t	Coef.	t
r 1_1	0.12151	3.91***	0.06125	1.97*	0.02656	2.01**	−0.0006562	0.07	−0.0115761	2.32**
r 1_1 xr 1_2	0.07608	1.91*	0.09592	2.71***	−0.0084587	0.43	0.01708	2.48**	4.2 E−05	0.01
r 2	0.02024	0.53	0.00333	0.1	0.00359	0.18	0.02612	3.3***	0.00703	1.82*
c 1_1	−0.0025814	−0.07	−0.0145407	0.31	0.00525	0.3	0.01198	0.75	0.01229	2.11**
c 1_1 xc 1_2	−0.1605137	−4***	−0.0745137	1.97*	−0.0400	2.67***	0.02300	3.14***	0.03163	3.13***
c 2	0.05563	1.73*	0.1079499	3.66***	−0.0299191	−2.3**	−0.0021488	0.28	−0.0028562	0.42
c 3	0.02325	0.71	0.11758	3.55***	−0.05030	3.83***	−0.000457	0.06	0.01682	3.64***
c 4	0.11015	3.82***	0.01906	0.63	0.04317	3.42***	0.00275	0.43	−0.0125245	2.06**
c 5	0.08104	1.89*	0.01515	0.3	0.03557	2.32**	−0.0156754	1.67*	−0.0124858	1.72*
c 6	−0.074814	−1.5	0.00776	0.11	−0.0425	1.59	−0.0119212	1.04	0.0137296	1.37
p 1	0.45502	5.33***	0.45911	5.53***	0.00818	0.18	−0.1211949	3.71***	1.4 E−05	0
p 2	0.07547	2.08**	0.01562	0.46	0.03776	3.1***	−0.0108677	1.56	−0.0225304	3.23***
p 3	0.06039	1.72*	0.04422	1.18	0.01276	0.86	−0.0037803	0.48	0.00683	1.49
loggdp	0.46913	11.57***	0.30363	8.39***	0.0791301	4.68***	−0.0251259	3.41***	−0.0285077	2.39**
ind	2.50472	7.16***	2.30529	6.12***	0.07974	0.5	−0.4299459	5.95***	0.31964	3.74***
ex	0.00262	1.35	−0.0014667	0.68	0.00184	2.37**	0.00026	0.69	−0.0004135	1.37
cons	1.99786	7.54***	1.91379	7.44***	−0.1461732	1.14	0.47059	8.46***	0.1477496	2.3**
Adjusted R-squared	0.9928		0.9918		0.6786		0.8853		0.9668	
Number of observations	322		322		322		258		324	
VIF test	2.47		2.47		2.47		2.53		2.45	
White test	248.16		227.82		150.01		173.51		232.85	
	0.0000		0.0000		0.0068		0.0001		0.0000	
Wooldridge test	100.33		35.145		47.848		3.913		21.473	
	0.0000		0.0000		0.0000		0.0666		0.0003	

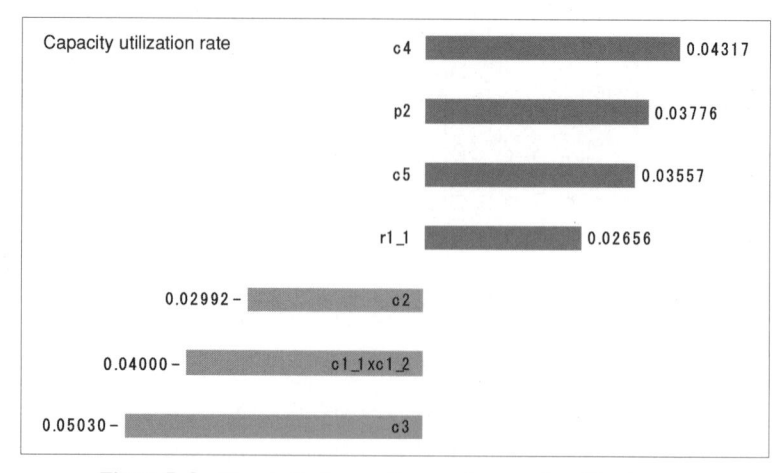

Figure 7. 3 Impact of reform options on the capacity utilization rate

impact. The explanation for this is the same as the above case.
Capacity Utilization Rate
In a capital-intensive industry, like electricity generation, the measurement of labor productivity needs to be supplemented by the measurement of capital productivity, which is represented by the capacity utilization rate. The impact of reform options on this indicator are reported in Table 7. 1, in column 5 and 6. The coefficients, which are statistically significant at 10% (at least), are graphically illustrated in Figure 7. 3.

Compared to hypothesis 3, regression results show a considerable difference. While the more competition options, like generation competition (C 4) and wholesale market (C 5), have a positive impacts t on capacity utilization, which is same to the hypothesis, all three restructuring options, C 1 to C 3, adversely impact the indicator. To explain this situation, the restructuring industry most likely resorts to the requirement of increasing reserve margin to ensure the stable supply. The share of non-government electricity generation, which positively influence to the net generation and installed capacity, is statistically insignificant in this regression. The positive impact may be exterminated due to the numerator of the capacity rate.
Electricity Loss
The regression results for electricity loss are shown in Table 7. 1 in columns 7 and 8. In general, it can be seen from the results that most of reform options take the negative sign, i.e. it leads to improvement in the electricity loss ratio. However, some other options increase of loss rate, for example, R 1, R 2, or C 1. Many coefficients are also statistically insignificant. To detect the problem, the statistics of

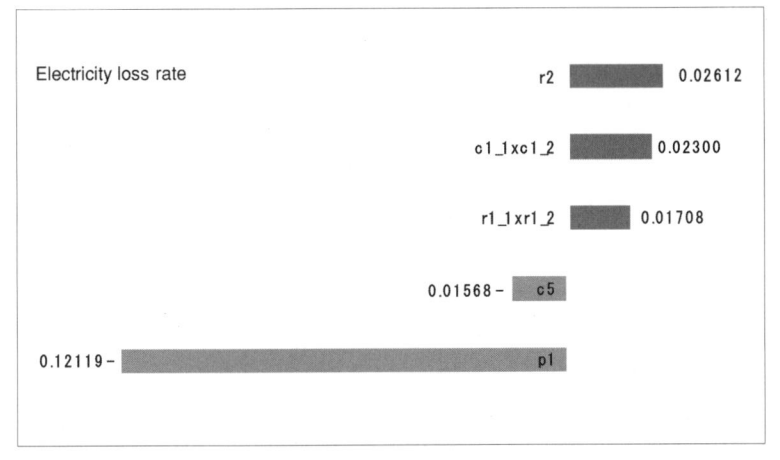

```
Electricity loss rate                     r2  ▇▇▇ 0.02612

                                    c1_1xc1_2 ▇▇▇ 0.02300

                                     r1_1xr1_2 ▇▇ 0.01708

                            0.01568 - c5

        0.12119 - ▇▇▇▇▇▇▇▇▇▇▇▇▇▇▇▇▇▇▇▇▇ p1
```

Figure 7. 4 Impact of reform options on the electricity loss

transmission and distribution losses should be looked at. Transmission and distribution losses constantly fluctuate. For developed countries like Japan or Korea, the electricity system has matured and the trend of transmission and distribution loss has generally declined. However, for some developing countries, such as India, transmission losses do not seem to decrease much despite ongoing efforts to form an integrated network nationwide. The distribution losses even increase with the expansion of the network to rural areas. For such reasons, the regression result may not show the real impact of reform policy on the industry. Figure 7. 4 extracts statistically significant coefficients. Private participation can reduce the electricity loss rate and creates a benchmark to reduce the loss rate to obtain benefits.

The Share of Nuclear Power Generation

The last two columns in Table 7. 1 show the estimated coefficients and t value of each independent variable on the share of nuclear power generation. Being consistent with the hypothesis, many of the more competitive and privatization options harm the share of nuclear power generation, for example, the competitive generation, wholesale market or selling of corporation shares to private owners. The impact of restructuring variables like corporatization, TPA, and especially the establishment of a regulatory body, although they act adversely to the hypothesis, can still be explained. The regulatory body may facilitate addressing nuclear development issues, implementing subsidiaries from government in order to strengthen nuclear policy. Based on these results, it can be considered that the more reform has been implemented in electricity industry reform, the more the attention that should be paid to regulation in order to maintain a sustainable energy policy, especially for energy

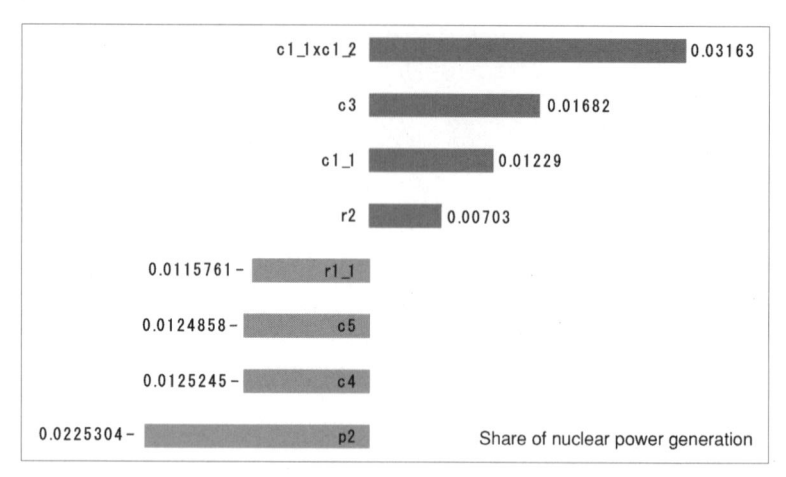

Figure 7. 5 Impact of reform options to the share of nuclear power generation

Variable	Value
c1_1×c1_2	0.03163
c3	0.01682
c1_1	0.01229
r2	0.00703
r1_1	-0.0115761
c5	-0.0124858
c4	-0.0125245
p2	-0.0225304

importing countries, such as Japan or Korea. Figure 7. 5 presents a set of statistically significant effects on the share of nuclear power generation.

Discussion

It can be seen from all five model estimates that the impact of reform options produce the same results (sign) as assumed in the hypothesis. The sign difference, while unique in the case of generation per capita and installed capacity, is repeated in other performance indicators. The reasons for that are the first two dependent variables did not vary much (in term of time series) in the period of analysis and they were also encompassed by GDP per capita. In contrast, variation among the three remaining variables is considerable and may cause inconsistent effects. Inconsistent impacts to the industry performance also mean that such reform options should be applied with caution on a gradual step-by-step policy.

As mentioned in Section 7. 1 and 7. 2, literature reviews and the rationales of the analysis, many previous studies have employed the aggregation method to examine the impact of reform in terms of three categories of regulation, competition, and privatization. So that, it may be not permitted to compare the regression results gained in this analysis with the previous ones.

In order to obtain several comparisons with previous studies, the effects of each reform option on the group categories should be combined. However, it should be noted this combinations unofficially allowed. As a result, discussion should focus on the following areas: (i) all of three aspects of reform have positive impact on generation and installation capacity per capita. These results are different from the conclusion reached by [Zhang et al., 2005]. (ii) Privatization options show an obvious impact on the performance indicators. However, it must be kept in mind that privatization here is only partial privatization.

The research does, however, have a number of limitations which are acknowledged by the author. To begin with, the sample is composed of most Asian countries, except those in Central Asia, which covers Kazakhstan, Uzbekistan, etc. The regression may be biased due to bad sample selection. There is no reason for this to be the case; however it can not be ruled out either.

Another limitation is that the paper has not explored the social and long-term developmental effects resulting from privatization and market liberalization in the electricity sector, which remains the subject of future research.

7.6 Predictions about the Probability and Shape of Change in Electricity Industry Reform: Group Modeling

It seems no ideal electric power reform model to be pursued by developing Asian countries exists. Chapter 6 of this book has shown differences in the current situation, policy, and the characteristics of the reform.

In designing institutional arrangements for the power sector in developing Asian countries, improvement in the economic efficiency of the sector is not the main target, as in Europe and USA. Instead, the three goals of "a stable supply of electric power through inviting private investment", "the stabilization of electric power prices (not price decreases)" and "the eradication of poverty and an increase in the rate of electrification, to be realized by rural electrification" should all be accomplished at the same time. To attain these goals, the government has to clarify the respective roles of government and the market, playing a development role in certain fields where market principles cannot be applied (such as investment in transmission, rural electrification, etc.), and establishing institutions which promote competition in the private sector.

The proper schedule setting for liberalization, as well as the proper design arrangement for institutions of the sector are prepared. It also means that a comprehensive strategy for electricity reform should be utilized. The essential steps and technical roles should include: establishment of transparent electricity market regulations; establishment of a regulatory organization which is independent and has sufficient man-power; unbundling of the power industry and establishing sufficient competition to keep market power under control; setting up an independent market operator and a power system operation; privatization of several electric power utilities, which are now oligopolies, and elimination of several of them in the course of competition and at the same time continuing the promotion of investments in power Greenfield projects. Each country is now in a transition period, moving towards success in privatization, and the issue is how problems occurring during this period can be handled in a flexible and timely manner.

Also in Asian countries, it should be a priority for all stakeholders, including not only the electric power sector itself, but also investors and consumers, have an awareness of being participants and playing active roles in operating within the system.

Considering the characteristics of each country, and in an effort to sketch a "Model to be pursued by the electric power sector in Asia", the following chapter of this book will introduce several verification analyses for group country representatives.

CHAPTER 8
VERIFICATION ANALYSIS: THE CASE OF FOUR ASIAN COUNTRIES, GROUP REPRESENTATIVES

Based on the results and discussions from Chapter 7, this chapter endeavors to develop several reform strategies for Asian countries. Obviously due to certain limitations, the empirical analysis done in the last chapter can serve as a foundation, rather than a final argument, in this chapter. Several case studies will be utilized as a verification and supplementation analysis.

8.1 Policy Implications for the Reform in Asian Electricity Industries

Results from Chapter 7 outline the impact of reform policies on electricity industry performance in the Asian electricity industry over the last 20 years. Due to limitations of cross-country data, the results may not truly show the real impact. There were also several unexpected results. These points were discussed briefly in the discussion section of the last chapter. However, despite limitations in the availability of data, the regression offers several interesting results. It should be noted the policy option most influenced to electricity reform is the policy to increase the share of non-governmental related generation, , and the second most affecting policy is the occurrence of legal framework for industry reform. One more notation is the adverse impact of reform options on the share of nuclear power generation.

Continuing such policies for future development is clear, all three aspects of regulation, restructuring and competition, and privatization should continue to develop in Asian electricity industries. However, which sequence is best? What is a priority? At what pace? How should it be done?

The policy implication for the next stage of the reform process depends not only on the impact on industry performance, as shown in Chapter 7, but also on the irreversible or reversible nature of policy with high or reasonable prices. Among the first considerations for any reform program is whether there is a logical sequence for reforms, and if there is, whether it is too costly to undertake them out of order. Early reforms should address the most important problems, and if possible, build momentum for future reforms and minimize risks of failure and policy reversal. Reversible and less risky reforms can be undertaken more readily than irreversible (or costly to reverse) and more risky reforms. Several irreversible reforms have the advantage of establishing commitment toward future changes, and privatization is often seen as one such reform. However, irreversible reforms require more careful design and assessment.

Regulation should be prepared as a requirement for reform, especially for regulatory framework, legal and function. If the functions are specified clearly, a regulatory body may not be established immediately, but should function in a clear and transparent manner. If priority is given more to competition it should be introduced gradually with at least 5 policies up to the highest competitive level, retail competitive market. Retail market may be implemented in advanced countries with mature electricity systems, such as Singapore. Privatization, as mentioned above has a positive impact on industry performance and investment, but is an irreversible reform and needs to be carried out step by step. Partial privatization might be best strategy for most Asian countries, except countries where electricity is already under private control. In terms of privatization methodologies, the role of non-government generation is important and still needs to be strengthened to attract more private investment for future demand. However, conflict between IPP development and the flexible to development of a wholesale market should be noted. Privatization of existing companies should start in the generation sector. The option of selling a minority share on the stock market should also be considered. More and more options may be used depend on country priority and status.

Another source of results used here is the classification of country groups in Chapter 6. Chapter 6 established four group countries based on the position of the reform process from countries with the old structure largely intact to countries in the second stage of reform. This classification is mainly based on the level of reform, but it also denotes the level of economic development.

Building a future plan for all of Asian countries is an ambitious task. To reduce a margin of the work force, while still ensuring that it is a compatible model for Asian countries, this book focuses on four case studies, which represent each of the four country groups. The countries selected are listed as follows: (i) Bangladesh, (ii) Thailand, (iii) Japan, and (iv) Singapore. In each case study will examine the facts of electricity reform to date, the most influencing factors, and predictions about the shape of change.

8.2　The Case of Japan

- **Industry Structure**

The Japanese electricity industry has been dominated by privately-owned, vertically-integrated electric utilities over the last half century. Regional monopolies are regulated under the Electric Utility Industry Law with an obligation to supply electricity. The Ministry of Economy, Trade and Industry (METI), formerly called the Ministry of Industry and International Trade, (MITI) regulates the electricity rate based on a fair rate of return.

At present there are 10 vertically integrated regional companies. The two dominant companies are Tokyo Electric (TEPCO) and Kansai Power. A third company, Chubu Electric, is also important. These three companies own about 60% of Japan's 216 GW of generating plants. The rest are owned by the six other interconnected companies, Kyushu EPC, Tohoku EPC, Shikoku EPC, Hokuriku EPC, Hokkaido EPC and Chugoku EPC. The other regional company, Okinawa EPC supplies Okinawa, but is not interconnected and owns less than 2 GW of plants. The remainder of the plants are owned mainly by two companies, the Electric Power Development Corporation (EPDC, 16 GW), trading as JPOWER, and Japanese Atomic Power Company (JAPCO, 2.6 GW), which build plants using new or innovative technologies and sells their output to the regional companies. A majority of EPDC was owned by the government with regional companies holding the balance of shares. However, in October 2004, the government and the electricity companies sold their shares and the company is now an independent generator. 90%

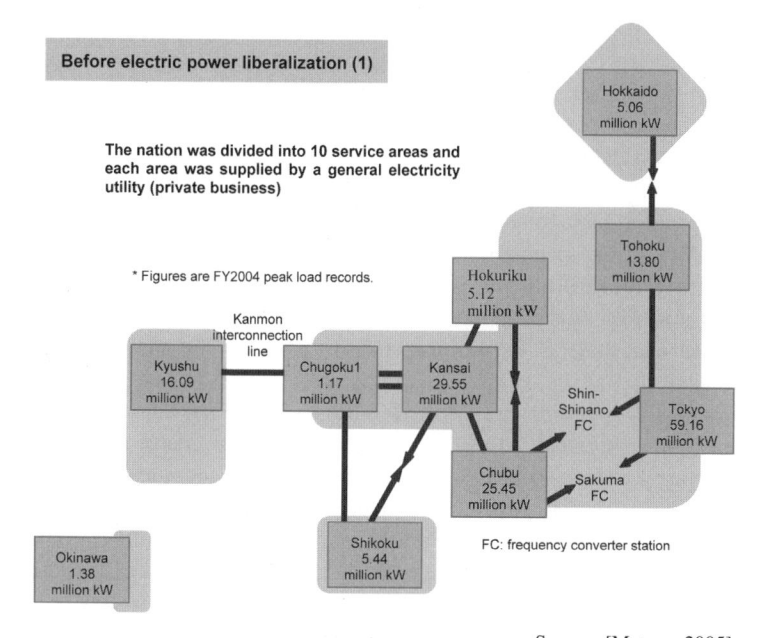

Source: [Matsuo, 2005]

Figure 8.1 Electric power utilities in Japan

of JAPCO's shares are held by the regional companies and JPOWER. In Japan there are 10 private electric utilities, all vertically-integrated from generation to retail supply. They are all regional monopolies, with their own franchise areas. Between generators, inter utility trade is carried out to ensure security of supply. There are three wholesale suppliers of power: the Electric Power Development Co Ltd (EPDC); the Japan Atomic Power Company, and the Joint-Venture Power Utilities.

- **Electricity Reform to Date**

Deregulation of the electricity sector only began recently. In 1995, the Electricity Utilities Industry Law, the main legislation covering the electricity industry, was amended as a result of a number of pressures, including: (i) the global energy sector reform trend, (ii) the comparatively high electricity tariffs in Japan; and (iii) the deteriorating load factor due to a sharp increase in demand in the summer.

1995 Amendment

The main provisions of the amendments made in 1995 to the Electricity Utilities Industry Law are as follows: (i) liberalization of entry for Independent Power Producers (IPPs), (ii) granting permission to allow special electric utilities to supply directly to their customers (Retail sales). The 1995 amendments of the Electric Utilities Industry Law are mainly characterized by the liberalizations of entry by IPPs. Although direct retail sales are now allowed for special electric utilities, they are limited to the suppliers with small capacities. The key amendment is that utilities can now conduct tenders for IPP investment in generation to cover short-term thermal power requirements. In 1996, 20 projects were accepted, with a capacity of 3,046.9 MW. In 1997, proposed projects amounted to 4,254 MW, while 16 projects were accepted for a total of 3,118.3 MW. In Japan, the electricity industry is owned completely by private companies; therefore the share of IPP doesn't serve as an indicator to show the role of private sector participants in the power sector. However, the process of competitive bidding revealed that the costs of IPPs were much lower than those of the existing utilities, increasing the pressure for further liberalization.

A variety of companies are involved in IPP ownership, including iron and steel companies, oil companies, and trading companies. The major source of fuel used by industries that generate electricity in excess of their own requirements (or which plan to build IPP capacity) is coal and oil. The steel industry uses mostly coal-fired generation, and the petroleum refining industries use mostly oil-fired generation. A number of firms in those industries are already auto-producers, possessing idle land on which they can construct generation plants, and having relatively easy access to fuel. Gas plays a relatively limited role in independent power generation. However, it may change drastically with the revision of the Gas Utilities Industry Law, which

allows the breakdown of the industry barriers between the electricity and gas sectors.
2000 Amendment
Even after the 1995 amendment, the Japanese electric industry has been waiting for more changes to take effect. The Program for Economic Structure Reform, which attempts to reduce costs to industry and pursue efficiency improvement, was adopted in December 1996. Subsequently, the Japanese Cabinet adopted "the Action Plan for Economic Structure Reform" in May 1997. The electricity industry has attracted increasing attention, as policymakers have come to realize that there is still more room for cost reduction and efficiency improvement through deregulation.

MITI has undertaken an inquiry into the Electric Utility Industry Council (EUIC), by establishing the Basic Policy Committee in July 1997, to determine the optimal structure of the electricity supply industry. It has been recognized that competition should go hand in hand with the requirement to meet public needs, such as universal service, maintenance of high reliability standards, energy security and preservation of the natural environment. In January 1999, the Committee released its report, and put forward recommendations for partial liberalization of electricity retailing. Subsequently, in May 1999, the Diet passed a bill to amend the Electric Utilities Industry Law. In the revised legislation, partial retail market competition was introduced. As March 21, 2000, extra-high voltage customers can select their supplier. This partial liberalization was expected to bring about a significant impact on the Japanese electricity sector, since 27.7 percent of total electricity demand would be affected. Under the amended law, power producers and suppliers (PPS) were established. PPS were allowed to supply power to contestable customers by using the power network owned by general electric utilities.
2003 Amendment (effective in 2005)
It was decided that all customers with a contract demand of 50 kW or more would be allowed to choose their suppliers by 2005 and a nation-wide wholesale power

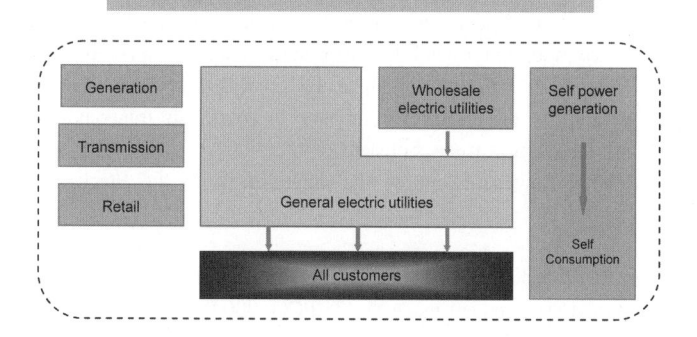

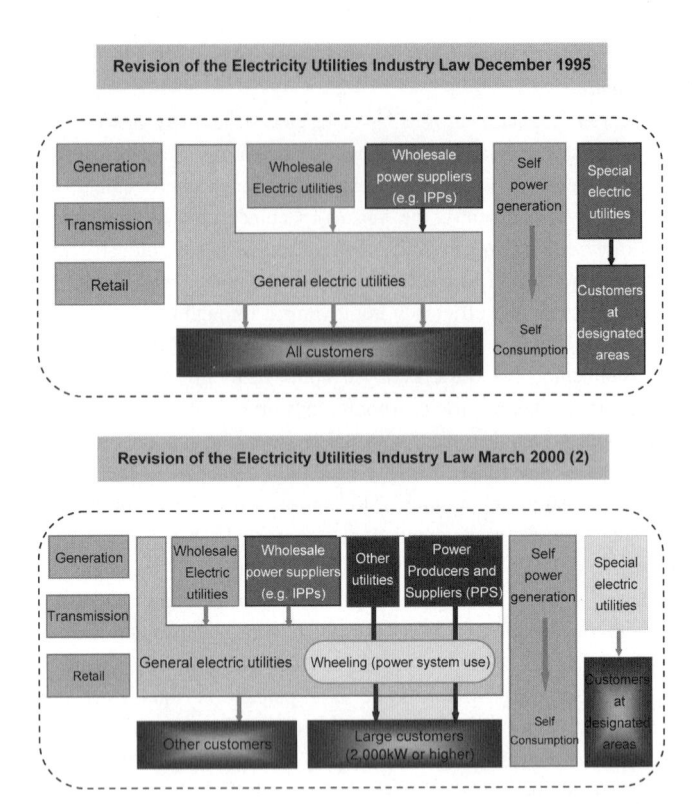

Source: [Matsuo, 2005]
Figure 8. 2 Structural changes in Japanese electric power sector

exchange (Japan Electric Power Exchange, JEPX) would be established by private initiative and commence operations in 2005 for trading day-ahead delivery and forward contracts. Vertical separation of electric utilities was unaffected. Accounting separation was required, and a neutral transmission system organization (later named the Electric Power System Council of Japan, ESCJ) was to be established to monitor the system operation of the electric utilities and to coordinate transmission planning 18. Transfer service charges were abolished.

The following flowcharts outline the amendments to the Electricity Utilities Industry Law in Japan.

- **Features of Electric Power Liberalization in Japan** (Based on the framework of new institutional arrangement made by [TEPCO, 2004])

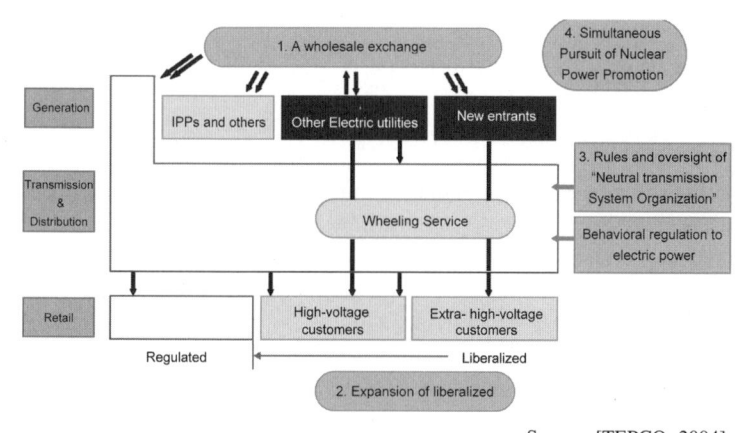

Figure 8. 3 Key points in the design of institutional arrangement

Source: [TEPCO, 2004]

Due to several conditions specific to Japan, such as a low self-sufficiency ratio in the energy supply, sharp demand fluctuations, long lead time for construction of facilities, and steady demand growth and the necessity of capital expansion to meet it, incumbent utilities should be maintained as a supplier with responsibility of supply security under the vertically integrated system from generation to transmission and distribution. The utilities are particularly relied upon for integral development and operation of generation and transmission facilities, as well as promotion of nuclear power.

Figure 8. 3 below shows the key points of institutional arrangement for EI reform in Japan.
- Japan Electric Power Exchange (JEPX). Participation in power exchange is voluntary. JEPX has 21 investors and operates as a private non-profit organization.
- Expansion retail liberalized market. It was decided that all customers with a contract demand of 50 kW or more would be allowed to choose their suppliers. Therefore, the share in electricity consumption of eligible (EHV/HV) customers is 63%. The review of complete liberalization is to begin in Apr. 2007.
- Rules and oversight of "Neutral transmission System Organization" (NSO). NSO is a self-governing organization operated by private entities (e.g. EPCos, PPSs), maintaining the process of neutrality. It is responsible for drafting and supervising basic rules relevant to the transmission and distribution divisions. ESCJ was designated as NSO in June 2004, which was included: general electric utilities, PPSs, wholesale power, suppliers, captive power suppliers and academic

experts. The government (i.e. METI) overseas the neutrality of the decision-making process, leaving the operation of NSO itself to the initiative of NSO members.
– Simultaneous pursuit of public interest and efficiency

- **Comment on the Reform Plan and Outline for Industry Future**

The Japanese electricity sector is on track for further deregulation measures. These are likely to occur at a steady and somewhat slow pace. The following section discusses several evaluations of the current status of EI reform in Japan, and then endeavors to predict the shape of the industry.

Privatization

The electricity industry is completely owned by private companies, privatization is not required. The calling for IPPs may have effects of increasing competition, but not privatization.

Unbundling and Competition

Competition is introduced with the term of 'partial liberalization'. This is a very cautious approach to introducing competition. The government and industry confirmed that they shared the objective of introducing competition without compromising public interests associated with electricity supply, such as energy security, environmental protection, universal service, and reliability of supply. Retention of vertical integration was deemed necessary, as it is was better suited to achieve these public interests.

Uncertainty remains over the future arrangement. First, retail access for small residential customers is not yet determined. The discussion is scheduled to begin in 2007. Small residential customers may not be aware of the benefits of retail choice, and few suppliers have interest in the residential market. Although the industry expressed its intention to proactively consider full liberalization, this is still open issue, as it is in many other countries as well. Second, it is not yet certain whether the market design of JEPX is satisfactory to market participants. JEPX started with a day-ahead spot market and a forward market, but a real-time or a balancing market has not been considered. JEPX has to motivate market participants to trade to increase trading volume since trading on the JEPX is not mandatory, and the potential for an illiquid market remains. In addition, there is a technical reason why an unconstrained nation-wide electricity market is unlikely to be feasible. The electrical system in Japan operates at two different frequencies: the Eastern region is served by three utilities operating at a frequency of 50 Hz; the Western region is served by six utilities operating at a frequency of 60 Hz. Inter-regional trade is very limited relative to the size of the national market because of limited capacity of frequency converters. Third, it is not yet clear whether the separation of generation and transmission accounting is sufficient to convince new entrants of neutrality in providing transmission services.

Partial liberalization was not prosperous for new entrants. In 2005, the share of sales by the PPS in the retail market opened for competition was only 2%. The number of new entrants is 23 operators (gas, oil, steel, paper manufacturing, trading firms, etc.), and total salable volume accounts for 1.8 million kW [Matsuo, 2005]. The only gains were seen in terms of commercial (EHV) customers, where the market share of electricity by PPS increased by 20.1%. This percentage is small compared to the 63% eligible as specified by law. By looking for the cause of the situation, we might discover the most important factor influencing electricity industry reform in Japan. Opening retail market, existing electricity companies, whose owns nearly entire power industry, have to open their network to new entrants, loosing their customers, and sharing their benefit to new entrants. Liberalization, therefore, is not attractive to these existing companies. In order to offer reform policy, a high degree of consultation between government policy-makers and private firms is required, and faces stiff legal and political hurdles.

The daily average trading volume on JEPX, inaugurated as part of electricity market liberalization, is barely flickering, at only one-20th of its target. The exchange was created by "power product suppliers" like Nippon Telegraph and Telephone Corp., Tokyo Gas Co. and 21 others, including the major established power companies, allowing other power companies to sell surplus power at cheap prices. The major utilities sell their surplus power to JEPX-member PPSs, which in turn sell it to contracted enterprises and factories. Trade, however, has been slow. The daily volume for spot deals has been about 1 million kWh, or 10 percent of the initial target. And the only forward-delivery deal was concluded on the first day of trading. Critics are blaming the major electric utilities, which sell their surplus electricity to startup suppliers but are also reluctant to lose customers.

Deregulation

Deregulation in Japan is fragile. Although electricity industry is obeyed by Electricity Law, which was amended three times in 1995, 2000, and 2003 to cover the reform policies, however these policies are far from comprehensive strategies for reform. The 1995 law adjustment concentrates in IPP invitations, the 2000 and 2003 said more about liberalization in retail sector. These adjustments ignore the most of essential issues of reform, in which the biggest drawback may be the establishment of an independent regulatory authority and capacity building of regulatory and supervisory institutions for electricity market. The role of an independent regulatory body in power sector reform is very important. Whether or not the reform is successfully performed depends on how effectively the independent regulatory body can carry out its duties. When the electric power sector is divided into power generation, transmission, and distribution, and the number of stakeholders including private power producers is increased, the role of the regulatory body will increase significantly. Minimization of market risk, establishment of transparency and governance in the market is essential. The important challenges are to enhance the practical ability of the regulatory body and to avoid situations where only a limited number of operators can use market power. Enhancement of transparency and governance in the market will help improve the investment environment, and consequently increase the number of market participants. Until now the regulatory system in Japan electricity was very complicated and a lot of overlap existed between many agencies. It should be noted that a variety of agencies have the some regulatory functions, including METI, FEPC, ESCJ, etc.

While there is an overlap in functions, a more serious problem of the regulatory system in Japan's power sector is the lack of true power. For example, ESCJ is designated as a neutral transmission system organization, that is a self-governing to secure fairness and transparency in the transmission and distribution sector. ESCJ has an important role in the process of transmission open access, but it is only a neural organization, a straw man in front of the existing powerful companies. No penalty mechanism exists for violation of the open access provision.

In conclusion, it can be noted that in economies with a large private sector involvement in the electricity supply industry, the reform strategies concentrate on setting regulatory framework and introducing competition. Competition, obviously offers more choice for customers, but it seems to contradict the benefits of existing straw man power companies. Substantive reform requires a high degree of consultation between government policy-makers and private firms, and faces stiff legal and political hurdles. From the author' point of view, there is an urgent need to establish an independent regulatory agency, which is legally-based and financially independent, has clear objectives, and is functional. Only in that environment can actual liberalization happen.

8.3 The Case of Singapore

Singapore has no natural resources. Energy needs are largely satisfied through use of electricity. The country pays great attention to ensure the electricity market is robust, competitive and secure.

State-owned companies continue to hold a monopoly over Singapore's electricity sector, although the restructuring and privatization process has begun. The three main generation companies, PowerSeraya, Senoko Power and Tuas Power, which are subsidiaries of Singapore Power, together generate 90% of Singapore's electricity.

- **Electricity Industry Reforms**

1995 Reform

1 Oct 1995 – Government decided to reform the vertically integrated electricity industry, facilitating competition in electric generation and retail. The Government of Singapore corporatized electricity and piped gas activities of the PUB (Public Utilities Board). Singapore Power Ltd. includes three generation companies (Tuas Power Ltd, PowerSenoko Ltd and Power Seraya Ltd), one transmission and distribution company (PowerGrid Ltd), and one Electricity retail company (SP Services Ltd (formerly known as Power Supply Ltd)).

Figure 8. 4 shows the electricity industry structure from 1 Oct 1995 to 31 Mar 2001.

On 1 April 1998, a wholesale electricity market via the Singapore Electricity Pool came into operation. PowerGrid Ltd, the Pool Administrator, operated the Pool to facilitate trading of electricity between generation companies and the retail company, in a competitive environment. Additionally, Tuas Power Ltd was floated in the Singapore stock exchange.

2001 Reform

In 2000, the Government announced further deregulation of the electricity industry to obtain full benefits of competition. The key restructuring initiatives to be implemented included separation at the ownership level of the contestable and non-contestable parts of the electricity industry, the establishment of an independent system operator and the liberalization of the retail market.

Energy Market Authority (EMA) was established on 1 Apr 2001. The Energy Market Authority takes over the responsibility of regulating the electricity and gas

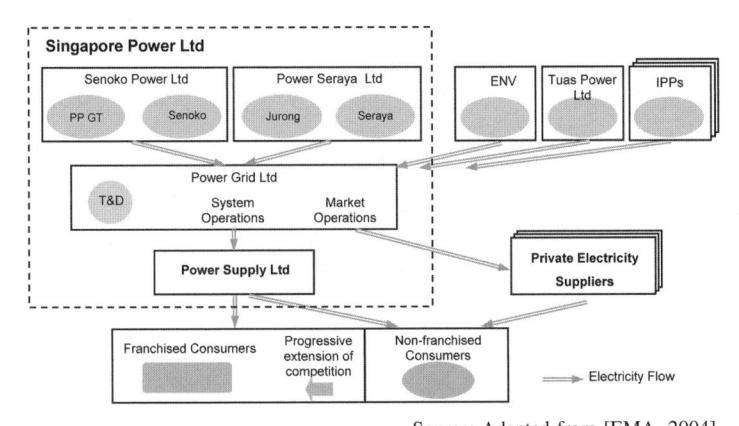

Source: Adapted from [EMA, 2004]

Figure 8. 4 Electricity industry structure from 1 Oct 1995 to 31 Mar 2001

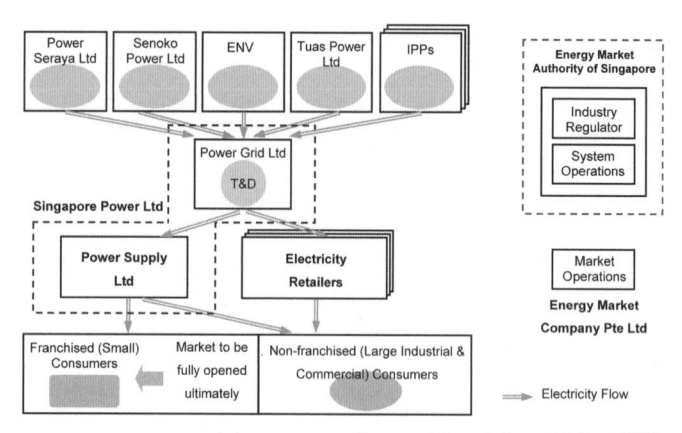

Source: Adapted from [EMA, 2004]
Figure 8. 5 Electricity industry structure from 1 Apr 2001 to 31 Dec 2002

industries and district cooling services in designated areas from the Public Utilities Board. In this year, the formation of the Energy Market Company (EMC) also occurred. The Energy Market Company (EMC), a joint venture company of the Energy Market Authority and the M-Co New Zealand, was formed to take over from PowerGrid Ltd as the Pool Administrator of the Singapore Electricity Pool. The divestment of generation companies was recognized in 2001. Singapore Power fully divested its generation companies, PowerSenoko Ltd and PowerSeraya Ltd to Temasek Holdings. The separation of ownership between the generation companies and the transmission and distribution company PowerGrid Ltd was carried out to enhance competition by ensuring a level playing field for all generation companies. Singapore Power (SP), i.e. PowerGrid, SP Services, is only allowed to participate in non-contestable businesses. Contestable businesses, e.g. retail and generation, have been divested by of SP.

Figure 8. 5 shows the electricity industry structure from 1 Apr 2001 to 31 Dec 2002.

2003 Reform
On 1st January 2003, the new electricity wholesale market commenced operation. About 250 consumers with a maximum power requirement of 2 MW and above became contestable.

Future Plan
The future plan of Singapore reform includes expansion of the liberalized retail market in order to ensure that benefits from competition in generation are passed on to consumers. Choices for Contestable Consumers consist of (i) purchase from

retailers or (ii) direct purchase from the wholesale market at spot prices by trading in the wholesale market or (iii) indirect purchase from the wholesale market through SP Services.

- **Comment on the Reform Plan and Outline for the Industry's Future**

Among Asian countries, Singapore is the one of the most developed countries in electricity reform.

Deregulation

In Singapore, a high level of reform has been seen in the comprehensive regulatory framework that exists, which covers both legal framework and regulatory body.

Unbundling and Competition

Especially in the aspects of unbundling and competition, reform seems to move quickly in Singapore. It can also be seen in the advanced model for the wholesale market. Figure 8.6 shows the structure of the wholesale market before and after 1 st January 2003.

The far-reaching reform also is present in the retail liberalization. In Phase 1, about 200 consumers became contestable as of 1 Jul 2001, with an average monthly consumption of 20,000 kWh. Phase 2 expanded the average monthly consumption of customers to 10,000 kWh. This phase began being progressively implemented from Dec 2003 and another 5,000 consumers became contestable by the end of 2004. Phase 3 will open retail liberalization to remaining consumers, which covers about one million consumers, mainly households. However, plans to extend the liberalization to residential users have been delayed until the end of 2006 in order to

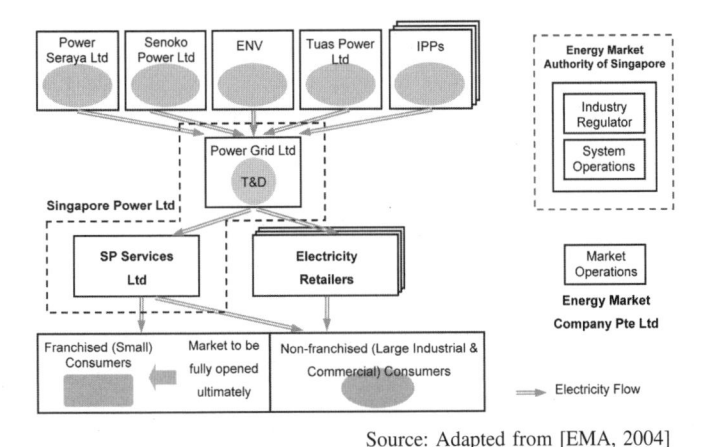

Source: Adapted from [EMA, 2004]
Figure 8. 6 Electricity industry structure since 1 Jan 2003

220   Part II

Figure 8. 7   Structure of wholesale market before and after 1 Jan 2003

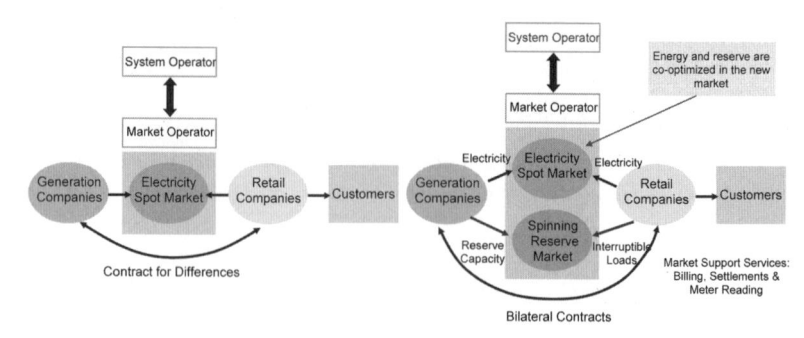

Source: Adapted from [EMA, 2004]

develop the retail market and the systems needed to ensure a smooth transition. In 2004, the EMA also began allowing large electricity consumers to sell their power off-take back to suppliers for a profit.

Privatization

Obviously, the level of regulation and competition reform in Singapore EI is at a relatively advanced level. However, the privatization reform is not realized yet. The privatization plan has been delayed several times. Therefore, the priority of reform strategy should now concentrate in privatization, promoting new investment, as well as selling assets. The process of selling assets may start soon. Following the end of the final phase of privatization, the primary retailers of electricity in Singapore will be Keppel Electric, Sembcorp Power, Tuas Power Supply, Senoko Energy Supply, Seraya Energy, and Marubeni International Petroleum.

Several issues need to be considered regarding the privatization process. For the pre-privatization, a re-evaluation of the value of assets is needed. The financial and operation data of assets have a great effect on asset sales. Therefore, the cost assessment work should be performed in the early stage of privatization in the electric power sector. Also, since the government does not guarantee the sale of power plants, except for power plants employed to supply base load, risk is high for off-takers and sales difficulties are expected. If government considers asset sales as the highest priority, they must guarantee or establish measures for the sale of power plants.

For post-reform, there are issues of supply reliability. Reliability has become a large concern for Singapore's electricity market. Although it is a mature market, it is characterized by overcapacity, the combined installed capacity of Senoko, Seraya, Tuas, and Sembcorp exceeds peak load demand by 80%. Singapore recently experienced five power outages in less than two years. As a result, a high-level

Energy System Review Committee has been established to evaluate options for improving the reliability of the market. Short-term solutions include the use of gas-fired generation plans that can operate with diesel in the event of a prolonged gas disruption (cogeneration plants), the installation of additional independent power producers, and the establishment of two gas sources for each power plant. LNG imports are a longer-term solution.

Another post-reform issue is technical risks relating the splitting of companies. Lack of consistency for planned operations among generation, transmission, and distribution poses a possible risk if vertically integrated public power utilities are split up, including: (i) establishing conformity between the development plan for generation and (load dispatching instruction and future) the development plan for transmission and distribution. The conformity between the transmission development plan and the production development plan would be necessary, (ii) optimization of the electric system operation. The optimization in generation, transmission, and distribution should be adjusted to reflect the entire operation. In addition, operation rules approved by the regulatory agency should be applied to the system. The operation rules and installation and improvement of the soft– and hardware to realize the operation are important tasks for the future.

8.4 The Case of Thailand

The current electricity industry structure is essentially a state-owned generation and monopoly transmission company, the Energy Generating Authority of Thailand (EGAT), sells wholesale to two state-owned distribution companies, the Metropolitan Electricity Authority (MEA) and the Provincial Electricity Authority (PEA). In 2004, EGAT produces around 50% of all electricity generated.

- **Electricity Industry Reforms**

The development of the electricity sector has depended to a great extent on the development of the economy as a whole. Therefore, the periods of liberalization in the power sector will have to be defined according to the overall situation of the Thai economy.

Before 1987, according to the Electricity Generating Authority of Thailand Act 1969, EGAT was the only entity allowed to generate and transmit electricity in the country; no individual or organization in the country could generate electricity, except for its own use. In Thailand, 1992 was considered as the first year of electricity industry reform. The government established the National Energy Policy Council (NEPC) and also passed a master plan for state enterprise sector reform, which provided a framework for restructuring and privatization of certain sectors, including energy. The EGAT Act was amended to end EGAT's monopoly on

222   Part II

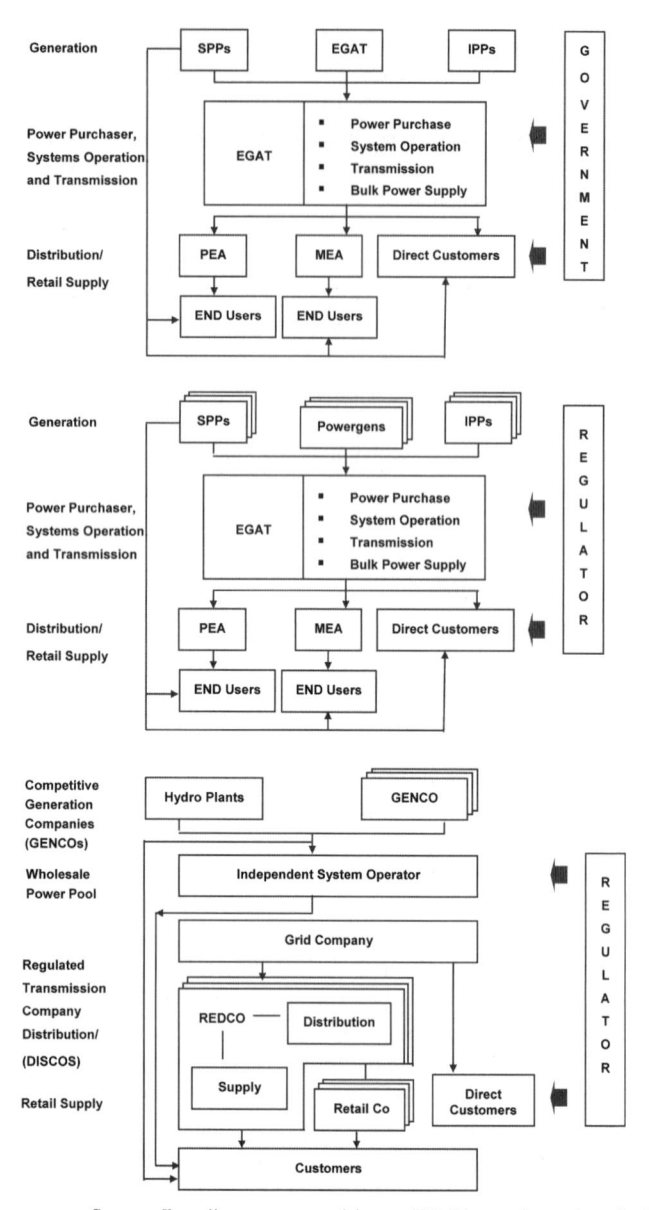

Source: [http://www.eppo.go.th/power/FF-E/pw-reform-1-main-E.html#4]
Figure 8. 8   Thailand's electric power sector in short-, medium- and long-term plan

generation and to permit the private production and sale of electricity. EGCO was created with EGAT as its parent and the off taker of power generated by EGCO's Rayong 1,232 MW Combined Cycle Power Plant and 824 MW Khanom Power Plant.

In 1994, EGCO was privatized and listed on the stock market. Also in this year, the first round of IPP solicitation was issued. EGAT would buy up to 5,800 MW of capacity from IPPs for the period 1996–2003. Power Purchase Agreements (PPAs) have been signed with 7 IPPs, with a total of 5,944 MW of electricity sold to EGAT. The cabinet approved a Master Plan for privatization in September 1998, detailing the privatization of state enterprises related to communications, the water supply, transportation, and energy. Privatization in the electricity industry was set to occur in three stages: (i) 1999–2000: to corporatize EGAT, yet keep it as a 100% state-owned enterprise, (ii) 2001–2003: to change EGAT into a holding company, with its business units established as operating subsidiaries, and launch the sales of shares in EGAT subsidiary generation companies, and (iii) beyond 2003: to establish a competitive power pool with retail competition in the electricity markets, and privatize all generation facilities (Figure 8. 8).

The first stage of this plan was completed successfully. However, during the second stage, privatization plans faced intensive opposition from EGAT's labor unions, the risk of higher prices, the risk of corrupt allocation of shares to cronies, and the risk of foreign control developing through the purchase of shares. The government backed down and announced the cancellation of EGAT privatization plans. Plans to liberalize the sector have been slowly abandoned, including the cancellation of the proposed power pool.

In 2005, the privatization plan was reviewed again. In May, the cabinet gave the green light to a long-delayed privatization plan for EGAT. It allows setting up EGAT PLC and then listing it on the Stock Exchange of Thailand. The Ministry of Finance will hold 75% of EGAT shares, foreign investors can hold no more than 5%, and the remaining 20% will be distributed to small-scale investors and the general public. To persuade labor union, the cabinet agreed to increase salaries by 15%. However, on November 15, the Supreme Administrative Court accepted a petition by 11 civic groups to halt EGAT's IPO pending a review of the legality of two decrees authorizing the privatization of the country's largest power producer.

- **Issues on the Reform Plan and Outline for Industry Future**

Many efforts have been made to reform Thailand's electricity industry. The reform has been clearly mapped out for a long time (1998). However, not many reform steps have been achieved and Thailand remains in the second country group with an unclear future. The electricity industry of Thailand is now confronted with two

Table 8.1 Proposals for future industry structure

Period	Planning	Regulation	Generation	Transmission	Distribution & Retail	Environment	Consumer Protection
Proposal I (The Power Pool)	NEPO	IRB	PowerGens, EGCO, IPPs, SPPs, etc.	Grid Company	Privatized MEA, PEA	Civil society, The constitution, etc.	IRB
Proposal II (Public Company)	NEPO	Not clear	EGAT, EGCO, IPPs, SPPs, REGC, etc.	EGAT	MEA, PEA	Not clear	Not clear

Source: [Nuntavorakarn, 2002]

options for their future, as illustrated in Table 8.1.

Obviously, in Proposal II the electricity industry structure remains the same. In order to successfully move to Proposal I, Thailand needs to cope with the issues that have caused failures in reform this far. Such causes come from regulation, competition, and privatization policy.

Regulation

It seems that the Thai government has put inadequate attention on the establishment of an efficient regulatory framework. In Thailand, an energy law has been drafted many times, but still remains unapproved. Regulatory functions are now taken by NEPO. To facilitate to reform, a regulatory body should be established as a truly independent entity to ensure fairness for both service providers and service consumers, and its decision-making must be depoliticized.

The regulatory body must take the extensive opposition of the labor union into consideration. Workers are often poorly informed on the methods of privatization and restructuring, and fear that they will lose their job security and all or some of their social protection. The regulator should open dialogue with the interested parties, such as labor unions, etc. The joint labor-management-government research group should formed for the decision making process of electricity industry reform.

Restructuring and Competition

There must be a distinct separation between the functions of generation and the transmission so as to create a level playing field. The rationale behind it is if a transmission system operator also owns a power plant, discriminatory treatment may arise with respect to, for example, the use of the transmission system and power dispatch orders. The Independent System Operator (ISO) should be formed, having no affiliation with any generation or transmission company to ensure fairness in the competitive generation process.

There should be adequate and real competition in the Power Pool. Therefore, it is essential to have a large number of power producers. Each producer must not possess an excessive proportion of production capacity compared with the overall capacity, which may induce exertion of market power. Hence, in restructuring the generation business of EGAT, it is essential to separate the existing generation facilities, at least, into 2–3 GenCos.

Promotion of several retailers is essential so that consumers can opt to receive services from several companies, which will enhance real competition in both price and service quality.

Privatization

A considerable share of existing non-EGAT generation is contributed partly by the IPP and SPP generation and partly by spin-off subsidiaries of EGAT. Privatization is considered as one of the main targets of Thailand's government in reform strategy. The first reform activity was the establishment of the EGCO comprised of spun-off EGAT power plants. However, in the period of reform that has lasted more than 10 years, after the establishment of EGCO, there was only one more successful privatization effort, which is the sale of the Ratchaburi Power Plant to set up RGCO. The government has put great effort into selling EGAT's shares, however, it has succeeded yet. The most recent effort to privatize EGAT although approved by the government was refused by the court.

In selling shares to the private sector, efforts should be made to allocate the shares as widely as possible to prevent any particular person from gaining market power. If the operation of the Stock Exchange of Thailand (SET) is efficient enough, sale of shares in the SET will be the most suitable and easiest way to raise equity. However, in certain cases, strategic investors may be required to hold a portion of the shares in order to assist in the development of businesses that are to be privatized in terms of, for instance, loan procurement and business expansion in the future. As for the concern of the employees and the general public about the takeover by foreign investors, setting conditions for the selection and joint venture can be a solution to this concern.

8.5 The Case of Bangladesh

- **Electricity Industry Reforms**

The power sector in Bangladesh is organized under the Ministry of Energy and Mineral Resources. The Government, through the ministry, wholly owns and supervises the Bangladesh Power Development Board (BPDB), the Dhaka Electricity Supply Authority (DESA), and the Rural Electrification Board (REB), which are its three executing agencies in the sector. Until fiscal year (FY) 1998, BPDB was responsible for all generation and most of the transmission in the country, and also for distribution in district towns, municipalities, and some rural areas, while DESA was responsible for distribution in the greater Dhaka area, including the capital city. The distribution of electricity in most of the rural areas of Bangladesh is the responsibility of the 67 rural electric cooperatives or palli bidyut samities (PBSs), which are organized, initially funded, and monitored by REB. DESA's (including DESCO) retail sale accounts for about 39% of total sales.

Initial Reform

The Government opened the power sector for private investment in 1992. In 1994, the Government of Bangladesh adopted a paper titled Power Sector Reforms in Bangladesh (PSRB). The PSRB outlined the reform process proposed to be followed by the Government to gradually remove the constraints in the sector through improvements in sector and corporate governance, introduction of competition, and public-private partnerships.

In 1995, power cell, an institution under the Power Division created in 1995 to design, facilitate, and drive reform measures. The Government, BPDB, and REB have taken steps to involve the private sector by (i) adopting a private power generation policy in October 1996, (ii) adopting a small power generation policy in 1998 to encourage distributed generation, and (iii) awarding contracts for about 1,800 megawatts (MW) of generating capacity on a build-own-operate (BOO) basis. As part of the unbundling process, in 1996, the Power Grid Company of Bangladesh (PGCB) was created to take over the transmission business from BPDB. It took over about 50% of the transmission assets and planned to take total control by the end of 2002, however, this plan has not been completed yet. Also in 1996, the Dhaka

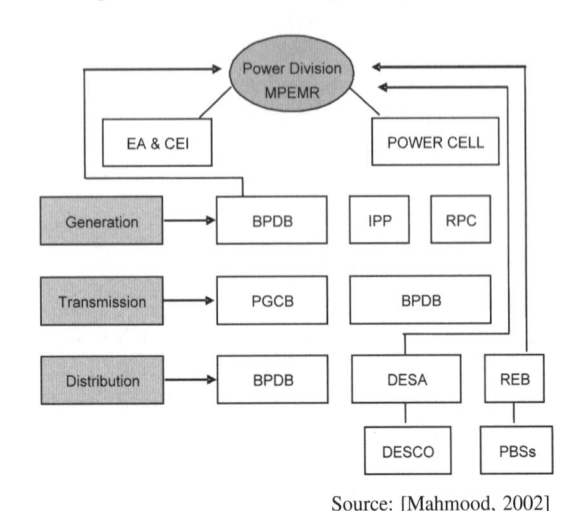

Source: [Mahmood, 2002]
Figure 8. 9 Present structure of the electricity industry of Bangladesh

Electric Supply Company (DESCO) has been created to take over a part of the Distribution Business of Dhaka City from DESA. By now it already took over the Mirpur area. The government has begun the corporatization process of several power plants of BPDB, DESCO, and West Zone Power Distribution, creating a manageable and competitive environment through bench-marking. The present structure of electricity industry of Bangladesh is shown in Figure 8. 9.

Future plan

The Bangladesh government has set long-term goals for the power sector (i) To make electricity available for all by 2020, (ii) To ensure a reliable and quality supply of electricity, and (iii) To provide electricity at a reasonable and affordable price.

The Government is committed to the Power Sector Reforms. The main components of the proposed reforms program are: (i) Segregation of power generation, transmission, and distribution functions into separate services, (ii) Corporatization and commercialization of emerging power sector entities, (iii) Creation of a regulatory commission, (iv) Private sector participation in power generation and distribution, (v) Introduction of a cost reflective tariff for financial viability of the utilities and promoting the efficient use of electricity, (vi) Development of demand management including energy efficiency measures to conserve energy, and (vii) Development of alternative/renewable energy sources [Mahmood, 2002].

The following tasks should be carried out for the generation sector: (i) Separation of all existing power generation units through a corporatized entity, (ii) Incorporation of future power plants and those currently under construction as independent companies, (iii) Generation projects be selected at least-cost option, (iv) Generation capacity be sought through a mix of public & private sources. Single buyer model should be adopted as the market structure. In the future, the multi-buyer model/competitive pool may be adopted when the industry structure is mature enough and market structure will be stable under a regulatory environment. The single buyer will be a public sector entity. Initially, BPDB will act as single buyer until a suitable alternative public sector organization is set up.

In transmission sector, the transmission network will be owned, operated, planned, and developed by a corporatized entity in the public sector. In the distribution sector, which is the priority area of the reforms, work should focus on the following: (i) The existing distribution system of BPDB and DESA should be transformed into a number of new corporatized entities, (ii) Private capital and management participation in distribution companies, (iii) The network and supply (commercial activities) business should be separated and private participants should be invited to participate in the supply business, (iv) The rural electric cooperatives (PBSs) under the REB should continue functioning and additional PBSs should be

228   Part II

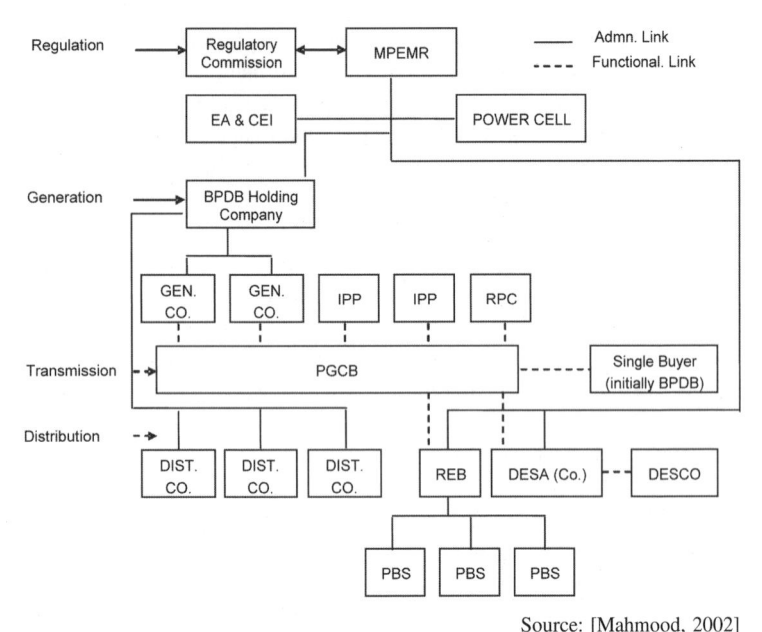

Source: [Mahmood, 2002]
Figure 8. 10   Electricity industry structure in Bangladesh after reforms

formed when required, (v) Introduction of consumer voice and organizational accountability in the form of citizen-client charters.

The planned structure of Bangladesh electricity reform is illustrated in Figure 8. 10.

- **Issues on the Reform Plan and Outline for Industry Future**

Bangladesh electricity industry reform started relatively early compared to other Asian countries. However, sector performance improvement was not the only driving force behind reform in Bangladesh. It was proposed under the pressures of international institutions, for example, the reform policy of 1994 was established under the guidance of the ADB. Therefore, reform policy has been implemented slowly and Bangladesh remains in the first group, which includes countries with a low level of reforms in Asia. The power sector in Bangladesh performed poorly through the late 1980's and 1990's. The major constraints in the sector were (i) lack of institutional capability; (ii) unavailability of long-term domestic capital for financing investments; (iii) limited foreign exchange debt service capability of the economy; (iv) poor management systems and procedures; (v) low employee commitment; and (vi) institutional weaknesses in governance, banking, law enforcement, and judicial processes, which are external to the sector but are essential for its proper functioning.

For the future plan of sector reform, a step-by-step policy still seems to exist. However, considering the performance of the electricity industry, the author does not attempt to criticize this policy. The priority of reform should be regulatory framework, corporatization and unbundling, and attracting private investment.

Regulation

Currently, the Government/Power Division functions as regulator to approve investment programs, to monitor performance of the public sector entities/utilities, to approve tariffs, and appoint an electrical adviser and chief electrical inspector. On 10 March 2003, the Parliament approved the ERC Bill, which lays the basis for the ERC for the electricity and gas sector. However, preparation of human resources and capacity building is required before the establishment of the ERC.

Restructuring and Competition

Although the introduction of competition in the EI has been stated in the government policy since 1994, it should be noted that competition here is limited to the generation sector only. Wholesale and retail markets are not considered in the short-term plan of reform. Priorities of reform in the Bangladesh EI are generation, transmission and distribution unbundling, and corporatization. The separation of all existing power generation units facilitates the establishment of corporatized entities, which can be applied to private management to improve its operation efficiency. Corporatization also should be pushed up in the distribution sector. Performance improvement in distribution functions would be at the top of the agenda for reforms. Considering system loss accounts for more than 30% of electricity generation, it cannot be over emphasized that unless and until the performance of distribution segment is improved, reforms in other segments, such as generation and transmission will not lead to significant improvement in the overall situation of the power sector, particularly with respect to consumer service standards and financial viability.

Privatization

Due to unavailability of the domestic capital market and the incomplete corporatization process, privatization of Bangladesh's electricity industry, in the short-term, cannot consider the option of selling shares or privatizing existing power companies. It concentrates on attracting private investments for new power plants instead. This is the proper strategy because Bangladesh is one of the developing countries which has the lowest electricity generation per capita (150 kWh per capita). It means that more investment is required to meet the demand. In addition, private capital and management participation in distribution companies is required to increase the rate of electricity access.

8.6 Generalization Issues for Asian Modeling in Electricity Industry Reform

Regulatory reform in the electricity sector, characterized by deregulation, competition, and privatization is beginning to become a common phenomenon amongst Asian countries. It several levels of reform development in Asian economies, from countries that have barely changed their structure to which hardly with a relatively advanced structure, can be seen. Differences are shown not only in the reform process but also in terms of the social fabric, political, legal, and financial framework, and the stage of economic development. Where reform i snot taking place in underdeveloped and emerging economies, the outcomes may differ from what economists and industry analysts in developed economies consider ideal. There is clearly no "one-size-fits-all" model for electricity supply industry regulatory reform. However, a number of possible models exist and each may be appropriate to a particular set of circumstances.

- Countries in the first step of reform process. Liberalization seems a long way off for such countries. Their electricity industries are still in the early stages of development both, in term of capacity, technology, and management. The priority of these countries may not be competition or privatization. The short-term objectives of such countries should be to strengthen state-owned companies, finish corporatization, and to attract investment to meet the increasing demand for economic development.
- Countries in the second step of reform. These include countries that have carried out several initial reform tasks, such as enactment of an electricity law, completion or near-completion of the corporatization process, preparation for reforms, such as practicing power exchange or preparing the sale of assets. Asian countries that belong to this group include Vietnam, Taiwan, Indonesia, Thailand, and India. For such countries, a proper step-by-step reform policy should be applied. At first, proper regulatory framework should be established and strengthened. The regulatory body should be given power to oversee fair competition and keep proper pace for privatization. Industry structure should be unbundled, introducing competition through single buyer, practicing in the power pool. The purpose of SB is to divide power generation, transmission, and distribution legally, and to conduct "equal" treatment for IPP and power plants of national electric power companies, and thus to promote competition. The next objective for this country group should be the preparation of favorable conditions for the wholesale market and the trial launch of a power exchange.

Acceleration of privatization is not a priority. Partial privatization is favorable. Especially, more effort should be focused on finding a new way to promote private investment to meet the increasing demand of electricity. Investment vehicles, important not only in this country group, demonstrate their essential role in most Asian economies. Under situations where government guarantees individual electric power projects be abolished, the PPA contract between the power distribution companies and IPPs becomes increasingly important. For this reason, since the off-taker risk of a power distribution company will be questioned, for promotion of private investment, establishment of a support scheme which ensures the payment by IPP such that the government provides its guarantees to the power distribution companies is required.

In order to attracting further investment, the most challenging point is how to deal with the two issues of finding the appropriate approaches to reconciling existing IPP contracts with emerging power markets and designing new IPP contracts better to facilitate subsequent integration into electricity markets.

[Woolf and Halpern, 2001] suggests the following solutions, which involve a combination of approaches:
- Facilitate voluntary renegotiation to encourage the IPP to participate in the market at least with respect to un-contracted capacity and energy and also ancillary services and congestion management; One instrument which may be considered in inducing IPPs to participate in the new market is to request conversion of a power purchase agreement into its financial equivalent, a so-called "contract for differences" (CfD), which would hedge the uncertainties surrounding prices in the new market.
- Adapt the market rules preferably for a transitional period to accommodate the IPPs to the extent possible without compromising the competitiveness and efficiency of the market and exempt them from the requirement of complying with rules and standards which do not affect reliability, particularly if these have cost implications.
- Appoint an experienced and skilled contract manager to manage the PPAs and minimize above-market costs.
- Consider the potential benefits of a contract buyout at a point in the future, once the market has settled down, on the expectation that there will be sufficient interest from suitably qualified bidders prepared to take a substantial degree of market risk without discounting the prices they are prepared to bid excessively.
- Establish a transparent and credible process of discussion and consultation with IPPs; treat all IPPs in the same manner in order to reassure IPPs, their lenders and investors.

One more measure for promoting private investment in the new environment of the industry structure is merchant plants. Merchant plants sell electricity for many unspecified customers via wholesale prices through a competitive pool market.

Merchant plants, independent from PPAs, have the merit that resources can be allocated more efficiently through the formation of prices reflecting ever-changing market trends. However, market risks (price risk and quantity risk) and exchange risks for investors are higher than those of traditional IPPs, because they cannot get the guaranteed cash flows such as in the PPAs. Since other risks such as political risk are not reduced, in order for business to be feasible it is essential, in addition to taking measures against the conventional risks for IPPs, that the wholesale electricity price stays stable at such levels that make investment feasible, and that there is a capital market where financing is available to Merchant plants. Introduction of a pool market is a precondition for establishing merchant plants.

Also, due to liberalization in the course of sector reform, a certain number of employees may loss their jobs. A comprehensive social policy that offers employment to those people would be needed. Subsidies in this area can be justified by securing transparency, defining the proper beneficiaries, and the beneficiaries' self-help efforts.

- Countries with the electricity industry controlled by private companies. In such economies, the reform strategies concentrate on setting regulatory framework and introducing competition. Competition, obviously offers more choice for customers, but it seems to be contradictory to the benefits of existing straw man power companies. Substantive reform requires a high degree of consultation between government policy-makers and private firms, and faces stiff legal and political hurdles.
- The most advanced in EI reform group. These countries under this group need to continue their selection of a competitive market. However, more efforts should be given on promoting privatization. New measures for asset sales should be considered. Sufficient assessment of costs for power generation, transmission, and distribution is required first. This financial and operational data has a great effect on asset sales. Also, since the government does not provide guarantees regarding the expected sale of power plants. If the government considers asset sales as the highest priority, they must make more guarantees or take other measures into consideration for the difficult sale of power plants.

Another notable issue is that in a privatized market, since rural electrification tends to be neglected, the central government would be directly responsible for rural electrification in off-grid regions, utilizing the government subsidies on the universal charge. For countries which have completed electrification, such as Singapore or Malaysia, this does not seem to be a problem, however this is not the case with other countries in this group, especially the Philippines, an island country most in need of distributed generators to meet the demand.

It can be concluded that while there is clearly no "one-size-fits-all" model for electricity industry reform, some common features for EI reform strategies for Asian countries can be seen. These include the simultaneous pursuit of public interest and efficiency, proper step-by-step policy, priority to promotion of private investment, establishing regulatory authorities, and expansion of liberalization. Also in Asian countries, it should be a priority that all stakeholders, including not only the electric power sector itself, but also investors and consumers, have an awareness of being participants and playing active roles in operating within the system.

CHAPTER 9
NUCLEAR POWER POLICY UNDER THE PROCESS OF EI REFORM IN ASIAN COUNTRIES

The model of reform of each country is different and depends on the current status, as well as their political orientation. However, in the reform process, there are several troublesome issues that every country must cope with, these include: social, environmental, nuclear power, and renewable, etc. The most serious concern should be given to regulatory risks involved in the process of liberalization that might harm the security of supply in the electricity industry, which is highly dependent on nuclear power. An important conclusion derived from the regression results in Chapter 7 is that while most of the reform policy options, in general, have positive impacts on electricity generation per capita and the installed capacity per capita, they show different effects on the share of nuclear power generation. The option of corporatization and establishing an independent regulator have positive impacts, while other options related more to competition and privatization have a negative impact. It means that the more reform has been implemented in electricity industry reform, the higher attention should be made on the regulation in order to maintain a sustainable energy policy, especially for energy-importing countries, such as Japan or Korea. In the framework of this book, Chapter 9 will try to address the issues of nuclear power under the process of EI reform in Asian countries. Policymakers have emphasized the need to make liberalization policy compatible with nuclear power development. The impact of reform will not be limited countries currently endowed with nuclear power, but also extend to countries that could potentially develop nuclear power.

9.1 Current Status and Future Potential of Nuclear Power in Asian Countries

- **Current Status and Future Potential of Nuclear Power**

Recently, Asia is the only region in the world where there has been significant growth in nuclear power and that currently has sizeable plans for the construction of new nuclear plants. In Asian economies including China, Chinese Taipei, Japan,

Table 9.1 Recent nuclear reactor additions and reactors under construction in Asia, as of April 2004

Economy	Number of operating reactors	Connected since 1995	Under Construction
China	44	63	61
Chinese Taipei	33	41	78
Japan	11	41	83
Korea	78	33	83
North Korea	44	52	72
Russia	22	37	39
India	11	30	44
Pakistan	41	45	71
Total	2.44	2.70	4.28

Source: [IAEA, 2004 a]

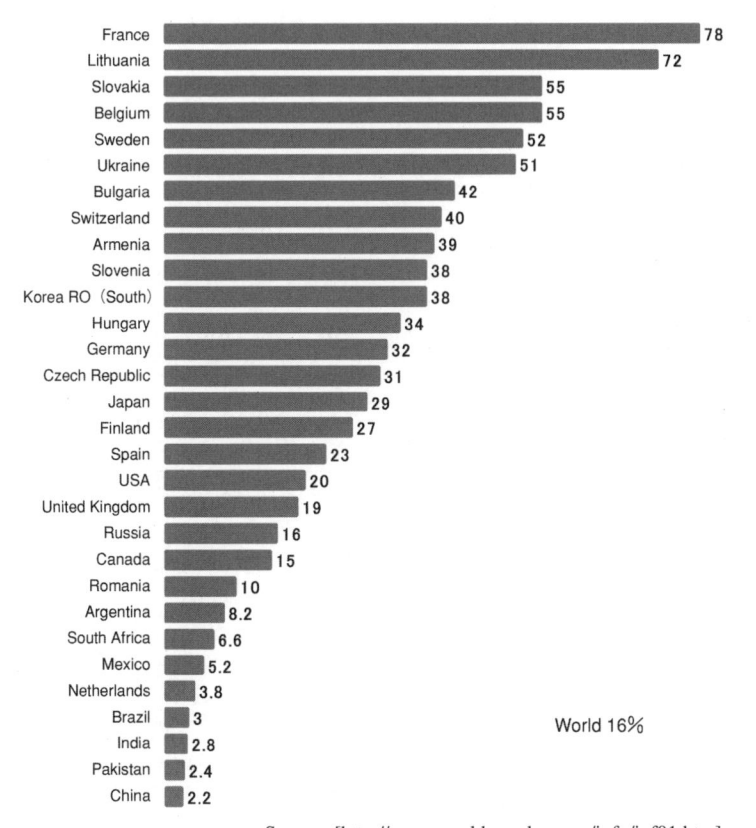

Country	Percentage
France	78
Lithuania	72
Slovakia	55
Belgium	55
Sweden	52
Ukraine	51
Bulgaria	42
Switzerland	40
Armenia	39
Slovenia	38
Korea RO (South)	38
Hungary	34
Germany	32
Czech Republic	31
Japan	29
Finland	27
Spain	23
USA	20
United Kingdom	19
Russia	16
Canada	15
Romania	10
Argentina	8.2
South Africa	6.6
Mexico	5.2
Netherlands	3.8
Brazil	3
India	2.8
Pakistan	2.4
China	2.2

World 16%

Source: [http://www.world-nuclear.org/info/inf01.htm]
Figure 9.1 Nuclear share in total power generation 2004 (percentage)

Korea, North Korea, India, and Pakistan, a total of 28 reactors have been put into commercial operation since 1995 and 25 more are under construction.

In addition to power reactors, there are many research reactors in other Asian economies, including Indonesia, Malaysia, Philippines, Thailand, and Vietnam. Figure 9. 1 shows the share of nuclear power in countries' electricity generation.

There are the three economies in Asia that have more than a 20 percent nuclear share in their total power generation: Korea, Japan, and Chinese Taipei. The nuclear share of generation in Japan was 27 percent in 2004, which is low compared the 30 percent share observed in previous years.

There are 12 reactors under construction in Asia as of 2006, accounting for an additional 6,704 MW of capacity. Table 9. 2 lists the number of reactors under construction and the official plans as of 2006 for new nuclear power plants in Asia economies. Figure 9. 2 also illustrates the projection of nuclear generation in those countries.

Table 9. 2 List of nuclear power reactors in Asian countries 4 January 2006

	Nuclear electricity generation 2004			Reactors operable Jan 2006		Reactors under construction Jan 2006		Reactors planned Jan 2006		Reactors proposed Jan 2006		Uranium required 2006
	billion kWh	% e	No.	MWe	No.	MWe	No.	MWe	No.	MWe		tonnes U
China	47.8	2.2	9	6,587	2	1,900	9	8,200	19	15,000		1,294
India	15.0	2.8	15	2,993	8	3,638	0	0	24	13,160		1,334
Indonesia	0	0	0	0	0	0	0	0	4	4,000		0
Japan	273.8	29	55	47,700	1	866	12	14,782	0	0		8,169
Korea DPR (North)	0	0	0	0	1	950	1	950	0	0		0
Korea RO (South)	124.0	38	20	16,840	0	0	8	9,200	0	0		3,037
Pakistan	1.9	2.4	2	425	1	300	0	0	2	1,200		64
Vietnam	0	0	0	0	0	0	0	0	2	2,000		0
WORLD	2618.6	16	441	368,386	24	18,816	41	42,707	113	82,220		65,478

Building/Construction = first concrete for reactor poured, or major refurbishment under way;
Planned = Approvals and funding in place, or construction well advanced but suspended indefinitely;
Proposed = clear intention but still without funding and/or approvals.

Source: [http://www.world-nuclear.org/info/reactors.html]

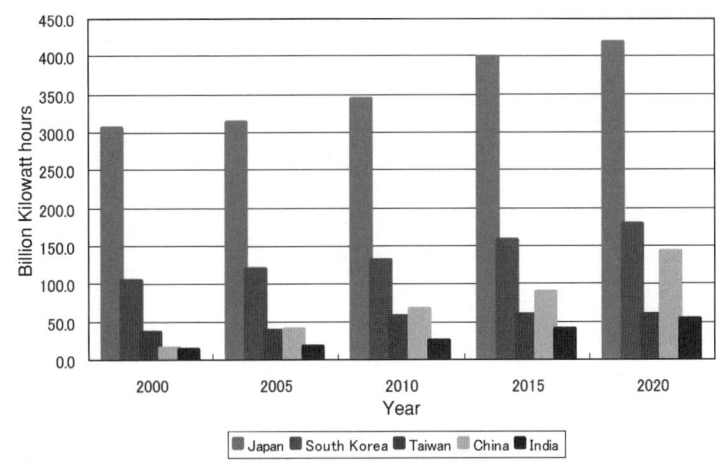

* Not shown: Pakistan, projected 4.3 GW by 2020, North Korea, projected 5.7 GW by 2020.
Source: [Energy Information Administration, 2002]
Figure 9.2   Nuclear generation projection for Asia

- **Drivers of Nuclear Policy**

The main drivers of nuclear energy are:
- Energy security in a context of scarcity or unevenness in the distribution of energy resources.
- High electricity demand growth and the need for energy source diversification.
- Sustainable development and the pressing need to reduce greenhouse gas emissions from power generation.
- Low nuclear plant generation costs as an incentive to maintain operation of existing nuclear fleets.

- **Nuclear Power in Japan**

Historical Development

Enactment of the Atomic Energy Law in 1955 introduced the promotion of atomic energy development and utilization toward peaceful objectives. In 1956, the Atomic Energy Commission was organized, working out "Long-Term Plans for the Research, Development and Use of Nuclear Power". Today, it is the basic program for the nation in regard to nuclear power development and utilization. The plan is revised and updated every five years.

In 1974, three basic laws for the promotion of electric power development were made into law; namely, the "Law for the Adjustment of Areas Adjacent to Power-Generating Facilities," the "Electric Power Development Promotion Tax Law," and

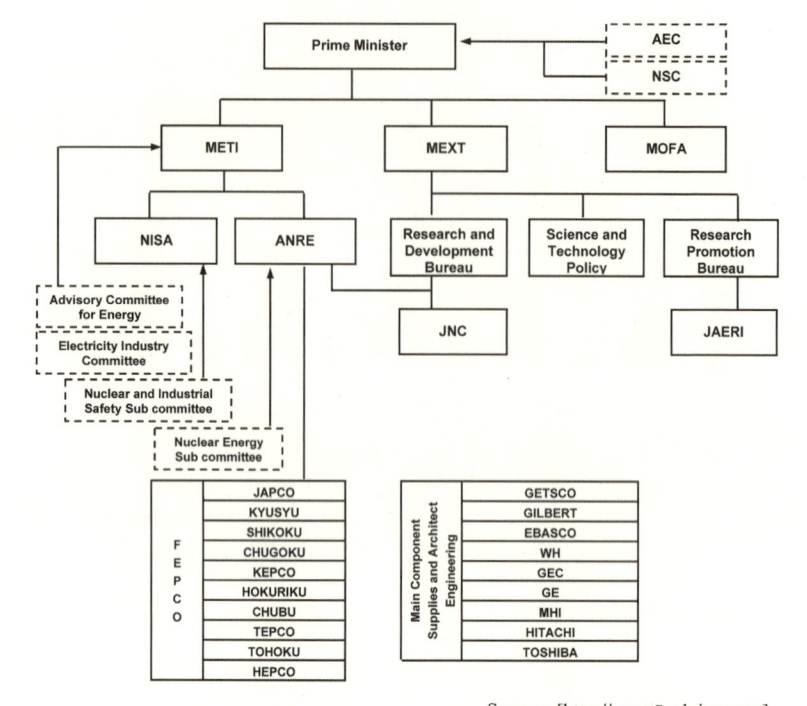

Figure 9. 3 Japan's organization chart in nuclear power

Source: [http://www.pub.iaea.org]

the "Special Account Law for Electric Power Promotion." These laws also advanced the appropriate siting of nuclear power stations.

In 1978, the Nuclear Safety Commission was formed as a separate entity from the Atomic Energy Commission. In 2001, the Nuclear and Industrial Safety Agency (NISA) was formed as a separate entity from METI to hold jurisdiction over matters of nuclear and industrial safety. Figure 9. 3 shows Japan's organization chart in nuclear power, comprising government regulatory authorities, electric power companies, and contracting engineers/suppliers.

Current Status and Trend
Japan, the largest producer of nuclear-generated electricity in Asia, is rated third-largest in the World in nuclear capacity (behind the United States and France). Table 9. 3 provides the data summary of the nuclear power plants in Japan. As of the 2004, the total capacity of nuclear power generation is 47,122 MWe. The total capacity of nuclear power generation plants under construction and planned plants as of 2006 are 866 MWe (one plant) and 14,782 MWe (twelve plants), respectively.

Table 9. 3 Nuclear power data summary, Japan

Reactors in operation	53
Nuclear installed capacity (gross)	47,122 MW
Reactors under construction	2
Total electricity generation	919.9 TWh
Nuclear generation	230.1 TWh
Nuclear generation share	25%

Note: Generation figures for 2003

Source: [IAEA, 2004 d], [JAIF, 2004]

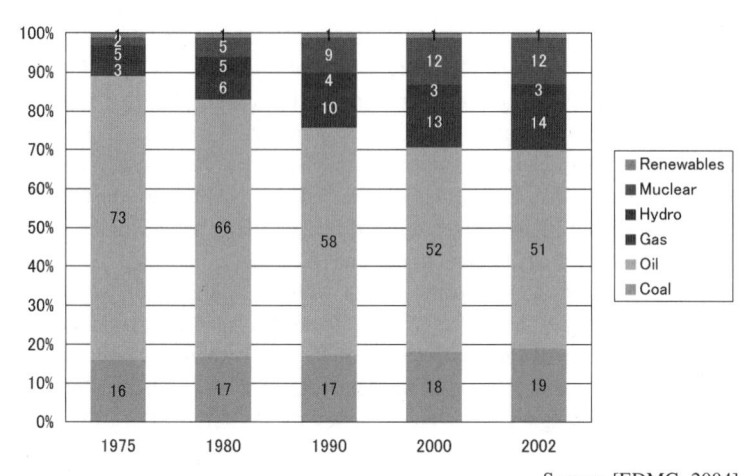

Source: [EDMC, 2004]

Figure 9. 4 Evolution of primary energy supply structure in Japan, 1975–2002 (percentage)

Nine of the major electric companies and the Japan Atomic Power wholesale company own and operate reactors in Japan. J-Power, the other wholesale company, does not own a nuclear reactor at present, but is planning the construction of its first advanced BWR type reactor to go online in 2012 at Ohma.

From 1985 to 2002, except for a couple of years, nuclear power was the largest contributor to electricity generation in the nation, averaging a 34 percent share between 1999 and 2001. Nuclear power plays a major role in Japan's present energy landscape and this condition is expected to continue in the mid-term future. Figure 9. 4 shows the evaluation of primary energy supply structure in Japan from 1975 to 2002.

Government Policy on Nuclear Power Development
Japan's heavy use of nuclear power comes as the result of a strategic need to reduce

its dependence on oil, gas, and coal imports. Nuclear power represents an important part of Japan's primary energy supply, at around 12 percent today, owing to nuclear energy's comparably reliable fuel supply together with its potential to help reduce emissions.

The Atomic Energy Commission of Japan (AEC) concluded a new "Framework for Nuclear Energy Policy", endorsed by the Cabinet in October 2005. Main points includes (i) to maintain or increase nuclear power's present share of 30–40% in Japan's total electricity generation beyond 2030 for stable energy supply and as a response to global environmental problems, (ii) steadily advance the nuclear fuel cycle, and (iii) aim towards the commercialization of FBRs.

METI's Long-Term Energy Outlook published in June 2004 foresees the need to construct 4 new reactors by 2010 and a total of 10 new units by the year 2030. The Outlook projects also that nuclear energy will account for 15 percent of the primary energy supply in Japan by 2030, up from about 12 percent today.

However, plans to construct more nuclear plants may be seriously impaired by a number of events that have greatly increased the public's mistrust in nuclear energy. A series of nuclear accidents has raised concerns about public safety and forced cancellation of plans to build at one site in Japan.

Now the government of Japan is faced with the dilemma of having nuclear energy as one of its few available options for countering fuel imports, and at the same time faces a crisis of public trust. Both the government and the nuclear industry in Japan recognize the need to regain the people's confidence as the only path towards formulating a nuclear power policy for the future. The Atomic Energy Commission of Japan in its White Paper on Nuclear Energy 2003 comment on the passage of new laws on nuclear safety and other reforms that will help prevent future accidents and avoid laxness or negligence in testing and inspections at nuclear installations. It also announces extensive public hearing activities, such as 'conferences for public participation and decision making in nuclear energy policy', to collect people's opinions and use it as a starting point towards a dialogue and a search for mutual understanding.

Moreover, to help make the siting of nuclear installations more attractive, the Japanese government has had a program to provide subsidies to the townships that are willing to host them.

R & D Organizations and Institutes
The Atomic Energy Commission (AEC), amongst other responsibilities, advises on R & D and revises the long-term program for the development and use of nuclear energy every five years. Government responsibilities for R & D are shared between the Ministry of Education, Culture, Sports, Science and Technology (MEXT) and the Ministry of Economy, Trade and Industry (METI). MEXT supervises the work of

the Japan Nuclear Cycle Development Institute (JNC), which was established in 1998, and also that of the Japan Atomic Energy Research Institute (JAERI), established in 1956. JNC is the main channel for the development of advanced reactors and establishment of the fuel cycle. In both there is close collaboration with the private sector, including shared funding on some projects.

- **Nuclear Power in South Korea**

Historical Development

Nuclear activities in Korea were initiated in 1957 when Korea became a member of IAEA. In 1959, the Office of Atomic Energy was established as a government organization in conformity with the global trend toward developing peaceful uses of atomic energy. The Atomic Energy Law was promulgated in the preceding year [IAEA web].

The Republic of Korea has carried out a very ambitious nuclear power program since the 1970's in parallel with the nation's industrialization policy, and has maintained a strong commitment to nuclear power development as an integral part of the national energy policy aimed at reducing external vulnerability and insuring against global fossil fuel shortage.

Current Status and Trend

Currently, Korea has one of the most dynamic nuclear power programs in the world. It is the APEC economy with the highest percentage of electricity coming from nuclear power (40 percent). It generated 130 TWh by nuclear means in 2003 and currently has an installed nuclear capacity of 16,716 MW coming from 19 reactors, the last of which began commercial operation only recently at Ulchin in July 2004. Another 1,000 MW reactor under construction at Ulchin was expected to begin operation by the end of 2004, while the construction of four other plants began between July and September of 2004. Additionally, another four reactors are in the preparation stage [APERC, 2004 b]. Table 9. 4 presents a summary of nuclear power data in Korea.

Table 9. 4 Nuclear power data summary, Korea

Reactors in operation	19
Nuclear installed capacity (gross)	16,716 MW
Reactors under construction	5
Total electricity generation	322.4 TWh
Nuclear generation	129.7 TWh
Nuclear generation share	40%

Note: Generation figures for 2003

Source: [IAEA, 2004 d], [KPX, 2004]

In Korea, nuclear-related activities are planned and carried out by various organizations, such as the Atomic Energy Commission (AEC), the Nuclear Safety Commission (NSC), the Ministry of Science and Technology (MOST), and the Ministry of Commerce, Industry and Energy (MOCIE).

Under the Atomic Energy Act, AEC is the highest decision-making body on policy issues and utilization of nuclear energy. The AEC is composed of nine to eleven members representing various sectors of the government, academia, and industry. The chairman of the AEC is the Prime Minister.

After the regulatory reforms in the electricity sector that brought about the break-up of the power generation segment of the Korea Electric Power Corporation (KEPCO), the nuclear power plants together with the hydro plants came under the control of the Korea Hydro and Nuclear Power Company (KHNP). KHNP is also responsible for the management of radioactive waste coming from all of its plants. Plans exist for the construction of centralized facilities for both spent fuel and low and intermediate level radioactive waste (LILW), but for the time being these materials are being stored on-site at each of the power plants.

Government Policy on Nuclear Power Development

Nuclear power continues taking a central role in Korea's long-term energy policy. The policy is based on the following tenets: (i) promotion of a policy to decreasing dependence on foreign sources of energy, particularly dependence on the Middle East for oil and natural gas, (ii) to consider nuclear power is an important part of the strategy to achieve sustainable development, and (iii) that nuclear policy supports the nation's industry. Owning a mature nuclear industry was originally intended to lessen the nation's dependence on foreign technology in the energy sector, but has of late been imbued with the added dimension of pursuing international export markets.

In order to achieve the objectives of the long-term nuclear energy policy, the government established a legal basis to formulate the "Comprehensive Nuclear Energy Promotion Plan (CNEPP)" every five years through the amendment to the Atomic Energy Act in January 1995. The CNEPP includes long-term nuclear policy objectives and basic directions, sector-by-sector objectives, budget and investment plan etc. The 1 st CNEPP was formulated in June 1997. As of July 2001, the Korean government formulated the second CNEPP, which includes a five-year implementation plan from 2002 to 2006, and the direction of the nuclear energy policy towards 2015 with a total of 28 nuclear units.

However, MOCIE has recognized that it will become increasingly difficult to secure sites and construct nuclear plants in the future. This factor could delay official plans for new plant construction, particularly cases that require the selection of new sites.

Role of the Government in the Nuclear R & D

The Atomic Energy Act stipulates that the Minister of Science and Technology shall formulate the National Atomic Energy R & D Program according to the sector-by-sector implementation plan. Major projects are being carried out currently and are funded by both the Government budget and the Atomic Energy R & D Endowment fund.

The Intermediate and Long term R & D Program covers 6 fields as follows:
- Nuclear reactor & nuclear fuel,
- Nuclear safety,
- Radioactive waste management,
- Radiation/radioisotopes application & radiation protection,
- Current issues related to the NPP, and
- Basic technology.

The R & D fund distribution by each sector is shown in Figure 9. 5.

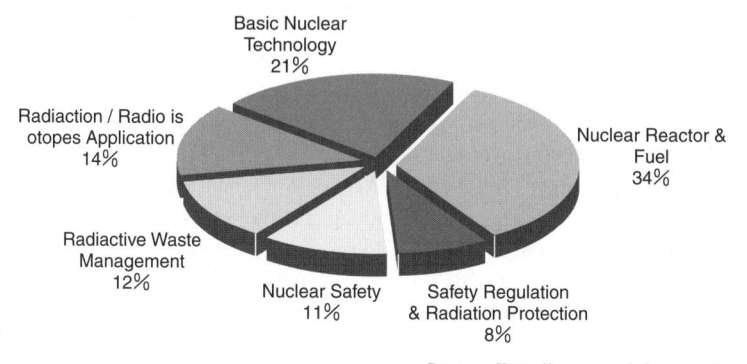

Basic Nuclear Technology 21%
Radiaction / Radio isotopes Application 14%
Nuclear Reactor & Fuel 34%
Radiactive Waste Management 12%
Nuclear Safety 11%
Safety Regulation & Radiation Protection 8%

Source: [http://www-pub.iaea.org]
Figure 9. 5   R & D fund distribution by sector (2001)

- **Nuclear Power in China**

Historical Development

In 1970, the former premier Zhou Enlai pointed out the necessity for the peaceful use of atomic energy and the development of nuclear power. This triggered the first step toward nuclear power development in China. In December 1991, the Qinshan nuclear power plant was connected to the grid for the first time. Thus, nuclear power generation began on the Mainland of China.

Current Status and Trend

Until now, nine nuclear units are in commercial operation, spread over three nuclear power bases: Qinshan, Zhejiang Province; Daya Bay, Guangdong Province; and

Table 9. 5 Nuclear power data summary, China

Reactors in operation	9
Nuclear installed capacity (gross)	6,988 MW
Reactors under construction	2
Total electricity generation	1,910.0 TWh
Nuclear generation	41.6 TWh
Nuclear generation share	2.2%

Note: Generation figures for 2003

Source: [IAEA, 2004 d]

Ling'ao, Guangdong Province. The total installed capacity of the 9 units is 6,988 MW. Two additional units are under construction in Tianwan, Jiangsu Province, and plan to enter commercial operation by 2006. By that time, there will be 8618 MW of installed nuclear power. Table 9. 5 presents the data summary of nuclear power in China.

In China, the National Development and Reform Commission (NDRC) is in charge of setting broad energy policy, energy planning at the central government level, and defining the participation of nuclear power in the electricity system. The State Commission on Science, Technology and Industry for National Defence (SCSTI) has administrative oversight of nuclear energy, and its subsidiary, the China National Atomic Energy Authority (CAEA), is responsible for managing the peaceful use of nuclear energy and promoting international cooperation. The China National Nuclear Corporation (CNNC), formerly the Ministry of Nuclear Industry and directly under the State Council, has responsibility for both civilian and military nuclear activities regarding nuclear weapons, power production, and waste disposal facilities. It also includes a significant research and development capability. Within the CNNC, the China Nuclear Energy Industry Corporation (CNEIC) is the organization in charge of China's uranium and enrichment services. Finally, the China Nuclear Engineering and Construction Group Corporation (CNEC) is the entity in charge of nuclear plant construction, nuclear engineering construction, and national defense engineering construction.

China's ambitious nuclear power development plan will make it the one of the world's largest markets for nuclear power development in the next 20 years. According to the latest economic plan, China will build 30 more units of 1,000 MW level nuclear power and will increase its installed capacity of nuclear power to 40 GW by 2020. This level of power will account for 4% of China's total installed power generating capacity. It also requires the construction of two 1,000 MW plants every year from 2004 up to 2020. The question now being asked is can China accelerate that growth rate to meet the even more ambitious pace of its new energy plan?

Nevertheless, the story of China's nuclear industry should be viewed in context. The nuclear industry is only one of many big stories related to China's energy market, including: (i) China was the world's third largest consumer of petroleum products in 2002, following the United States and Japan, (ii) Coal makes up the bulk, 64 percent of China's primary energy consumption, and China is the largest consumer and producer of coal in the world, (iii) The largest project under construction, by far, is the Three Gorges Dam, which, when fully completed in 2009, will include 26 separate 700-MW generators, for a total of 18.2 GW, (iv) The Chinese government is in the early stages of formulating a fundamental long-term restructuring of its electric power sector ... " Given the previous information, China's nuclear power target seems completely feasible.

Government Policy on Nuclear Power Development

China's Tenth Five-Year Plan (2001–2005) incorporates nuclear energy as a part of China's strategy to guarantee energy security, one of its major goals. The Plan calls for a policy that includes a moderate development of nuclear power generation in order to fulfill domestic electricity consumption.

In March 2005, China's Premier Wen Jiabo called for the rapid development of the nuclear power industry. The key points of Premier Wen's announcement were:
- Overall planning and rational layout;
- Maximizing domestic manufacturing of nuclear power plants and equipment, with self-reliance in design and project management;
- A self-sufficient nuclear fuel supply. Present capacity is limited and therefore to meet such goals more fuel making facilities will have to be built in the near future;
- Encouraging international cooperation;
- Set quality and safety as priorities.

Research and Development

In view of the massive development expected in China's nuclear industry in the future, the Chinese government believes it is in their best interest to develop domestic capabilities in all areas of nuclear technology, including the construction of facilities and the provision of nuclear services. Much has been accomplished already through international collaboration in research projects and through joint partnerships in the construction of nuclear power plants. According to the Institute of Nuclear Energy Technology, China basically has the technical capability to build power plants on its own. At the moment, one goal of the nuclear industry in China is to increase the level of domestic content in all nuclear power equipment from 50 percent in 2003 to 70 percent in the following 3–4 years.

- **Nuclear Power in Chinese Taipei**

Taiwan has few domestic fuel resources. For Taiwan, energy security is viewed as not only essential to economic growth, but to survival.

Taipower is Chinese Taipei's state-owned power utility that controls nuclear and hydro power plants in Taiwan. Currently, it has 6 reactors in operation at 3 plants, which came into commercial operation between 1977 and 1985. Installed nuclear capacity totals 5,144 MW and accounted for 21 percent of total generation in 2003. Table 9.6 provides the data summary of nuclear power in Taipei.

There are two reactors under construction at the plant at Lungmen, referred to as the fourth nuclear power plant scheduled for commercial operation in 2006 and 2007, respectively. The units utilize General Electric's advanced boiling water reactor (ABWR) and will have a capacity of 1,350 MW each.

The current Taiwanese government, however, has been cool to nuclear growth. This has caused construction delays for the last of Tawain's nuclear power plants to date, Lungmen. When Lungmen is completed, there are no known plans to build any more.

Table 9. 6 Nuclear power data summary, Chinese Taipei

Reactors in operation	6
Nuclear installed capacity (gross)	5,144 MW
Reactors under construction	2
Total electricity generation	173.8 TWh
Nuclear generation	37.4 TWh
Nuclear generation share	21.5%

Note: Generation figures for 2003
Source: [IAEA, 2004 d], [TAIPOWER, 2004]

- **Nuclear Power in India**

Current Status and Trend

India is an interesting case in nuclear power plant development. Although the share of nuclear now accounts for a relatively small amount of the total generation, however, it is estimated to hold one fourth of the world's thorium reserves. Thus, India has an advantage in nuclear development.

By the end of 2005, it had 15 reactors in operation, totaling 3,310 MW and contributing 3.3 percent to the total electricity production. It has a further 7 reactors under construction with a total capacity of 3,420 MW [NPCIL web, 2004].

The Department of Atomic Energy (DAE) was established on August 3, 1954, under the direct charge of the Prime Minister via Presidential Order. The Nuclear Power Corporation of India Ltd. (NPCIL) is a public sector organization under authority of the DAE that is responsible for design, construction, commissioning, and operation of nuclear power plants using boiling water and pressurized heavy water type reactors in India.

Nuclear Policies

In view of the limited fossil fuel availability within the country, the relevance of nuclear power in meeting the short and long term needs of our energy was recognized from the initial stage of development. From the very beginning, as a long term strategy, the Nuclear Power Program embarked on the three stage nuclear power program, linking the fuel cycle of Pressurised Heavy Water Reactor (PHWR) and Fast Breeder Reactor (FBR) for judicious utilization of our limited reserves of uranium and vast thorium reserves. The long term objective emphasized self-reliance and thorium utilization. The PHWR was chosen due to its extensive research and development facilities which cover diverse areas for supporting technology absorption.

India also has impressive plans for nuclear expansion. Under its 3 Stage Plan, a total of about 25 additional reactors are planned to be built between the years 2010 and 2020, which would increase the total installed nuclear capacity in India to around 20,000 MW and would represent a 25 percent share in total electricity production in that year. The nuclear capacity target is part of the national energy policy.

Nuclear power generation and related fuel cycle activities are under the authority of the Central Government. NPCIL, a wholly owned company of GOI and DAE, is responsible for setting up and operating the nuclear power plants.

Currently, there is no equity participation by the private sector in the area of nuclear power generation. The possibility of joint ventures with the public and private sector is being explored. This is essentially with a view to attracting investment in the nuclear power sector for capacity addition. This, however, will require amendments to the Atomic Energy Act, 1962.

- **Nuclear Power in Pakistan**

In Pakistan, nuclear power makes a small contribution to total energy production and requirements, supplying only 2.4% of the country's electricity. Total generating capacity is 19,540 MWe.

The Pakistan Atomic Energy Commission (PAEC) is responsible for all nuclear energy and research applications in the country. The Pakistan Nuclear Regulatory Authority is responsible for licensing and supervision.

In 2005, an Energy Security Plan was adopted by the government, calling for a huge increase in generating capacity to more than 160,000 MWe by 2030. This includes plans for lifting nuclear capacity to 8400 MWe, and 900 MWe of this by

2015. The government has announced plans to build two further Chinese reactors of 600 MWe each.

Under the current policy, an open and competitive electricity market is to be developed in Pakistan by 2010 at the latest. The focus of this policy is to increase private sector participation in electricity market and reduce dependence on public sector. All fossil fuel based power projects will be developed by the private sector. Nuclear power remains in the public sector. Because all activities in the electricity market are to be under the license of NEPRA, nuclear power plants, although are part of PAEC, they are registered and will be operating under the rules and regulations given in the generation license issued to them by the NEPRA.

- **Nuclear Power in Indonesia**

Research in atomic energy started early in Indonesia, in 1954, with the creation of the State Committee for Radioactivity Research. Today, Indonesia has 3 nuclear research reactors in operation and is planning another 10,000 KW test reactor.

A feasibility study of the installation of a nuclear plant was concluded in 1996 just before the Asian financial crisis, when all plans for nuclear power were deferred indefinitely. More recently, in 2002, the National Long-Term Energy Planning 2000 −2025 study included the possibility of a nuclear power plant in operation by 2016, and the latest National Energy Policy for 2004–2020, included nuclear energy as one possible alternative for the electricity generation fuel mix. Nuclear power is seen as a tool to achieve an optimum energy mix in terms of cost and impact on the environment, and to relieve increasing demand on fossil energy. Several areas have been identified as potential sites, with five of them pre-selected on the island of Java. Upon governmental decision 2005, a first reactor could be operational by 2016 [APERC, 2004 b], while, 5% of the total electricity for Java-Bali grid is expected be generated from nuclear energy by 2025.

Presently, all nuclear installations are owned and operated by the government via the National Nuclear Energy Agency (BATAN). This agency, formed by the Act No. 10 of 1997, is a supreme executive in charge of promoting the application and the research activities in the field of atomic energy in Indonesia.

- **Nuclear Power in Malaysia**

Malaysia is well endowed with conventional energy resources such as oil, gas, and coal, as well as renewables such as hydropower and biomass. Malaysia has implemented a five-fuel energy strategy optimizing the use of oil, natural gas, coal, hydropower, and renewable energy to achieve a balanced energy supply mix and reduce the nation's high dependence on oil and gas. For the power sector, this has meant substituting natural gas plants for coal-fired units and promoting the use of renewable energies wherever possible. According to government officials, nuclear power is currently not an option for Malaysia. Malaysia has at present one research reactor in operation.

- **Nuclear Power in Philippines**

The Philippines indigenous energy reserves are relatively small. It relies on imports for 60 percent of its total energy supply.

In the Philippines, atomic research began in 1958 with the creation of the Philippine Atomic Energy Commission, and with the nation becoming a member of the International Atomic Energy Agency. Construction of the 650 MW Baatan Nuclear Power Plant was initiated in 1977. However, due to reasons of political instability, as well as safety requirements, that power plan has not been put into operation yet.

In May 1995 a Nuclear Power Steering Committee was created by executive order to reanalyze the nuclear power program. The long term Philippine Energy Plan 1996–2025 included 2,400 MW of nuclear plants as part of the total 102,424 MW required for the year 2025. The more recent Philippine Energy Plan 2004–2013 has no mention of nuclear power for electricity generation, but the option is preserved in the long-term plan for consideration after 2020 [APERC, 2004 b]

The Philippines shut down its only research reactor in 1988.

- **Nuclear Power in Thailand**

Thailand is highly dependent on energy imports, particularly oil. Thailand has no policy for the time being for development and utilization of nuclear energy. A nuclear plant project was initiated by the Electricity Generating Authority of Thailand (EGAT) in 1966. A site was approved in the early 1970s, however the project was cancelled in 1977 in view of the prevailing global opposition and from the Thai public [APERC, 2004 b]. In 1997, a study on nuclear power was reinitiated, and at one point plans existed for the construction of one reactor within 10 years, with 5 more to follow afterwards. The project was deemed unsuitable after the Asian economic crisis and progress remains stagnate. The high growth expected in demand might revive interest in the evaluation of nuclear plants for the future.

Thailand has one research reactor in operation and another one under construction.

- **Nuclear Power in Vietnam**

Vietnam is endowed with fossil energy resources such as oil, gas, and coal as well as hydraulic resources suitable for power generation. However, the government of Vietnam foresees major and increasing difficulties in meeting electricity demand after the 2015 without resorting to either electricity imports, LNG, or coal imports for power generation, or development of nuclear power. The Ministry of Industry plans to introduce nuclear power plants in the near future, citing the urgent requirement to provide a stable supply of energy for socio-economic growth sparked by rapid development. Nuclear plants are seen as a way to diversify energy options to increase energy security, while contributing to environment protection.

Viability studies looking into all aspects of nuclear power began in 1995. These studies included plant site selection, reactor technology options, radiation safety, waste disposal, and nuclear law. Important steps towards deciding whether to construct nuclear plants have been made. One step is the identification of three suitable sites in PhuocDinh and VinhHai, both in NinhThuan province, and HoaTam in PhuYen province. Another step, set to begin in March 2003, is the elaboration of a nuclear law, related codes, and standards with cooperation from Japan and the International Atomic Energy Agency. The studies are now focusing on the construction of 2 to 4 units of 1,000 MW each between 2017 and 2020, and a decision on whether to go ahead with the project should be made before 2008.

Vietnam currently has a research reactor in operation.

9.2 Effects of Electricity Reform on the Development of Nuclear Power

The advent of restructuring and deregulation in many Asian countries and the rest of the world is another important factor to consider in the analysis of the future development of the nuclear industry. Deregulation of electricity markets is a trend expected to be followed by many economies. Nuclear power cannot remain isolated from electricity market competition. Both existing and future nuclear power plants will be affected by competition.

In following section, the impact of electricity industry reform on nuclear power generation is discussed under the following headings: (i) energy policy issues, (ii) operating nuclear power plants, (iii) safety and safety regulation, (iv) new nuclear plants, (v) nuclear liabilities and insurance, (iv) research and development activities.

- **EI Liberalization Brings Transparency of Policy Measures**

An important change in the framework for nuclear power development under competitive markets will be the clearer separation of commercial and government decision-making. Under non-competitive supply systems, governments had a number of mechanisms available to pursue public policy objectives without any easily identifiable public or private expense. These expenses could be passed on quietly and diffusely to electricity ratepayers.

Competition in electricity markets, and the increased policy transparency it brings, could be an important means to erase the mistrust in government supports. To the extent government financial support for nuclear power continues, especially for the construction of new plants, it will become more explicit for both market participants and the public. Owners of power plants in competition with nuclear power plants will not hesitate to bring to the attention of the public and competition authorities those support measures that they believe to be unfair or not justified by public benefits. The perception of nuclear power as a fully democratic and market-based choice will be strengthened. Competition could, in this respect, help to strengthen public acceptance of nuclear power.

- **Positive Impact on the Performance of Current Nuclear Power Plants**
Performance of Nuclear Power Plants
The impact of electricity market deregulation on the performance of nuclear power plants is expected to be positive. Increased competition in a deregulated market should bring about cost reductions through reductions in staffing, increased productivity, and higher availability factors, thereby improving economic performance.
Plant Life Extension and Power Plant Upgrades
There is substantial interest in nuclear power plant life extension and nuclear power plant power upgrade because the cost of life extension as well as plant upgrade is expected to be much less than that of building a new power plant of any kind. Competitive electricity markets will likely increase the incentive for life extension and upgrades of nuclear power plants, particularly for plants that are economically competitive, although there may be some concern about adequate return on costs in the long term.

- **Little Impact on the Nuclear Safety and Regulation**
The argument that competition will lead to lower levels of nuclear plant safety has been advanced by, among others, nuclear plant worker trade unions due to liberalization. Under liberalization, management decisions at nuclear companies may over-emphasize short-term economics. This is not true, since nuclear plant safety is assured by independent authorities who are, in principle, outside the control of nuclear plant operators and independent of short-term political influence. Nuclear safety regulators are required to ensure compliance with regulations, regardless of the consequences on the companies affected by their actions. The introduction of competitive electricity markets should therefore have little impact on the application of existing nuclear safety regulations by safety authorities.

- **Negative Impacts on New Nuclear Power Plant**

Under the prospect of more competition brought about by electricity market restructuring, nuclear projects face increased investment risk that further deters construction of new projects.

Nuclear power projects have relatively higher capital costs and longer lead times. Because of competitive electricity pricing, recovery of initial investment in higher capital projects takes longer. Thus, highly capital-intensive nuclear projects place more capital at risk and tie that capital up for longer periods of time, putting them at a disadvantage against projects that can be amortized more rapidly. Heavily capital-intensive projects also magnify the effects of overruns in construction costs or in time schedules due to the interest paid on capital.

Longer lead times characteristic of nuclear projects present other disadvantages in a liberalized market, aside from requiring more annual interest payments on capital. The financing of any power project requires anticipating the price of electricity after the end of construction and at the beginning of commercial operation, so that estimations can be made for the recovery of costs. A nuclear project with longer construction time has to predict prices over longer periods of time which introducing additional uncertainty.

- **Increase the Requirement of the Nuclear Liabilities and Insurance**

In a monopoly, electric utilities or regulators assign risks and liabilities as they see fit, often to captive customers. As the electricity sector moves toward liberalization, risks and liabilities will inevitably be reallocated.

Competition will put pressure on power generators to clearly identify and quantify future economic liabilities of nuclear power plants, and to include them in electricity prices. In principle, the growth of competition in electricity markets poses the same financial risks on both nuclear and other power generators; the only essential difference is the degree of these risks, which may be affected by the size of investment. Nuclear power, however, has some specific liabilities and associated risks. Decommissioning and waste management liabilities may be the most important of the various economic risks of nuclear power in competitive electricity markets. Another risk of nuclear power is the slight possibility of a large nuclear accident.

Concerns associated with decommissioning and waste disposal include: the adequacy of funding provisions to meet current estimated target costs, the accuracy of the target costs themselves, and the adequacy of regulatory requirements for ensuring sufficient funding.

- **Declining Research Activities**

Government funding of R & D for nuclear power has declined and this trend is expected to continue as electricity market deregulation increases. Utilities may tend to reduce R & D expenses in order to reduce costs and their efforts will likely focus on applied research aimed at performance enhancement.

Competitive markets will lead to short-sightedness in research and a drop in long-term research projects. The question in the nuclear field is particularly difficult to analyze since governments historically have spent enormous amounts of money that were never borne directly by electricity ratepayers. Nuclear research institutes already face many challenges, even without changing electricity markets. Many involved in nuclear research apparently fear that the difficult environment caused by decreases in government research and development funds will be compounded by "inadequate" money from utilities.

It is clear that competition leads companies to re-orient their research and enhance its effectiveness. Utilities will tend to reduce research into generation technologies and instead focus on improvements to plant operations and in-plant technologies. "Public good" research projects with no commercial benefits will be dropped.

9. 3 British Energy's Experience regarding Nuclear Power Issues in Liberalized Electricity Industry

British Energy (BE) operates 15 of Britain's 31 nuclear reactors, including 14 advanced gas-cooled reactors and Britain's newest reactor, a single pressure water reactor, completed in 1995, employs 5,200 staff and provides 20% of Britain's total electricity supply. BE restructuring was completed in January 2005 and BE relisted on LSE (FTSE 250). Current BE share price approx £4.65 (€6.80) November 2005.

- **Nuclear Liberalization in Britain**

Margaret Thatcher's Conservative Government came into power in 1979. Energy market liberalization, designed by Energy Minister Nigel Lawson, began in 1982. The Energy Act in 1983 set out the framework to introduce gradual competition to energy supply markets.

British Gas was privatized 1986, British Petroleum in 1987, National Power in 1990, and National Grid in 1990. Nuclear was the last sector of the UK energy market to be privatized. British Energy was privatized in 1996 through a public offering. Currently, BE stock has raised £2.1 billion (€3.1 billion) since its initial public floatation in 1996. However, the government had to grant £450 million (€ 658 million) in emergency state aid to BE in 2002. At the year of 2002, BE stockholders lost 97% of the value of their share.

- **What Went Wrong at BE?**
Homeowners and businesses customers were allowed to choose their electricity supplier from June 1999. Severe NETA competition forced wholesale market prices down to the marginal cost of generation. Price reductions of 10–15% were expected, however electricity prices have actually dropped 40% since 1998. Meanwhile, BE was locked into expensive reprocessing contracts with BNFL, equivalent to 25% of BE operating costs. At the current wholesale price of £16 per megawatt hour, British Energy, producing at £19, makes a £3 loss. This activity that hardly justifies British Energy's operation, not to mention paying shareholders handsome dividends.

In 2002, the UK government was forced to grant emergency state aid to BE. BE, therefore, was financially restructured. The government now owns a majority of the equity and has effectively renationalized BE.

- **Finding the Causes of BE failure and of Nuclear Power in the UK**

The causes of BE failure consist of: (i) the ability of nuclear accidents to deliver shocks to energy supply markets, damaging confidence, (ii) lack of experience in the financing and building of new nuclear reactors in a liberalized energy market, (iii) high capital costs for building nuclear plants mean that the nuclear option is extremely risky, (iv) it would lock the government into a 100-year energy lifecycle (10 yr build + 60 yr operate + 30 yr decommission).

In the environment of liberalization, the preferences of retail and wholesale customers should be included. For household retail preferences, a relatively small amount of nuclear power is sold to UK household consumers, roughly 10–15% of electricity mix compared with UK average of 20.6% nuclear. Mainly, this is because British Energy does not have a vertically-integrated retail supply outlet business, i.e. BE does not sell directly to households, and market research shows that 40% of household consumers are strongly anti-nuclear [Jackson, 2005]. Regarding industry wholesale preferences, most UK nuclear power is sold to large industrial consumers who need reliable baseload generation. Industrial users consume roughly 40% of the nuclear electricity mix. British Energy sells wholesale directly to industrial users, the most natural market for nuclear. However, as mentioned above, due to severe competition wholesale market prices dropped 40% since 1998. Thus, the environment of liberalization has exposed BE many difficulties.

- **The Future of Nuclear Power in the UK**

It is important to note that despite economic problems, BE continued to reliably supply power in Britain. Nuclear is an excellent baseload supplier. Nuclear remains an important source of low carbon energy and must remain a viable option for companies serious concerned about global warming and climate change.

"Tony Blair today indicated publicly for the first time that he will support building new nuclear power stations to meet Britain's future energy needs" [Jackson, 2005].

Therefore, nuclear power seems to have a bright future under liberalized market on one condition: Market interventions are necessary for the economic development of nuclear power in a free market.

9.4 Impacts of Industry's Reform and Mechanisms for the Promotion of Nuclear Power in Asian Countries

Recently, there seems to be little interest in nuclear power development in world due to the serious public opposition regarding the fear of a nuclear accident. Liberalization, while increasing the performance of existing power plants, also raises the uncertainties of nuclear power concerning safety, investment costs, and waste disposal. Meanwhile, Asia is the only region in the world where there has been significant growth of nuclear power in the recent past and that currently has sizeable plans for the construction of more nuclear plants. As mentioned above, the growth of nuclear power in Asia comes from many drivers, both political and economic. However, to gain in the dynamic development of nuclear policy Asian governments must provide specific mechanisms for nuclear power promotion. Following sections try to find impacts of EI reform to nuclear policies in Asian countries and then discuss several mechanisms to promote nuclear power.

- **Impacts of Industry's Reform in Asian Countries**

Regression results from Chapter 7 have shown that while restructuring policy options positively impact the share of nuclear power generation, competitive market options, which include both generation competition and wholesale market, harm nuclear power development. Coefficients are significantly negative.

Figure 9.6 illustrates the pattern of nuclear power share in the fuel-mix of several Asian countries in the period of 1985–2004. A reduction in the trend of nuclear power is seen in South Korea and Taipei. Presently, in Taipei, there are no new nuclear power plant candidates. The share of nuclear power in Japan, however, was stable thanks to the sustainable policy of Japanese government. Only in 2003 the share of nuclear power dropped. The cause of the drop was that TEPCO had to check the condition and safety of its nuclear power plants as a result of a revelation of an inappropriate management in the safety records of self-imposed inspections[21].

Section 9.1 has shown the bright future of nuclear power plants in China, with an impressive plan to build a 2×1000 MW nuclear power plant every year until

---

21 http://www.tepco.co.jp/en/press/corp-com/release/03031101-e.html

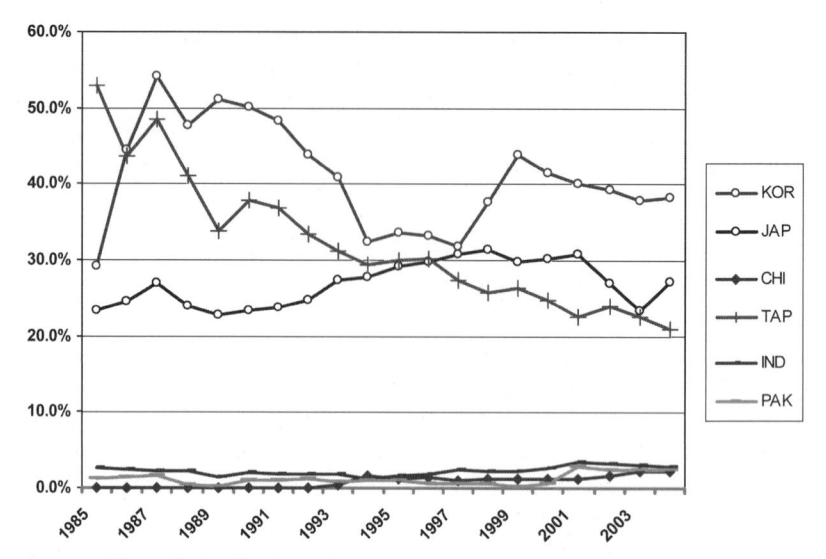

Figure 9. 6 The share of nuclear power generation in Asian countries

2020. The future also seems bright for India. It is obviously that a liberalized environment can harm to nuclear development. However, because the nuclear power issue is strongly influenced by political will, there is still hope for the further development of nuclear power in Asia, especially in countries where the development of nuclear power is aggressively pursed by government with the support of international corporation.

- **Promotion of Nuclear Power in Asia**
Strengthen Regulatory Authority
The regulator must fully understand the economic conditions of the competitive electricity market and the range of competitive pressures facing the operator. The empirical analysis in Chapter 7 shows the important positive impact of a regulatory body on nuclear power development in the environment of industry liberalization.

Under liberalization, regulatory authority needs to deal with the requirement of reducing nuclear power risks, in which the most important includes (i) risk of nuclear accident through safety regulation and (ii) risk of lacking payment for plant closure and waste disposal expenses.

− Safety Regulation

There must be independent and powerful nuclear safety authorities, which is able to act independently of short-term influence by both nuclear utilities and political

authorities. There is a real danger that safety regulators, sensitive to influence from either side, could let safety levels drop in a newly competitive market. Adequate regulatory independence is also a pre-requisite under any electricity market system, however it is even more so in a competitive one.
– Regulations on Ensuring Adequate Provisions for Decommissioning and Waste Disposal Expenses
Ensuring adequate reserves under any electricity market system has two important components: (i) establishing the technical requirements for plant closure and the corresponding sum of money needed to meet those requirements, and (ii) setting the financial arrangements for accumulating that money over the lifetime of plant operation. This includes managing any funds held in trust.

The first component is one that must be done by nuclear regulatory authorities. They must define exactly the standards, methods, and timing of plant closure. From these definitions they and the plant owner can estimate the amount of money needed to close the plant. The nuclear industry believes that, with clear definitions, costs are reasonably well known, despite variability among different plants in different countries of up to a factor of six in estimated plant closure costs.

The second component is primarily a financial oversight function. There must be an agreed way to collect and manage money over an extended period of plant

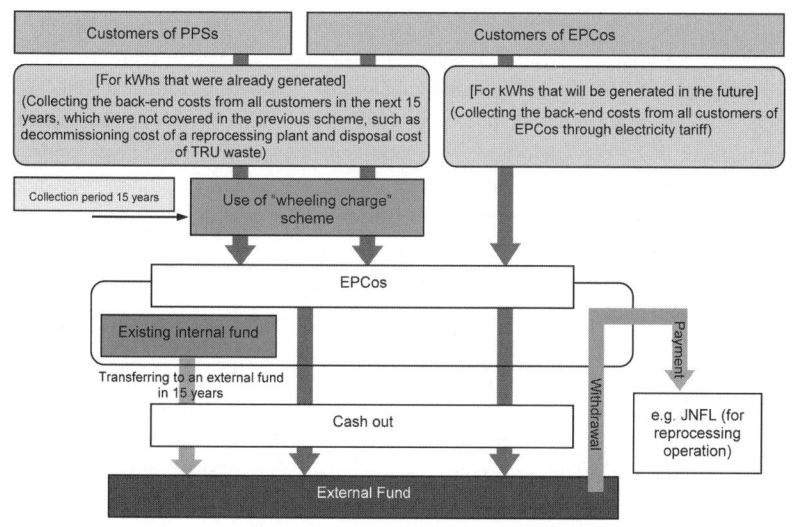

Source: [TEPCO, 2004]

Figure 9. 7 Scheme of back-end funding

operation for meeting future liabilities. To reduce risk, a decommissioning and waste disposal funding system should be established by law. Two examples of such funds are the decommissioning fund that has existed in Taiwan since 1989 and the new Back-end fund (Figure 9. 7) established in Japan in May 2005.

Research and development of nuclear technology has perhaps been the most visible component of government support. In the nuclear power field, the R & D requirement is quite enormous, not only in terms of the amount of money required, but human resource capability. Governments have also built specialized facilities, such as enrichment plants, fuel reprocessing facilities, and radioactive waste facilities.

Under the environment of liberalization, there also remains a role for government funding of research programs that provide an element of "public good", however, other participants such as co-operation with the private sector and on an international level should be included. Co-operative arrangements between industry and government, such as those developed in the IEA Technology Implementing Agreements, can help combine private and public research efforts. International co-ordination of nuclear research through organizations such as the NEA will take on added value from the introduction of competition.

Making Full Use of Economic Competitiveness of Nuclear Power
After all other issues are dealt with: nuclear safety, waste disposal, public opinion, and proliferation, in the end nuclear power will only be viable if it can be competitive in today's electricity markets. An often discussed topic is that the internalization of the cost for externalities into competing forms of electricity generation can make nuclear power projects more attractive to undertake. The establishment of carbon trade practices or implementation of carbon taxes could put nuclear power and renewable energy technologies in a better level of competitiveness. Efforts to promote the inclusion of nuclear projects under the Clean Development Mechanisms are expected to continue. However, all these measures are policy dependent and it is not possible to predict when and to what extent they can be adopted. For nuclear power to be a valid alternative fuel source for the future, it has to be economically competitive on its own.

[APERC, 2004 b] has made several conclusions about the economic competitiveness of nuclear power. The unanticipated positive performance of nuclear plants in markets undergoing restructuring has made governments and utilities re-analyze the prospects of nuclear power as a viable source of electric power. Nuclear plants currently in operation are economically attractive. With fuel prices that are lowest compared to fossil fuel alternatives and operation efficiency records that continue to improve, nuclear power plants have become the least operating cost option for base load capacity in many places. They also offer predictability in fuel costs which fossil fuel plants cannot provide. However, as noted, the competitiveness of nuclear power rests on the ability to reduce investment costs, which should include increasing decommissioning costs. Capital costs have also come down thanks to advanced reactor models which promise lower investment costs and shorter construction times, but the nuclear industry still has to demonstrate that it can finish projects on time and under budget, and new licensing processes, in those economies where they have been implemented, have to be tested. The recent construction record in APEC economies in Asia is on the right trend towards bringing investment costs and construction schedules under control.

Providing Policy Options to Offset Investment Risk

To offset investment risk, there four types of financial policy can be considered: loan guarantees, accelerated depreciation, investment tax credits, and production tax credits.

Loan guarantees for a part or all of the borrowed funds for capital could allow the borrowing rate on that portion of the debt to be as low as the risk free rate. Accelerated depreciation schedules can help nuclear owners be exempt from profit tax. Two popular accelerated depreciation schedules are 7 years and expensing. A refundable investment tax credit allows an owner to apply the credit to the income earned from other assets if the credit is larger than the tax on the asset, for instance, if the nuclear plant operated under a loss for the first few years and thus had no tax obligation. One more policy option is production tax credits; however, that policy can help cash flow only after the plant has been built and does not reduce near-term money requirements during construction.

Moreover, internalizing the cost of externalities into competing forms of electric generation can make nuclear power projects, which to a large extent already include them, more attractive to undertake. The establishment of carbon trade practices or implementation of carbon taxes could put nuclear power and renewable energy technologies in a better level of competitiveness.

Government Subsidies for Locating New Nuclear Power Plant

Finding sites for new nuclear power plants is one of the troublesome issues in nuclear development due to the strong opposition from the public media. To help make the siting of nuclear installations more attractive, the government should have a program that provides subsidies to the townships that are willing to host nuclear power plants. It could be upgrading the infrastructure system, an environmental improvement program, or social healthcare funding.

PART III.
MULTIDIMENSIONAL ELECTRICITY INDUSTRY REFORM – A STRATEGIC ANALYSIS FOR THE CASE OF VIETNAM

Electricity industry reform has been assessed through the standard prescription in Part I, and an empirical analysis based on econometric modeling in Part II for a more concentrated scope of Asian countries. Now, for the time for the most pointed and deepest analysis, the case of multidimension reform in Vietnam's electricity industry.

According to the results from Chapter 6, Vietnam is classified in the second country group, however, compared to other countries in the same group, it is slightly behind. Therefore, in the case of Vietnam, most reform policies will have to be considered when designing reform strategy. Part III of the book endeavors to, first, examine industry reform to date, and then apply policy implication findings from Part I and Part II to establish comprehensive strategies for electricity industry reform in Vietnam, which cover most reform aspects, such as the establishment of efficient regulatory framework, restructuring of the Electricity of Vietnam, introducing competitive market, implementing privatization to addressing regulation mechanisms for some most sensitive social-benefit issues. Such challenging tasks will be solved, respectively, in Chapters 10 to 14.

This part, to some extent, uses information and some results from [Mai and Nomura, 2005]. However, the focus is to bring a more detail and comprehensive analysis of multidimensional reform in Vietnam's electricity industry.

CHAPTER 10
ANALYSIS OF ELECTRICITY INDUSTRY REFORM TO DATE IN VIETNAM

10. 1 Historical Development of Vietnam's Electricity Industry

Vietnam's electric power industry was founded in 1954 with a small inheritance from the French. From a historical standpoint, the institutional structure of Vietnam's EI development can be divided into two periods: (i) 1954–1995: Government Entities, and (ii) 1995–now: Public Corporation.

- **1954–1995: Government Entities**

From 1954 to 1995, the organization of the electricity industry in Vietnam naturally changed several times; however, most of these changes were in terms of the renaming or the reallocation of divisions, not actual institution-wide structural changes.

Industry structure includes the following: ministerial divisions, power companies, electricity departments, and transmission departments (Figure 10. 1). The Ministry has controlled both the governmental and business management, in addition to directly controlling three regional power companies. These companies managed

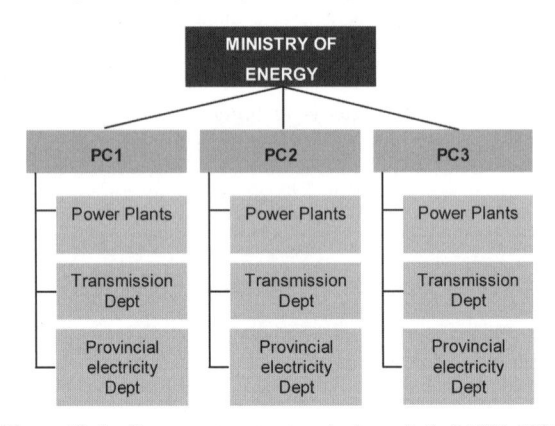

Figure 10. 1 Power sector structure in the period of 1954–1995

their own power plants, transmission departments, and provincial departments. There is no inter-regional connection between these companies.

Operation of the electric industry followed central planning mechanisms: the Ministry was responsible for setting an annual 5 year plan. Power companies must obey and carry out the plans. Power companies operated with non-profit objectives, it was not necessary to determine financial indicators; the main objective is to comply with given planning indicators. Most of the capital investment for power system development was injected through government & foreign grants, especially from Soviet Union.

This model seemed to be good in the period of war, when all country's resources should be concentrated for the struggle for independence. Power companies operated with non-profit objectives.

Vietnam began its transition to a market based socialist economy in 1986 under the reforms of "Doi moi". According to [Dung, 1996], in December 1986 "The 6 th Congress defined this first phase of the transitional period from capitalism to communism, with the main concept being the acceptance of diversified financing, provision, and market mechanisms throughout all sectors". Undoubtedly, the reforms freed up the economy to domestic and foreign private sector investments, resulting in rapid economic growth and an improvement in the average standard of living, particularly in urban areas. The high economic at the same time raised the demand for power and, of course, required more and more investment for additional power capacity. Meanwhile, grants from socialist countries ran out and capital injections from the government became limited. In 1995, these factors forced the electricity industry to change to a publicly corporate business structure.

- **1995–2004: Public Corporation**

In 1995, Electricity of Vietnam (EVN) was established as a state-owned corporation that operates in the areas of generation, transmission, distribution, and the sale of electricity. That was one of the initial efforts made by the Vietnamese government to separate the state from business management. The Ministry of Industry (MoI) is responsible for policy and oversight of the sector.

Under the EVN there are two operation blocks: (i) dependent accounting units including power plants, transmission companies, and (ii) independent accounting units including distribution companies, consultant companies, and mechanical companies. The industry structure is shown in Figure 10. 2.

The establishment of EVN marked a profound revolution in terms of organization and management in order to raise the sense of initiative in production and business, and create a pre-condition for the vigorous and dynamic development of Vietnam's power sector. Up to the end of 2004, the total installed capacity of the

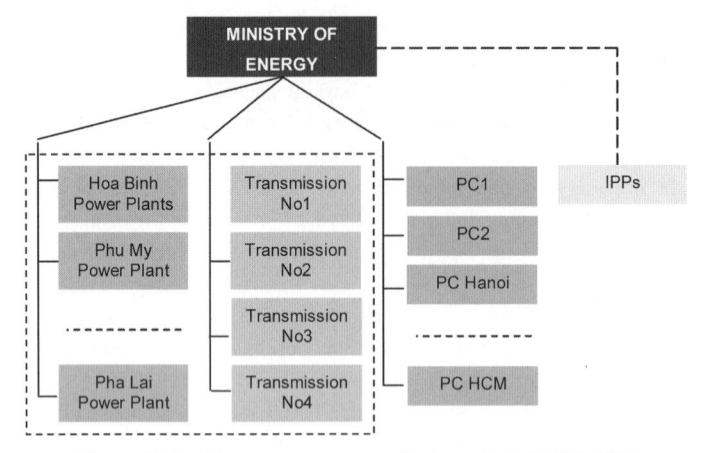

Figure 10. 2 Power sector structure in the period of 1995–2004

country reached 11.340 MW, the generation output gained 46.2 billion kWh increased from 4,500 MW of installed capacity and 14,617 GWh of electricity generation in 1995. Generation capacity in Vietnam in 1995 was about 4,500 MW. Expansion and new construction of transmission lines, distribution lines, and substations were also carried out. In 1995, about 50% of all households were electrified. This rate reached 87% in 2004. System loss was improved from 21.4% in 1995 to 12.1% in 2004.

In 10 years of development, the EVN has operated effectively, shown sustainable growth, and is financially viable. The annual growth rate of net profit is 12.35% and the SFR (self-financial ratio) is around 30%, which satisfies financial covenants imposed by international finance institutions. EVN also succeeded in mobilizing a huge capital of VND 78,000 billion for investments. Book value of total electricity network assets in 2003 increased six times compared to the value in 1995.

The vertically state-owned monopoly structure seems to be a conformable choice in the transitional period from a centrally planned economy to a market-oriented economy. The bundling structure between generation and transmission facilitates for economic dispatch of the power system, matching a good balance in the fuel mix, and good coordination between investment and operation activities. In addition, it should not be forgotten that this structure allows EVN to fulfill social objectives, such as electricity for poor people, rural electrification, etc.

However, at this juncture, the vertical monopoly type of structure has started to show its weak points, which contributes to pushing electricity industry toward a reform. The weak points are: (i) Unclear operating objectives. There is a controversy over the two main functions: to ensure its financial viability and to fulfill universal services on behalf of the Government. Being subjected to universal services also caused a lack of transparency in the EVN's finances. (ii) Passive operation of subsidiary units. Dependent accounting units operate as cost centers and do not consider their profit. (iii) A low rate of return: reduced from 12.3% in 1996 to 6.3% in 2002, which was slightly lower than the bank's annual interest rate of 7–8%. (iv) Low productivity rates in comparison to other ASEAN countries, and (v) non-transparency of pricing policies, not cost-based and strongly cross-subsidized.

10. 2   The Need for Electricity Industry Reform

Why the reforms should be done? What do people hope to gain from reform benefits? In the Part I and II of this book such questions have been answered from a world-wide and Asian scope. The reform pressure have influenced to the electricity industry in Vietnam. The drawbacks of monopoly structure mentioned above pushed industry to a turning point in 2004. The following section discusses the needs for reform in Vietnam in terms of both internal and external factors.

Internal Factors

Not limited to the drawbacks of the existing organization model mentioned before, from 2005 and in the years to come, the EVN will be confronted with more many challenges, including:

- Dangerously low capacity reserve margin – Presently, the peak demand stands around 6,400 MW and installed capacity is approximately 8,900 MW and available capacity is about 8,650 MW. The actual degraded value of the generation capacity is about 6,500 MW during the dry season. This extremely low reserve margin cannot ensure the reliability and security of the power system. New capacity is urgently needed to avoid power shortage in the future.
- Continued rapid demand growth and huge capital requirements – In the period of 2000–2005, the demand growth rate varied in a high range of 14–15%. The demand is forecast to continue to grow at 12 to 13% annually in the 2006 to 2010 planning period. By 2010, the total power demand is forecast to be 82 billion kWh in 2010 [EVN, 2004]. In other words, approximately 1,000 MW of new capacity will be required each year for the next ten years in order to meet the forecast demand. Satisfying the demand is essential for the nation's economic development objective. Based on EVN's estimates, the total requirement for capital investment in the power sector is about US$ 14.8 billion from 2003 to 2010. According to another preliminary estimation in Master plan number IV[22],

---

22 The final draft of electricity power development master plan number VI was submitted to Prime Minister in 2006.

the total requirement of investments from 2006 to 2025 reaches an ambitious total of US$100 billion.
- Growing gaps in financing of sector development – In the revised Master Plan on Vietnam Power Sector Development approved by the Prime Minister on 21 March 2003, the Government can finance only approximately 70% of the capital investment requirements for the power sector development. Approximately US$ 6 billion in capital investment will be required from non-Government sources in the period from 2003 to 2010. It will be a great challenge for EVN to attract such a large amount of non-Government investments to the power sector.
- No clear legal and regulatory framework – It would be difficult to attract non-Government sector (domestic and foreign) investors to Vietnam's power sector if there is no clear legal and regulatory framework to protect their investments and ensure full cost recovery and a reasonable return on their investments.
- Inadequate revenue realization and cost recovery – At present, the average electricity tariff yield is approximately US 5.6 cents per kWh, less than the estimated long run marginal cost of US 7.0 cents per kWh. The financial viability of the power sector cannot be achieved and sustained without appropriate tariff increases.
- Conflict of national uniform electricity tariff with market-based reform – Due to different customer density, the mix of customers, and geographical circumstances, and the cost of providing electricity is inherently different in different regions. EVN could not move forward in forming limited liability power companies at the provincial and district levels due to the national uniform tariff. This is because that the limited liability company could only charge its electricity services at the national uniform tariff regardless of the costs of providing services, at the EVN level, the national uniform tariff requires EVN to provide cross subsidies among regions (power companies) through internal bulk power transfer pricing. The cross subsidy practice does not encourage power companies to cut costs and improve efficiency. The national uniform retail electricity tariff is in conflict with the Government's economic reform and the cost-based pricing principle.

External Factors and the Role of International Financial Institutions

As mentioned in Part II, one of the most remarkable characteristics of the EI reform in Asian countries is it has been pushed under the strong initiatives of international financial institutions, such as the World Bank and ADB. In Vietnam, since the middle of 1990s, EI reform has always included as one of the objectives of the loan or key policy and institutional reforms supported by the projects.

The World Bank project of CR 2820-VN – Power development project (1996) supported to rationalize power sector institutions, commercialize operation of sector entities, initiate appropriate legal and regulatory framework, and initiate private sector participation in generation.

The World Bank project of CR 3034-VN – Transmission, Distribution, and Disaster Reconstruction Project (1998), besides to strengthen, rehabilitate and expand the high voltage transmission network and distribution systems to meet demand, reduce losses, and improve reliability in Vietnam, also objected to support power sector reform and restructuring through: (i) the separation of "transmission" and "generation" functions; (ii) the implementation of regulatory reforms; (iii) the introduction of tariff reforms; (iii) increasing operational and management autonomy for distribution companies and creating a commercialized organizational structure, and (iv) equitization of one or more operating units of the distribution companies on a pilot scale.

Encouraging continued implementation of power sector reforms is one of the key policies supported by Rural Energy Project I and II, in 2000 and 2004 respectively.

ADB has the same strategy to incorporate power sector reform in project loans. The Power Distribution and Rehabilitation (LN 1368-VIE of June 1995), Central and Southern Vietnam Power and Distribution (LN 28187-VIE of November 1997) both consisted of the objective of implementing sector reorganization and institutional reforms to ensure efficient use of resources.

In additional, the WB and ADB have provided several technical assistances that are directly relevant to power sector reform. These TA projects are summarized below.
- The World Bank TA on Preparation of Electricity Law (1997–2000) (executed by the Ministry of Industry)[23]. Under this TA, the Electricity Law Working Group was able to prepare and revise the draft Electricity Law. After more than 20 revisions, the law was finally approved in 2004 and has been enacted since June 2005.
- The ADB TA 2888-VIE: Improvement of Power Sector Regulatory Framework, (1998 to 2000). Under this TA, the Consultant provided extensive training to the Electricity Law Working Group and other industry stakeholders on the regulatory reform of the power sector, and tariff principles and methodologies
- The ADB TA 3763-VIE: Vietnam Road Map for Power Sector Reform carried out by the PA Consulting Group. This is the most relevant study regarding power sector reform. This TA assists the Government of Vietnam to define a clear and more detailed vision of the desired future power sector and market

---

23 At the same time, WB also provided a TA on Electric and Gas Sector Reform (the Executing Agency was Ministry of Planning and Investment).

structure 5 and 10 years ahead, and establishes a roadmap for power sector reform. It should be noted that the road map recommend by the TA did not cover complete reform strategy; it seems to give more priority to market development rather than privatization. However, it is true that the recommendations from this TA have constructed the basic ideas for further policy development in sector reform. This book makes use of such recommendations, especially in the section on regulation reform.

Besides pressure from international financial institutions, several other factors have a strong influence on EI reform in Vietnam. The process of joining WTO or AFTA can be included in these factors because they require an open economy and facilitation of competition. In the last stage, the provisions of AFTA will be enacted in 2006, while finalizing the negotiation process to join the WTO. These conditions place more pressure on industry reform.

10.3 Reform to Date in Vietnam's Electricity Industry

This section summarizes the progress made to date in implementation of power sector reform in Vietnam in such aspects as government policy, private participation in power generation, and EVN restructuring activities.

- **Government Policy on Reform**

The Government's power sector policy and reform objectives were first articulated in the Power Sector Policy Statement issued by the Ministry of Industry on December 1, 1995. In 1997, the Government updated its Power Sector Policy Statement, in which: (i) to transform the power sector enterprises into efficient commercially run entities that are financially strong and creditworthy, and have management autonomy in operations, (ii) to form a consistent legal basis to facilitate commercial sector operations and Government oversight of the sector, (iii) to use cost-based principles to set retail prices to promote efficient use of electricity, and efficient internal transfer prices to motivate the distribution companies to operate efficiently and purchase power economically, (iv) to diversify financial resources (including foreign investors) in the form of debt and equity financing to supplement the public sector investments, and create an environment to foster competition in project financing, and (v) to increase rural access to electricity and facilitate electricity conservation.

The Government Action Plan 2001 to implement the decision made at the third session of the Communist Party Central Committee, IX Tenure also had great influence on electricity industry reform. It included: (i) Restructuring and improving the efficiency of state corporations, (ii) Building several strong economic groups by adjusting and adding a model Charter or organization and operation of State Corporations (as issued in Decree No. 39/CP), promulgating a Decree on the

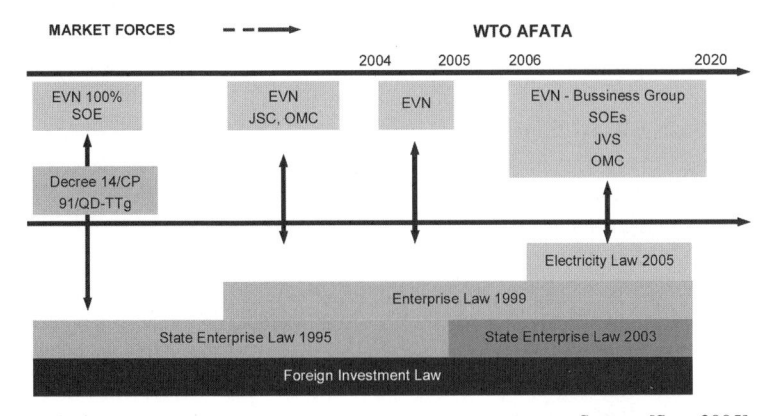

Source: [Son, 2005]
Figure 10. 3 Legal framework of electricity industry reform

transformation of the current structure of State Corporations into the parent-subsidiary model, and drawing up a plan for establishment of economic groups.

From 1999–2005, three laws relevant to electricity reform were enacted, including Enterprise Law 1999, Competition Law 2001, State owned enterprise Law (revised) 2003, and the most important one-Electricity Law enacted in 2005 (Figure 10. 3).

For the new stage of reform, [Institute of Strategy, 2004] said: (i) the continuation of reform strategy that includes developing a regulatory framework, adopting cost-based tariffs, and promoting private sector participation in the power supply, (ii) speeding up the unbundling and equalization process in public utilities such as telecommunications, the electricity supply, the water supply, gas, etc, (iii) increasing transparency in public utilities, (iv) stabilizing and reducing the electricity tariff, and (v) developing the electricity and gas markets.

- **IPP in Vietnam**

The EVN Annual Report 2004 says that non-EVN installed generation capacity as of December 2004 was 2518 MW, and that generation in 2004 was 6,363 GWh. The share of IPP in installed capacity is 22% and in generation 13.0%.

Operated under a BOT scheme, there include two combined cycle gas turbine power plants, Phu My 2-2 of 715 MW and Phu My 3 of 717 MW, and the Candon hydropower plant with a capacity of 72 MW. The developer of Phu My 2-2 is a consortium comprised of EDF International with 56.25% interest ownership, Sumitomo Corp. with 28.125%, and TEPCO International with 15.625%. Phu My 3's developer consists of BP with 33.3%, SembCorp Utilities of Singapore with

33.3%, Kyushu Electric Power with 26.6% and Nisho Iwai with 6.7%. Songda construction corporation, a domestic developer invested in the BOT Candon scheme.

Power plants being operated by the foreign IPPs in February 2004 included (i) 3 units of 125 MW oil firing thermal units in total 375 MW by Hiep Phuoc, (ii) 2 units in total 13 MW by Amata, (iii) 2 units in total 72 MW by VeDan, (iv) 1 unit of 12 MW by Bourbon, and (v) 4 units in total 24.8 MW by Nomura.

Hiep Phuoc power plant, which is the largest IPP plant in Vietnam at present, commenced operation in 1998. The company was established by a Taiwanese investor, the CT & D Group, as an enterprise with 100% foreign owned capital. Amata power plant is operated by an IPP established as a joint venture between Vietnamese investors and foreign investors. Vedan and Nomura IPP companies are wholly foreign invested companies.

Additionally, there are many small hydro IPPs which are operated by domestic investors.

IPPs greatly contribute to meeting the high growth rate of electricity demand in Vietnam. It also illustrates the success of EI reform policy to attract new investments from other participants. However, one point needs to be recognized is that most domestic IPPs are not funded by private investment, instead, most of domestic investors are non-EVN state companies such as VinaCoal, Songda construction corporation, etc. This means that idle domestic capital still has not been mobilized.

- **EVN Restructuring Activities**

In order to corporatize and commercialize the EVN, a series of works have been initiated (Figure 10. 4). Since 1998, EVN has continued to strengthen the corporate finance through issuing a bulk supplied tariff with distribution companies, and an internal generation price system with power plants. Several efforts were made to separate universal services and bring transference to pricing; however, both are not successful as yet. Since 1 st January 2005, EVN has launched an internal trial market for power plants based on production cost bidding mechanisms. However, since no actual cash flow has been transferred between EVN and power plants, the operation of market has a very limited effect.

In the past couple of years, EVN has taken initiatives to study the feasibility for creating new single-share limited liability companies (so called One Member Limited Company, OMC) and joint-stock limited liability companies. Creation of single-share and joint-stock limited liability companies will increase managerial autonomy and improve the efficiency and accountability of such companies. It should be noted that the State will not be responsible for the debts of the limited liability companies.

The trial establishment of one-member limited liability companies includes two generation companies Can Tho and Phu My, and two distribution companies, Ninh

EVN Restructuring

Figure 10. 4 EVN restructuring activities in 2005

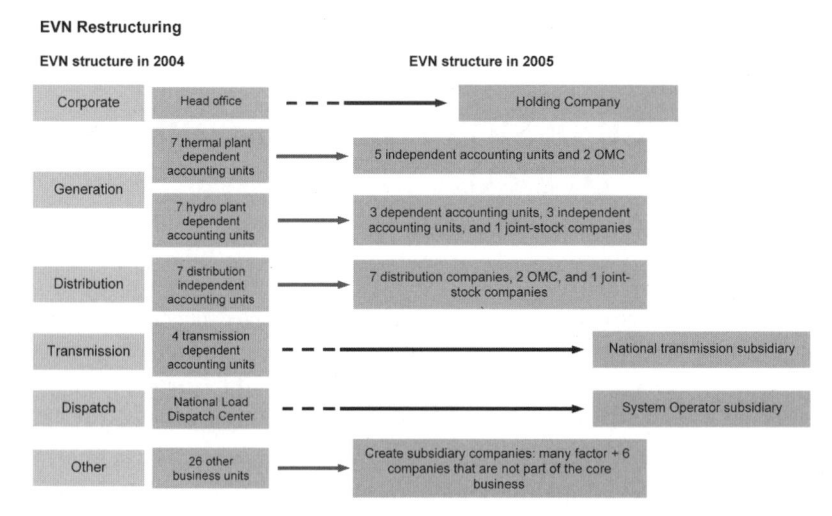

OMC: One Member Limited Companies

binh and Hai duong. The following companies transferred to a limited liability structure and were listed on the stock market: two generation companies, Vinhson-Songhinh and Phalai, and one distribution company, Khanh hoa. According to Decision No. 12/2005/QD-TTg, dated 13th January 2005, EVN also transformed 7 of its power plants into independently accounting units.

At the end of 2005, the EVN's core member organizations including the pending restructuring consists of (i) Generation sector: five generation dependent accounting units, three independent accounting units, one one-member limited company, and four joint-stock companies (two of the companies were in the process of constructing power plants), (ii) Transmission sector: four regional transmission companies, (iii) Distribution sector: seven regional power companies and two one-member limited companies, (iv) National dispatch center. In addition, EVN has a number of non-core member units, including manufactures, consultants, etc.

CHAPTER 11
A PROPOSED DESIGN FOR EI REFORM IN VIETNAM – OVERALL DESIGN

The analysis of electricity reform to date in last chapter provides one more foundation for design reform strategy in forthcoming years. The following chapter will try to incorporate experiences drawn from worldwide EI reform, Asian-based analysis, as well as Vietnam circumstances, to shape the future of electricity industry in Vietnam.

11.1 Considerations for Policy Selection of Electricity Reform

There is an urgent need for power sector reform to help resolve all of the issues described above. However, "There is no method of easily solving the problem of annoying the power industry" [Joskow and Schmalensee, 1985].

In order to discover the appropriate strategies for reform, many issues need to be considered. First, we list the points related to the electricity industry in general. Then, the special issues related to the Vietnamese context will be discussed.

Electricity bears several special characteristics which make it different from other commodities. These characteristics include: (i) the inability of storage, (ii) the need for supply and demand to match at all times, (iii) the lack of substitutes, (iv) it plays a vital role in modern society, (v) it is a standard product, and (vi) long periods for power plant construction are required.

Generally accepted prerequisites for a competitive market include the following: (i) a stable and steady increase in the demand for electricity, (ii) sufficient installed system capacity, and (iii) financially and technically credible market participants. For developing countries, another important prerequisite is a stable macro-economic and political environment.

The electricity industry is an essential infrastructure for economic development; even a trivial failure in industry reform can affect the economy. Therefore, the government must be careful in launching its liberalization policy, especially more and more care needs to be given to irreversible policies such as privatization. Moreover, reform is not a "free lunch". For many products, the costs of competition may indeed be low compared to the benefits gained from actively participating in a competitive market, however, for the electricity industry the balance is very different. Costs of competition are various and often very high. For example, the cost for IT systems and software in England's NETA, was US$2 billion.

Additionally, the experience from far-reaching reform industries shows that there are many issues arising from organizing the EI as a publicly owned monopoly. However, it is far from clear whether breaking the industry into privately owned companies operating under open market conditions will necessarily solve these problems. In many cases, changes introduce a new set of problems, no less intractable than the old problems (refer Chapter 5 Section 5.2 for more detail discussion of reform's risks).

For the case of electricity industry in Vietnam, there are some more considerations such as: (i) the issue that most concerns the Vietnamese government is ensuring a stable electric supply for economic development. The government has no intention of loosing control of the electricity industry, at least in the near future; (ii) the supply-demand imbalance, low capacity reserve, and power shortages still exist. As a result, the power market must be effectively regulated to maintain stable market conditions; (iii) hydroelectricity counts for a significant proportion of the power system and seasonal factors often affect generation scheduling; (iv) the retail electricity tariff set by government, is nationally uniform for each customer category, and incorporates significant cross-subsidies and is not cost reflective. The national uniform tariff set an obstacle to EVN in forming limited liability power companies at the provincial and district level. This is because that the limited liability company could only charge its electricity services at the national uniform tariff regardless of the costs of providing services; (v) the introduction of a market is using precious resources and will require capital to install appropriate infrastructure; and (vi) the legal and regulatory framework for reform in Vietnam is insufficient. Given that the structure and commercial arrangements in the power sector are evolving rapidly and the participation of the private sector is relatively new, the legal and regulatory framework has not developed adequately to respond to these changes, (vii) the capability of the domestic banking system and stock market is very limited. Quotations of state-owned company shares in the stock market were launched in February 2005. The strength of investor demand, both domestic and foreign, has not been examined yet. The process of asset re-evaluation has faced a lot of difficulty, likely to lead to the loss of public money.

Under such conditions, it is too risky to choose the stretched reform strategy, which was considered successful in Argentina and several other Latin American countries.

11. 2 Expected Benefits and Potential Risks

To establish reform strategies, the benefits and risks should be known in advance.

This section will summarize[24] the expected benefits and potential risks of reform in Vietnam.

Expected Benefits

Substantial benefits can be expected from the reform program. All the stakeholders including consumers, the Government, EVN, and private investors will benefit from the reform program. The key benefits are summarized below.

- Preventing potential power shortage. Once a sound and clear legal and regulatory framework is established and adequate tariff levels are set to attract new investments, non-Government investors will have more confidence and be more interested in investing in Vietnam's power sector. With the reform program, there is a better chance for achieving the development of over 3,600 MW of new generation capacity anticipated to be invested by non-EVN entities between 2003 and 2010. Thus, the power sector will have adequate generation to satisfy the forecast demand and support the nation's economic development.
- Generating proceeds for Government. Once the EVN is restructured and the new limited liability companies have improved their financial performance, these companies can generate proceeds for the Government by listing on the stock exchange (equitization).
- Improving efficiency and minimizing cost increase of electricity services. By introducing competition to the generation market and implementing economic regulations on transmission and distribution services, we can expect efficiency improvement in the power sector and lowered cost increase in new power supplies. There will be pressure for generators to improve efficiency and productivity and cut costs so they will be able to price their power competitively and still be profitable. The cost increase of electricity services in the future will be less than the increases without a reform program.
- Improved system reliability and power quality. When a new grid code is established and a new regulatory authority is in place to enforce it, improved reliability of the power system, such as less frequent power interruptions and faster power restoration, can be expected. In addition, the power quality (voltage fluctuations) will be improved with the enforcement of the grid code.
- Better customer service and consumer protection. We can expect better customer service and consumer protection resulting from the creation of an independent regulatory authority. One of the key functions of the regulatory authority is to protect consumer's interests by establishing and enforcing customer service performance standards and resolving consumer disputes.
- Wider access to electricity and more affordable electricity to the poor. Continued

---

24 Adapted from [ADB, 2003].

implementation of the Master Plan for Rural Electrification will increase the rural electrification rate. It is expected that 95% of the population will have access to electricity when the Phase III of the Master Plan for Rural Electrification is completed in 2010. With tariff reform, appropriate tariff subsidy to the poor will be implemented to make electricity more affordable to the poor.
- Electricity price stabilization and reduction in the long run. Smaller increases in the price of electricity and reductions in the long run via power reform are expected. However, it should be pointed out that the electricity tariffs for certain classes of consumers (e.g., domestic class tariff) will not be reduced immediately after the reform. This reason is this class of electricity tariffs is currently subsidized. Tariff reform towards cost- and market-based pricing will require tariff levels to be raised initially. These initial tariff increases are necessary for full cost recovery and attraction of investments to the power sector in order to prevent power shortages and sustain the sector development. After initial increases in the tariff level, efficiency gains from reform will help slow future price increases and may reduce the prices in the long run.

Potential Risks

Power Sector reform, like any other business program, will have its own risks. Chapter 5, Section 5.2 covered several reform failures including economic, environmental, and political contradictions. Hereby, this section will specify some major risks for the Vietnam situation. Fortunately, many potential risks can be minimized by proper market structure design, market rules, and market regulations. The potential risks are summarized below:
- Escalating power supply price in the spot power market. In a competitive spot power market, market forces determine the power prices. If there is not an adequate power supply in the market, the market structure is not designed to prevent market domination, market rules are not properly designed, or market surveillance on anti-competitive behavior and mitigating measures is not effective, the spot power market could see the power prices going up rapidly. This was what happened in the California power pool in 2000 and 2001. In Vietnam, more attention should be given to the risk of violation of electricity price because one of the characteristics of Vietnam's electricity system is the limitation of reserve margin. This in turn makes system operation more difficult. As noted before, however, this risk can be minimized through sound market rules, effective market regulations, and surveillance. International experience, such as the power pools in the United Kingdom, Australia, New Zealand, Singapore, South America, and North America, has shown that a properly designed power pool can substantially reduce power supply prices.
- Regulatory risk. In a restructured power sector, an independent regulatory

authority, under the Government, plays a vital role to ensure the fair and efficient operations of the power market. If the regulatory authority fails to establish and implement fair and transparent regulations to attract investments to the power sector and protect consumers, the benefits of power sector reform may not be realized. Only by establishing a sound and transparent regulatory framework and creating an independent regulatory authority, can it minimize the regulatory risk.

- Financial market risk. One of the important prerequisites for attracting investments to Vietnam's power sector is a strong and healthy financial market. A crisis in the financial market, such as the 1997/1998 financial crisis in Asia could derail the success of power sector reform. It should be also noted that Vietnam's power sector competes with other sectors in Vietnam and similar sector in the neighboring countries for private investments. If the power sector in Vietnam fails to improve and maintain its competitiveness, it will be unable to attract the required investments to the sector. An appropriate strategy for legal, regulatory, and tariff reforms may help attract private investors to the power sector.
- Economic growth risk. The power sector development and reform strategy is based on the forecast of rapid demand growth of approximately 15% per year in the next ten years. If the forecast economic development and power demand growth does not materialize, then the invested fixed assets (generation, transmission and distribution capacity) will be underutilized. As there will be less electricity sale volume (kWh) to share the fixed costs than the forecast, the per unit electricity price (dongs per kWh) will go up. To minimize this potential risk, it is recommended that the EVN monitor economic growth closely and adjust, on a regular basis, the Power Sector Master Plan accordingly.
- Fuel supply risk. In Vietnam's power sector, approximately 40% of the generation is thermal power. Fuel cost accounts for 60 to 70% of EVN's total annual generation operating expenses. Presently, the price of fuel (coal, natural gas and oil) sold to EVN's power plants is fixed by the Government. Power sector reform and performance could be adversely affected if fuel prices substantially increase or if fuel supply shortage occurs. It is strongly recommended that the Government embark on fuel sector reform in parallel to the power sector reform. Thus, there will be a coherent, comprehensive energy sector reform to achieve overall benefits of economic efficiency, financial viability, and social development for the nation.

11. 3 Overall Design for Electricity Industry Reform

Starting later than many reformed countries, Vietnam gains the advantage using information based on those countries' experiences. Making use of these experiences may allow Vietnam's EI reform to evolve at a quicker pace. Among many valuable lessons, two principal lessons may be (i) Step-by-step policy. In Part II of Chapter 8, a step-by-step policy for electricity reform was chosen for Asian countries in the forthcoming years. The considerations for Vietnam's case also illustrate that this policy is the best practice. "Slow and steady wins the race". This policy guideline needs to be obeyed in all aspects of proposed reform design, and (ii) Multidimensionality of reform policy. Balance between economic efficiency and social benefits should be maintained.

Reform Components and the Multidimensionality

Following the ideas from Part I and II, the fundamental features of reform strategy proposed to electricity industry should include: restructuring the industry and developing a market system, privatizing state-owned companies, introducing a system of regulation, and devising mechanism to address specific social and environmental issues. Such main aspects, in turn, are divided into several detailed policies. The detailed roles of several major policies were empirically evaluated in Part II.

Sequencing

Among the first considerations for any reform program is whether there is a logical sequence to the reforms, and if so, whether it is costly to undertake them out of order. Early reforms should address the most important problems and, if possible, build momentum for future reforms, and minimize risks of failure and policy reversal. Reversible and less risky reforms can be undertaken more readily than irreversible (or costly to reverse) and more risky reforms. The irreversible reforms require more careful design and assessment.

Privatization is reversible only at high external cost, such as a diminished reputation among foreign investors, and poorly designed privatization can complicate subsequent reforms. Structural choices, such as the degree of vertical or horizontal integration, can also be costly or difficult to reverse. Resources need to be carefully prepared for the introduction of a competitive market, such as capital and also human resources to establish and operate market. Moreover, smooth market operation requires a complete system of rules. It is widely accepted that regulation should go first, establishing policy framework and a powerful regulatory body to prepare and overseeing reform processes. Figure 11.1 below illustrates the sequencing of reform dimensions.

Speed

Obeying the process of step-by-step policy, the speed of reform should be at a moderate level. In terms of the entire reform process, the moderate speed of reforms demonstrates the existence of "a transition period" in which the industry may have

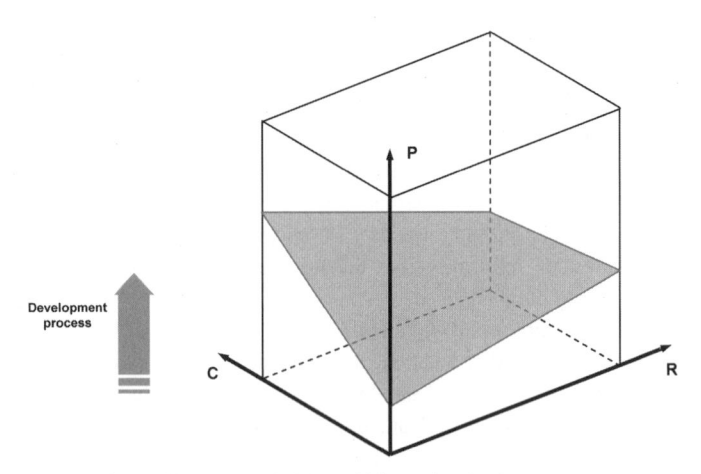

Figure 11. 1 Imaginable multidimensional reform strategy

time to establish effective regulation, to prepare to meet general requirements, and implement generation competition, and prepare the necessary conditions, including rules, stakeholders consultation, and negotiation for privatization.

The moderate speed can be seen also in each reform dimension. In regulation reform, the government still maintains an important role. The "independent" regulatory agency might be set at first under the ministry in charge. Regulatory mechanisms to ensure social benefit must remain or even be strengthened.

For the competitive aspect, there are several options for market and organizational structures in power sector reform, including: (i) a phased transformation starting with a single buyer market model, followed by a wholesale competition model and then a retail competition model, (ii) moving directly to the wholesale competition market to allow generators to sell electricity to distributors and large consumers, and (iii) moving directly to a competitive power pool. Obeying the Electricity Law of 2005, phased transformation is the appropriate path to follow. This option also matches the step-by step policy process.

In terms of privatization, the moderate speed can be seen as partial privatization, applying management incentives from private sector rather than a massive program of privatization.

Relation between Reform Dimensions
In conclusion, the basis of design is to ensure the multidimensionality of reform, as well as to obey step-by-step policy, to give priority to establishing effective regulation, introducing competition, and carrying partial privatization with caution.

Figure 11. 2 synthesizes the relation between reform aspects. The detailed

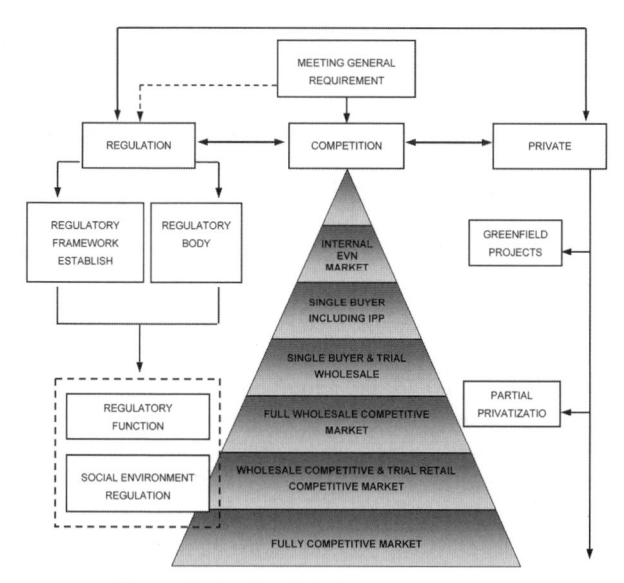

Figure 11. 2 Relation between reform aspects

description of each will be presented in the next three chapters, respectively.

CHAPTER 12
REGULATION REFORM IN VIETNAM'S ELECTRICITY INDUSTRY

Chapter 4 Part I has shown that the more liberalized the electricity industry, the higher the requirement for regulation is. Proper regulatory framework is one of the preconditions that must be met for the reform policy. Regulation is defined here as the creation and enforcement of rules that promote the efficiency and optimal operation of markets as well as the devising mechanisms to address specific social and environmental issues. The role of the regulator is to maximize those social benefits taking into account social values and the need to maintain financial viability and promote economic efficiency of the sector in the long run. This chapter endeavors to establish an effective regulatory framework and mechanisms to deal with social and environmental issues in the context of electricity industry reform in Vietnam.

12. 1 Building an Effective Regulatory Framework

Before 2005, in Vietnam there was no meaningful distinction between policy-making and regulation. In fact, the government of Vietnam sets policy and regulations through decree at either the ministerial level (MoI) or Prime Ministerial level. Several regulatory functions, such as license granting, system planning, pricing, etc, is spread between EVN, MoI, MoP&I and MoF. While these institutional arrangements can work in the absence of competition or greater private sector participation, sector reform will require changes.

To meet that requirement, the framework for regulation in the electricity industry needs to be outlined with a note for the distinction between policy-making and regulation. The recommendation of [ADB, 2003] in regulatory framework is shown in Figure 12. 1 as follows.

According to this recommendation, there is a clear distinction between the three components of a regulatory system, which includes a policy maker, a regulator, and an industry player. The Policy maker is responsible for setting high-level policy directions of the Government. The Regulator functions as the interpreter and enforcer of those policy directions consistent with its defined objectives. Figure 12. 2 shows the relationship between such components.

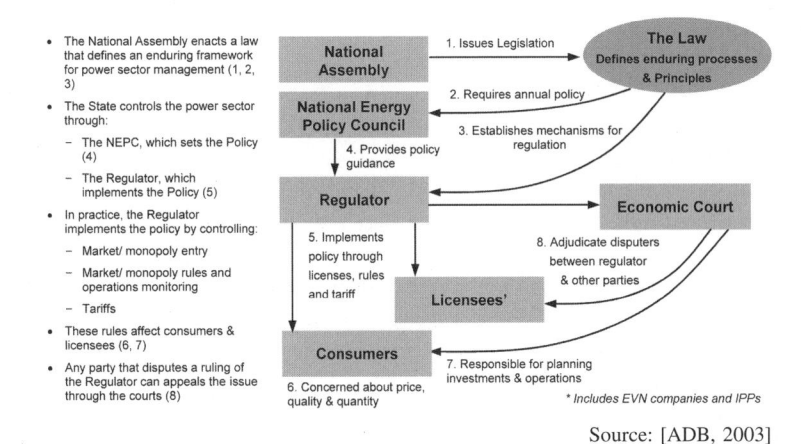

CHAPTER 12   281

Source: [ADB, 2003]
Figure 12. 1   An institutional framework for separation of functions

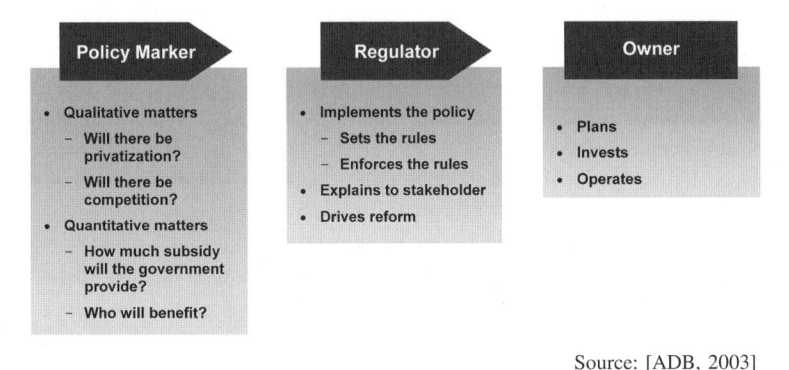

Source: [ADB, 2003]
Figure 12. 2   The distinction between policy making, regulation and ownership

The author of this book has the same viewpoint of separating the role and the objectives of each component, however, doesn't support the proposal of establishing a national energy policy council. To avoid overlapping responsibilities between government organizations, the roles and duties of the policy maker can remain with Ministry of Industry. The task should be done explicitly defining the content of power sector policy by law. It can prevent the extension of government policy making into regulatory or operational matters and policy making.

Chapter 4 Part I mentioned two dimensions of regulatory reform, which cover the regulatory government and regulatory instruments. In Vietnam, the regulatory reform may include establishing a comprehensive legal and regulatory framework

and creating a regulatory authority.
Establishing Legal and Regulatory Framework
Results from the empirical analysis conducted in Part II shows the positive impact on sector performance through establishment of an electricity law. The over-arching document in the regulatory architecture for Vietnam's electricity market is Electricity Law, which was enacted in July 2005. Under the Law, a series of sub-law documents need to be created, such as: the decree of guidance on implementation of the Electricity Law, the Prime Minister's decision on a road map for market development, the decision on an establishment of a regulatory authority, and other implementation documents. These documents are fundamental for the reform process and need to be issued as soon as possible in the beginning of transitional period.
Create a Regulatory Authority
Regulatory authority is the linchpin role in a liberalized industry. Regulation is defined here as the creation and enforcement of rules that promote the efficiency and optimal operation of markets. The role of the regulator is also to maximize social benefits, taking into account social values and the need to maintain financial viability and promote economic efficiency of the sector in the long run. In developing countries, which are just as hurried for investments as Vietnam, the role of the regulator becomes increasingly important due to its support of new investment. One conclusion from a survey of 48 international investors in the power sector recently conducted by the World Bank highlights four critical items that emerge as potential "deal breakers": a strong legal framework that protects investor rights, consumer payment discipline, availability of government or multilateral agency financial guarantees, and regulatory independence. The existence of a market carries little weight with investors.

In October 2005, a regulatory body has been established within Ministry of Industry (according to the decision No 258/2005/QD-TTg).

As stated in Chapter 4 Part I, the regulator needs "coherence, independence, accountability, transparency, predictability and capacity". Such requirements should be considered as a guideline for the establishment of a regulator in Vietnam in terms of location, characteristics, and functions.

Supported by Article 65 of the Electricity Law, the regulatory body in Vietnam can meet the requirement of coherence. Independence is critical, however the regulator remains a government agency. The regulator should ultimately be established as an independent agency of the Government of Vietnam. However, this cannot occur overnight. To ensure the capacity characteristic of the regulator, the proposed approach entails development of regulatory capacity within an existing ministry, and once the Electricity Law has been enacted and completion of capacity building for regulatory staff, separation of the regulatory unit from that ministry and establishment as a separate authority. The Electricity Law was enacted in 2005, however, several more years needed for the spin-off of the regulator from the MoI. The regulator might be a part of ministry until the electricity industry is ready to move into a wholesale market.

In the liberalization market, players, especially private investors, do not want to play a football game with a referee who is also a member of the opposing team, so even a regulator under the umbrella of Ministry has to be established with a clear set of functions which includes: (i) supporting the policy maker in analyzing and advising policy options and (ii) developing rules and licenses. To implement power sector policies, the regulator must develop rules that will govern sector operations, and licenses that will obligate operators to comply with these rules which include the rules that govern the monopoly segments of the industry, the rules that govern competitive segments of the industry and the technical codes, principally dealing with network operation, (iii) conducting public hearing and disseminating information. In the interests of transparency, the regulator will be authorized and be obligated to conduct public hearings to finalize and apply these rules and review license applications, provide information to consumers about their rights and obligations under these rules, and serve as a last resort for consumer complaints, (iv) Issue licenses. The regulator should be the sole licensing authority for power sector licenses, (v) Monitor and apply rules. Once the licenses in operation, the regulator must monitor compliance with the relevant rules and codes. In competitive parts of the industry this is primarily to ensure that there is no excessive market power or collusion. In monopolized parts of the industry, this would focus principally on price and service levels. It will be necessary to review tariff applications submitted periodically by the licensees in accordance with the tariff code, (vi) Enforce rules and resolve disputes. If a licensee does not comply with the rules, the regulator must enforce the rules, possibly leading to sanctioning of the licensee. Again, this would be conducted through a public hearing process in accordance with prevailing laws and rules, and last, but not least (vii) Provide support services. According to Article 66 of the Electricity Law, it seems that several necessary functions do not exist yet, for example, conducting public hearings and the dissemination of information, this should be clarified in other by-law documents.

Moreover, due to the movement of industry structure, the regulator's functions also need to change. A liberalized market increases the competition regulation, while reducing the scope of regulation in a monopoly. Two main regulatory functions which change according to industry structure will be discussed in the next sections. Another very important regulatory function missing from Article 66 is related to devising of mechanisms that deal with social and environmental issues in the context of electricity industry reform, which will be specified in Section 14.4.

The independence and transparency of a regulatory body has to be shown through appointment and dismissal of commissioners and funding for its operation. The regulatory agency is not meant to be representative of the players, but independent of the players. The regulator should be independent in terms of financial matters. However, under the current structure, when the regulator has been newly established, the operation budget may be granted from state budget. However, in forthcoming years, it needs to be collected from regulatory fee.

The Structure of Regulator

The regulator can be single- or multi-sector. Based on the advantages and disadvantages of each structure mentioned in Chapter 4, in the context of Vietnam's energy sector, it is appropriate to choose a regulator for two close sectors: electricity and gas. The major argument for this decision is the electricity and gas regulator has better position to guard against the distortions created by inconsistent regulation of two utilities. Petro Vietnam sells gas to the EVN as fuel for electricity production. It can also generate electricity and then sell it to the EVN, so gas and electricity, in some extent could be used as substitutes for each other. Multi-sector structure also allows sharing fixed costs, scarce talent, and other resources.

Regulatory Objectives and Instruments

Most regulatory regimes try to meet multiple objectives and the government must try to address these objectives simultaneously. The regulatory challenge then becomes the need to consider various instrument combinations as a way of simultaneously meeting primary and secondary objectives or at least minimizing the need to face socially and politically difficult tradeoffs. In the case of Vietnam, the primary trade-offs are also found in the consideration of the following objectives: sustainability and efficiency, efficiency and fairness, and sustainability and fairness.

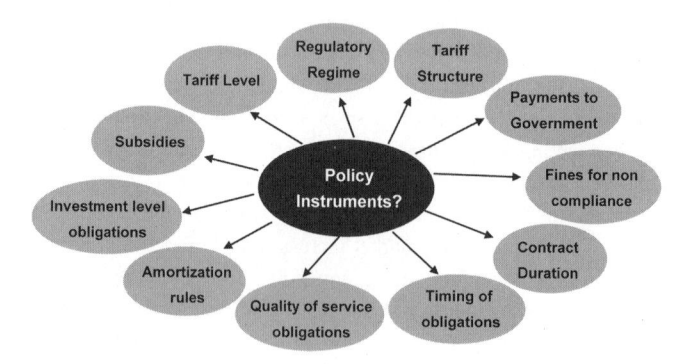

Source: [Estache et al., 2003]
Figure 12. 3 The main policy instruments for regulators

To achieve any combination of regulatory objectives, the regulator can pick from a wide set of specific instruments. Figure 12. 3 provides a visual description of this diversity. The instruments listed in this figure can be aggregated into three broad categories: regulatory regime, contractual obligations, and tariff level and design. The following sections of this book will discuss several principle regulatory instruments in the case of the electricity industry in Vietnam.

12. 2 Mechanisms to Regulate Price

Electricity pricing principles consist of: (i) promoting economic efficiency and fairness, and attracting investment to the sector; (ii) achieving economic efficiency by setting electricity prices as closely as possible to the long run marginal cost of providing electricity services; (iii) separating electricity prices at generation, transmission and distribution levels to promote efficiency, accountability and transparency; and where and when tariff subsidies are necessary, they should be explicit and transparent with a specific timetable for gradual removal.

Regulating price is hardest for the regulator in the first stage of reform because there is a need to regulate price in all segments of the electricity supply chain. The requirement for regulating price will be reduced according to each phase of market development (Figure 12. 4). The range of regulated prices includes generation price for vesting contracts, transmission fees and retail regulation tariffs. This part will discuss and find the proper mechanisms for the regulation of prices.

Mechanism for Regulating Prices

Chapter 4 discussed two alternative mechanisms for regulating prices: cost-plus and price caps. Cost-plus regulation is better for sectors with large investment requirements and countries with a weak commitment capacity, while price cap

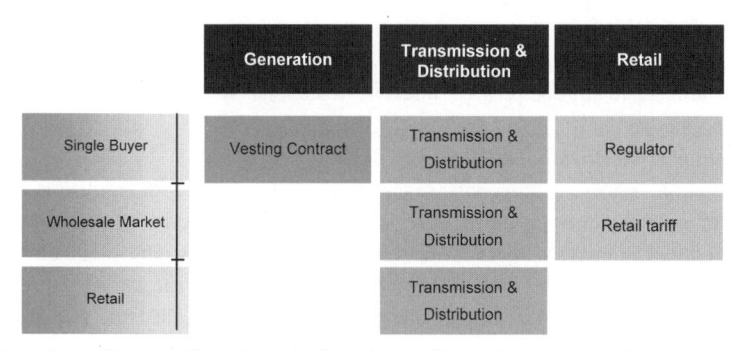

	Generation	Transmission & Distribution	Retail
	Single Buyer	Vesting Contract	Transmission & Distribution	Regulator
	Wholesale Market		Transmission & Distribution	Retail tariff
	Retail		Transmission & Distribution	

Note: In retail competitive phase, in fact, there still remain several regulated customer groups, such as small residential customers or subsidized-target groups.

Figure 12. 4 The scope of regulating price in different phases of market development

regulation is better for industries with excess capacity supported by institutions with strong commitment powers. The principal short-term power sector issue in Vietnam is the need for investment in new generation, as such, the cost-plus mechanism should be utilized without objection. It permits the firm to earn sufficient revenue, including a fair return on its investment. This mechanism may be an appropriate choice under the single-buyer phase. When the industry structure transfers to a more competitive model, price-cap regulation may be preferable.

Regulation in Generation Market

This is the segment of the EI most susceptible to the introduction of competition, either through competition for the market in the form of IPPs or real competition through a power pool. The internal "transfer pricing system" applied in the EVN can now be easily adapted to the requirements of the single buyer market where EVN-owned generation plants and IPP's sell power to an independent single buyer. The role of the regulator here is not setting a particular fee level, instead the regulator provides the guideline for calculating generation price for the vesting contract. In the single-buyer phase, this task is very critical due to the percentage of electricity trade under vesting contracts normally accounts for 80–90% total electricity.

Regulation for Bulk-supply Tariff

The bulk power supply tariff will change during power sector restructuring. In the single buyer market, the bulk power tariff to the distribution companies will be the weighted average power costs of the single buyer's PPA's plus the single buyer's overhead costs. In the wholesale competition market, the power supply price of a distribution company will be the weighted average prices of the power from the company's bilateral power contracts and the spot power market. In the retail competition market, power supply prices are negotiated between the buyers (eligible consumers) and the power suppliers.

The role of the regulatory body is shown in regulating single buyer's overhead costs. The mechanism that should be used is the rate of return for the EVN. In wholesale and retail competitive markets, regulation for this price will be not necessary anymore.

Regulating Transmission Tariff

Transmission is natural monopoly segment which requires tight regulation in any phase of market development. The regulating transmission price should allow the transmission company to recover the full costs of providing and maintaining the transmission network. The total cost to be recovered in any year, called the revenue requirement, will be subject to regulatory approval. This is called the rate-of-return regulation.

The role of regulator in the transmission sector also can be seen in the approval method used to allocate shared network costs among system users. At present, transmission costs are internalized by the EVN. The EVN does not distinguish the costs of providing transmission services to different distributors connected to its grid so, in effect, charges a uniform transmission tariff. To avoid price shocks and to allow the reforms to focus on establishing and developing the institutional capabilities of the new business organizations, the method of a postage stamp tariff and connection charge is recommended and should be initially implemented, especially in the phase of a single buyer. By then, the regulatory body will consider whether to introduce more cost-reflective transmission tariffs. As noted above, price shocks should be avoided so cost-reflective transmission tariffs need to be phased in over several years. The principle of regulating transmission tariffs can be applied to distribution segment when the retail market is implemented.

Regulation of End-use Tariffs

The end-use tariff will change during power sector restructuring. When the retail competitive market is implemented, the range of regulated end-use price will gradually be reduced step by step. At first, the high voltage level and large customers can negotiate prices directly. Then, range of eligible customers will increase. However, to ensure social benefits, even in a completed retail market, there still remain several regulated customer groups, such as small residential customers or subsidized-target groups. The role of the regulator is limited only within that kind of special customer group.

12.3   Mechanisms for Long-term Planning and Investment Incentives

With the valuable lessons learned from the California electricity crisis, long-term planning and investment incentives will be very important issues in the reform process. This section discusses how system planning and development functions will evolve as various phases of competition are introduced.

Current Planning Process

Under the current configuration of the sector in Vietnam, EVN drives much of the planning process subject to MoI approval and financial approval by Ministry of Planning and Investment (MoP&I). Through technical approval and financial resources, EVN then constructs power plants. Figure 12.5 outlines the current planning process.

In this case, decisions on the amount, locations, types, and timing of investments in new generation have been made by EVN with approval from state public utility commissions. It is the responsibility of utility companies to assure that enough generation capacity is available and usually there is a centralized generation plan. Under this type of organization, underinvestment is less preoccupying, since the vertically integrated firm is vested with the mission to meet all consumers'

288   Part III

Source: [ADB, 2003]
Figure 12. 5   The current planning process

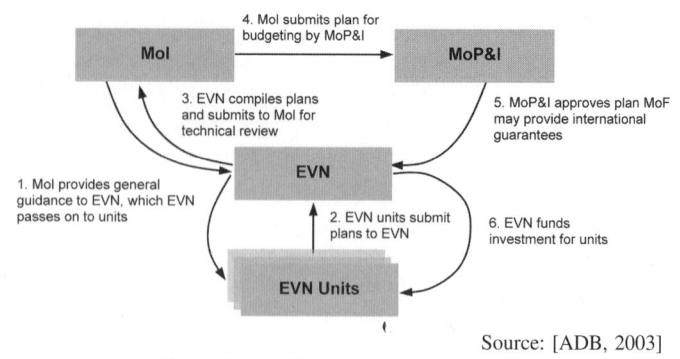

demand. As such, there will be sufficient incentives for the firm to undertake any necessary investments.

System Planning under the Single Buyer Market
EVN, as a single buyer, retains its pre-eminent role for the development of new generation capacity. In order to meet the high growth rate of demand, EVN will also likely retain capacity set-asides that will not be subject to the normal competitive tendering process, at least in the initial stage of the single buyer market. As shown in Figure 12. 6, the process starts with the single buyer forecasting load and planning new capacity needs as a part of the power sector master planning process in Vietnam. The regulatory authority/the MoI/the Prime Minister will then approve the forecast and plan, and supervises the procurement of the new facility according

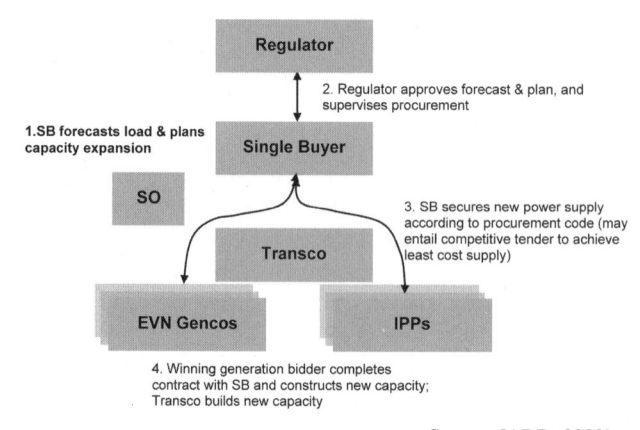

Source: [ADB, 2003]
Figure 12. 6   System planning under the single buyer market

to the terms of the least cost planning and bidding code. However, policy of set-asides for EVN in the early stages of the single buyer would not fatally compromise reform and the introduction of competition. The Government, especially the policy-making agency, must ensure that this is only a temporary phase. Otherwise, reform may stall without delivering its full benefit. Moreover, the rate-of return regulation also gives the incentives to private investors.

System Planning under Wholesale and Retail Competition

Wholesale and retail markets cannot function unless the players in the market are responsible for their own planning. If they are instructed where, when, or what type of plant to build, they may well reject the risk they would subsequently bear trying to sell the output of the plant into the market on terms not of their choosing. Based on information from System and Market Operators, it is up to investors to assess consumer demands and build generation accordingly. They assume all risk for these plants, and retain all rewards. The role of the Regulator is to ensure that all market participants play by the rules.

Meanwhile, the Transmission Company would also be planning and constructing its own capacity. However, because transmission is a monopoly, this planning will be subject to regulatory oversight. In addition, the Regulator will also monitor the behavior of generators to ensure they do not collude or engage in prohibited practices. This process is shown in Figure 12. 7.

However, according to [Zhang, 2005], it is insufficient to offer electricity investment either in generation or in transmission only by market competition mechanisms. Solutions for that include market design solution with demand bidding, capacity payments or capacity market and also the active role for ensuring the system adequacy of regulator. The regulators transferred the responsibility of

Source: [ADB, 2003]

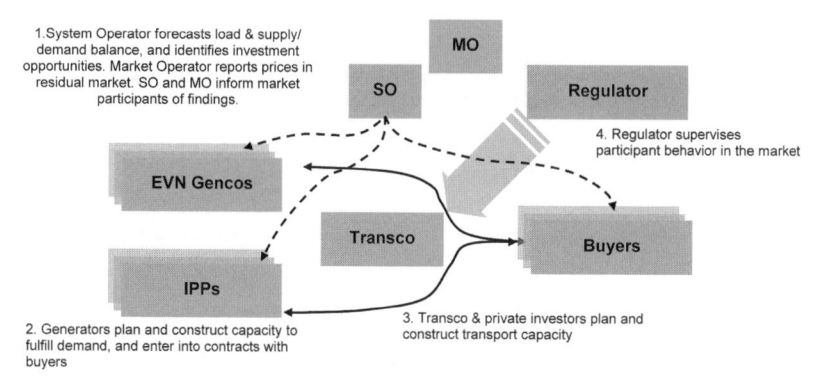

Figure 12. 7 System planning under wholesale and retail competition

290   Part III

expansion of generation capacity additions to the private sector in a market environment, but the government still keeps an indicative role to strategically direct the generation capacity expansion. The government would have to provide adequate signals. One example of regulatory framework is auctioning reliability contracts based on financial call options [Vazquez et al., 2002], and both the price and allocation among different plants are determined through the competitive mechanism. In this way, the income of generating companies can be stabilized and this actually provides a clear incentive for new generation investment.

12. 4   Mechanisms to Address Social and Environmental Issues

Chapter 4 stated that the reform has been designed to mainly address the economic and particular financial concerns, with insufficient consideration for social and

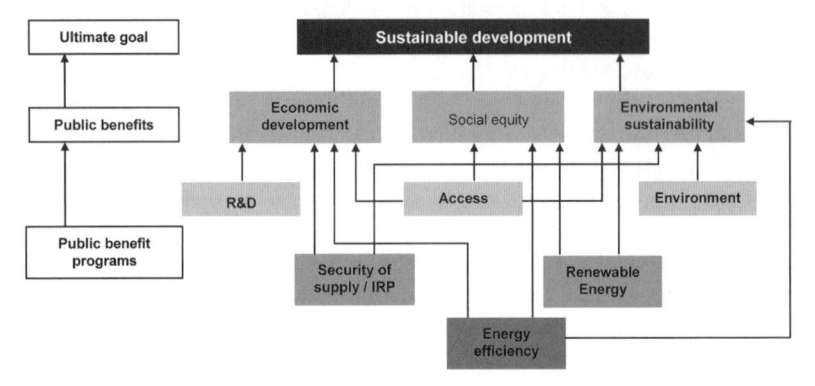

Source: [Eberhard, 2003]
Figure 12. 8 Public benefit consideration in EI reform

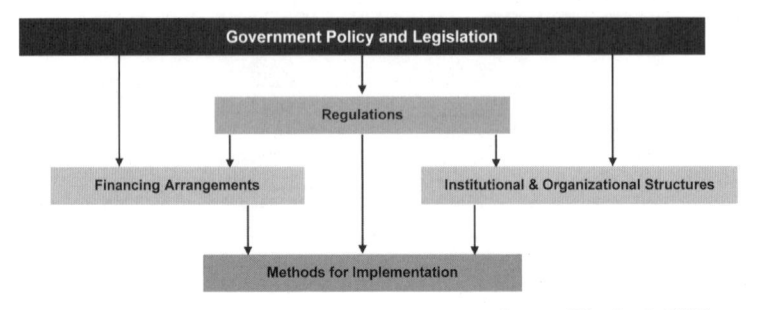

Source: [Eberhard, 2003]
Figure 12. 9 Policy, regulatory, financing and institutional mechanisms to advance public benefits

environmental issues. It has also shown the important role of the government's intervention to deal with such of issues. Figure 12. 8 illustrates the scope of public benefit issues. A reformed power sector would be separated from the Government's social support program and can be run on a commercial basis.

International experience shows that public benefits can, in principle, be promoted in any electricity market structure with the requirements of policy clarity, regulatory instruments, financing mechanisms, and implementation agencies and systems (Figure 12. 9). The following section will endeavor to address several major mechanisms may be appropriate for the EI reform in Vietnam.

- **Access to Electricity for the Poor and Rural Areas**

Rural electricity supply is usually very costly to develop and operations often suffer from low revenues because connection ratios are low and energy consumption from such customers is low. In the short-term, the electricity supply in rural areas often cannot recover the financial costs of supply from customers. As a result, distributors in the restructured electricity sector will generally not be able to commercially expand electricity in rural areas. To achieve the Government's objectives of rural electrification, special assistance or subsidies will be required. The rural electrification program in Chile demonstrates that if appropriate policy is set in place, the electrification program will develops well.

The legislation for rural electrification in Vietnam is provided for in Article 60 and 61 of the Electricity Law. The Law states that "the State shall adopt policies to provide investment support for construction of transmission lines from the outlets of electricity meters to households entitled to social policies and meeting with exceptional economic difficulties as certified by local people's committees. The State support policies cover: support in investment capital, support in interest rates on investment capital loan, and tax preference".

However, the Law does not clearly mention sources for subsidies. In principle, such subsidies should be borne by the government. However, in Vietnam, as well as many developing countries, state budgets are usually scarce and it is not easy to receive allocations. The most feasible method is to use a funding system. A social benefit fund should be established to support social benefit programs which are not limited to rural electrification, but also include universal service policies, for example, service to low-income customers or the promotion of energy efficiency, the development of renewable energy, and other research and development activities. The social benefit fund can be mobilized from various sources: (i) obligated contribution from power companies participating in the market, (ii) a social benefit surcharge applied to the kWh electricity sold, (iii) subsidies from the state budget, and (iv) grants from international institutions, energy tax, carbon tax, etc.

There might be a disagreement over the use of government intervention through social benefit funds because it may lead to market distortion. Although this may be true, the argument is that we need to strike a balance between economic efficiency and social equity. However, to minimize the market distortion, subsidies should be considered as the last choice only and targeting subsidies should be improved. The purported rationale for such support mechanisms is to ensure that essential services remain affordable to poor segments of society. Yet many subsidy programs involve almost no targeting: price structures do not discriminate between rich and poor people, so everyone benefits. As an alternative to traditional subsidies, direct subsidies have been proposed using various targeting mechanisms. These alternative mechanisms have several advantages: they are transparent, explicit, and minimize distortions in the behavior of the utility and its customers.

The following procedure or subsidy mechanism could be used to finance access and enable distribution companies to establish viable rural electrification projects. The scheme is independent of ownership considerations, thereby opening up options for encouraging the private sector and cooperative initiatives in distribution. It includes: (i) Proposals made by distributors on rural electrification since they are closest to local needs and situations, (ii) the development of a "business case" considering estimates on electricity sales and revenues (based on the prevailing tariff structure at the time), capital cost, operation and maintenance cost, administrative cost and other customer related cost (metering, billing etc.), (iii) the results of the business case are made available to the Government (including central, provincial, and local governments), the public and the Regulator, (iv) after an appropriate amount of time, the results are discussed via public hearing to review the proposition and receive comments, (v) the final decision will rest with the Government, (vi) the government will fund the capital cost of the specified rural electrification scheme(s), (vii) the government provides subsidies as "contribution in aid to the construction and operation", (viii) the rate base used to calculate the regulated return to the distributor would exclude the amount invested by the Government.

- **Environmental Issues**

How can reform influence the relationship between electricity and the environmental? [Bernow et al., 1998] argue that, "while there are potential environmental benefits from restructuring, environmental threats appear larger". On the one hand, the future looks relatively bright if gas combined cycle plants replace coal-fired generation, leading to lower CO_2 emissions. On the other, it is argued that capital-intensive renewable energy and end-use efficiency, which tends to have a longer pay-back period, will suffer due to the higher cost of capital, reflecting greater levels of perceived risk in the marketplace. Thus, to ensure sustainable development for the electricity industry, what does the regulator need to do? For example, the methodology to work out the true cost of power from the kind of power plant: the old dirty thermal power plant, the hydro power plant, etc.

In the past, in addition to the improvement of quality and efficiency of operation and business, the EVN endeavored to protect the environment. Power plants under the EVN strictly carried out supervision of environmental pollution and reported those supervisory results to the local environmental management authorities. As per the existing power plants, EVN allocated a given amount of the budget for periodic maintenance and repair of environmental pollution treatment equipment in order to ensure good operation and optimal equipment efficiency. The EVN also finalized the substitution of high efficient electrostatic dust filters for wet dust filters at thermal power plants to resolve pollution caused by dust emitted from power plants. It also implemented several sewage treatment projects.

As per new investments, EVN seriously performed environmental impact assessments for projects and advance mitigation measures in order to minimize adverse impact on the surrounding environment, as well as the community. Under the current economic and technical conditions of Vietnam, the EVN made great efforts to comply with the prevailing stringent environmental standards via enhancing operational efficiency of dust filters and desulphurization equipment, using low NOx creating nozzles. EVN has invested in combined cycle technology in order to increase the productivity of units and considerably decrease CO_2 gas emission.

In 2005, the EVN expedited research of reusing ashes emitted from coal fired thermal power plants as an additive to rolled compacted concrete and cement. This research will significantly contribute to better environmental conditions at power plant ash dumps. In parallel, the seeking and proposal of projects to be developed under Clean Development Mechanism (CDM) and renewable energy projects were continuously encouraged by EVN.

However, negative impact on the environment due to electricity industry reform has been recognized as well. Since the power sector has been opened to private participants, the first IPP, Hiepphuoc, an imported oil-fired thermal power plant utilizing second-hand technology, which has been in operation for thirty year. To mitigate investment risks, new investors try to make use of cheap technology, even if it is harmful to the environment. The same thing happens with many small hydro power plants invested in by non-EVN companies. Most of them use outdated Chinese turbines with low efficiency. The Naduong thermal power plant, invested in by Vinacoal, uses low quality coal causing problems with sulfur dioxide, dust, etc.

In terms of the EVN, thanks to strict regulations, threats to the environment seem to be decreasing. However, the EVN continues to maintain several old dirty coal-fired power plants, which should be retired in coming years, for example Ninhbinh and Uongbi. Moreover, the policy of equitizing several power plants has threatened to the environment since power plant owners may cut environmental protection costs to gain more profit.

Thus, under the vertically-integrated structure, non-market intervention, such as command and control or forced retirement of plant, still have the greatest applicability in improving the environment. However, obviously such policies can only show their influence in the regulated section of the industry, for example the EVN. Under liberalization of the industry, others mechanisms need to be developed in order to attain the goal of sustainable development. The regulator should give attention to new market-based environmental policies such as taxes and fees (energy tax, carbon tax, etc), energy trading and the promotion of CDM projects. Revenue from taxes and fees contributes to the environmental fund whether separate or merged with public benefit funds mentioned above, which can be used to support environment-friendly technologies, promote energy efficiency, and develop renewable energy.

CHAPTER 13
RESTRUCTURING AND MARKET DEVELOPMENT IN VIETNAM'S ELECTRICITY INDUSTRY

Chapter 2 Part I has covered four basic structures of the electricity industry. The empirical analysis in Chapter 7 Part II shows that a competitive market does not always have a positive impact on industry performance. It seems that there is no detailed guideline on how to introduce a competitive market. The process of market development is normally different country by country and influenced much by the country's political and economic specifications. Chapter 12, in an effort to establish the general design for electricity industry reform in Vietnam has selected step-by-step policy for all dimensions of reform. The following chapter will discuss policy implications regarding the restructuring and competitive market development dimensions of reform in Vietnam. The issue of introducing a competitive market has been raised recently by Vietnam policy makers in different documents, such as the Electricity Law and the Prime Minister's decision No 26/2006/QD-TTg. Instead of simply following such plans, the writer will put more effort on expounding the arguments, as well as the structure and operation scheme of the market development process.

13. 1   Market Development Standpoints in Vietnam's Electricity Industry

The legal background for market development in electricity in Vietnam is outlined in the Electricity Law enacted in 2005. The Law states that "to build up and develop the electricity market on the principle of publicity, equality, fair competition with the State's regulation to raise efficiency in electricity activities; to ensure the legitimate rights and interests of electricity units and electricity-using customers; to attract all economic sectors to participate in activities of electricity generation, electricity distribution, electricity wholesaling, electricity retailing and/or specialized electricity consultancy." [Article 4 Chapter 1].

The Law also provides that the electricity market needs to be formed and developed through three stages: (i) The competitive electricity generation market, (ii)

The competitive electricity wholesale market, and (iii) The competitive electricity retail market [Article 18 Chapter 4].

One essential principle that must be ensured when designing the electricity market in Vietnam is "reliability". Reliability here means more than just the reliability of the electricity supply. Market development needs to be implemented gradually, avoiding shocks and mitigating negative impact on economy, society, and environment. The development process should start with restructuring of the EVN. The initial phase of reform should be the period when general prerequisites for restructuring of the market, involving the establishment of fundamental legal, regulatory, and commercial frameworks, take place. The process of EVN restructuring started several years ago. Recently, in 2004 and 2005, a series of events happened, marking a new step in the restructuring process. This was the trial establishment of one member limited companies and the initial public offering several of units, in particular the establishment of regulatory authority has greatly contributes to the restructuring process of the electricity industry in Vietnam (for detailed information please refer to Chapter 10). Now is the time for new industry

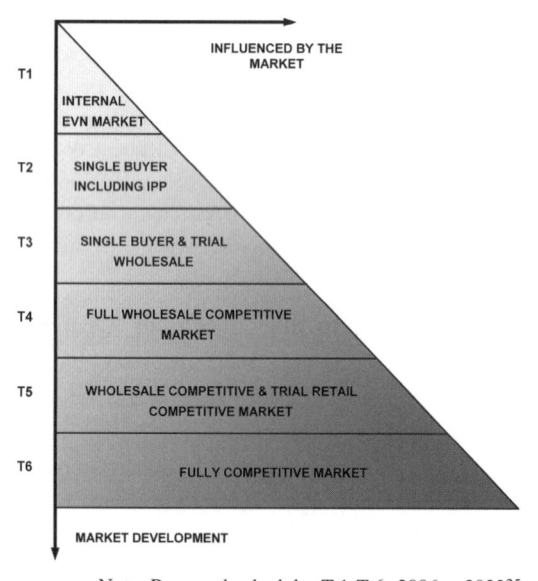

Note: Proposed schedule: T 1-T 6: 2006 – 2022[25]

Figure 13. 1 Market transformation phases

---

[25] According to the prime minister's decision No: 26/2006/QD-TTg dated on 26 January 2006.

structure.

A market-oriented electricity industry is a long way off in Vietnam. In the absence of most of preconditions, development of the market structure should follow step-by-step policy, however the three-phase development strategy provided for in the Electricity Law is appropriate. Moving to a more competitive market structure phase by phase can be implemented if and only if the identified preconditions are met. Moreover, to mitigate the risks accompanied with each step of opening market, to facilitate sustainable development, each of the three periods should be divided into two steps: trial and full competition. Figure 13. 1 shows the phases of transforming the existing market structure to the desired market structure in the future. Each phase of the development will be discussed in following sections, in which priority will be given to first stage-single buyer development, which is directly concerned with electricity industry reform in Vietnam.

13. 2   Competitive Electricity Generation Market

Model Selection and Arguments

The generation market can be seen as a "transitional period" toward the creation of the truly competitive market. The existing structure of EVN is transformed without any anxiety into the competitive power generation market.

In a competitive generation market, the popular model for the industry is the single buyer model. The level of competition in this model is very limited especially in terms of the electricity buyer. However, for the initial period of market development, limited competition is recommended because: (i) it does not require much change in industry structure, (ii) it promotes rapid investment and expansion, (iii) it facilitates system balancing, system reliability, and (iv) it keeps the important role of state-owned enterprise. The important role of the state-owned company should be continued because in Vietnam, as well as in many developing countries, national electric companies are centers for skill development and good employment practices. This company is responsible for securing stable operation of the new power structure, and for preparing the market design. Through state-owned national companies, in the transitional period, government support may be provided through investment in special technology, for example, nuclear power development, through bearing universal services, or solving environmental issues. The existence of a national power company can also provide a stable environment for attracting foreign investment.

The weaknesses of the single buyer structure can be able to overcome by selecting a single buyer's organization or adding market mechanism in electricity trading. In fact, several organization options can be proposed for a single buyer, including: (i) creating a new, non-profit government-owned single buyer company,

(ii) creating a new independent national transmission company that has a single buyer subsidiary and a system operator subsidiary, and (iii) through an EVN holding company owning three subsidiary limited liability companies: National transmission company, a single buyer, and a system operator. However, the last organization option, remaining part of the EVN, might be more appropriate for the argument of the single buyer mentioned above, especially in two point: this organization has the least dramatic changes in comparison to the existing situation and the EVN, as the parent company, can help strengthen a single buyer's financial credibility necessary to convince new investors to sign PPA's.

The second approach is to modify the single buyer structure by adding an electricity power pool. This variant of the single buyer model allows not only an increase in competitiveness, but also the facilitation of the next phase of market development. Thus, it should be recommended and depending on the scope of the pool can implemented in both steps of the competitive generation market: the internal EVN market and a fully competitive generation market.

Internal EVN Market
– Structure and operation (Figure 13. 2 a)

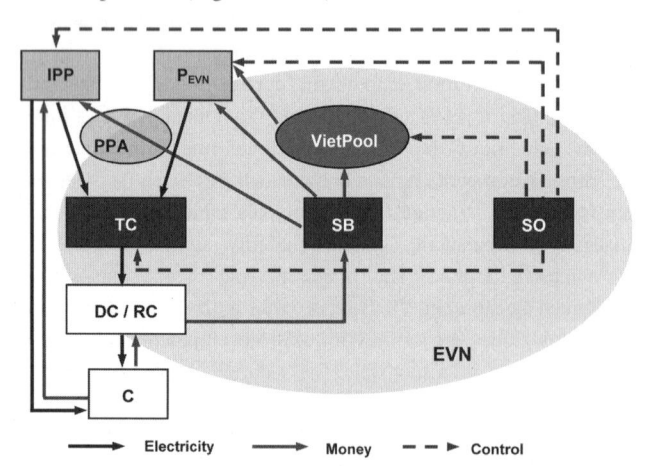

IPP = Independent power producer P_{EVN}= EVN power plant VietPool = Power exchange
TC = Transmission company   SB = Single buyer SO = System/market operator
DC = Distribution company   RC = Retailer C = Customer

Figure 13. 2 a   Competitive generation market step 1: an internal EVN market

Under internal single buyer structure, three major components of a competitive generation market exist: a national transmission company, a single buyer, and a system operator. All are subsidiary limited liability companies owned by EVN as a holding company. The single buyer is responsible for procuring new generation, buying power from generators, and then reselling bulk power to distribution companies. System operator is responsible for power control, scheduling and dispatch, and procuring ancillary services to ensure power system reliability. The existing National Load Dispatch Center can be transformed into the new national system operator. A grid code will be established to guide the operation of the national system operator. The transmission company owns and maintains transmission networks and provides open and non-discriminatory transmission service to all users. This can be done by merging the four existing EVN transmission dependent accounting units.

Regarding the internal EVN pool, most of the power plants belong to the EVN or are mainly directed by the EVN (one-member limited companies or joint-stock companies), except three hydro stations Hoabinh, Trian, and Yaly. These are multipurpose hydro power plants. The operation of these plants can be for purposes besides economic operation, such as flood control. Excluding these hydro power plants is also confirmed by the Electricity Law provision which states that the "government monopoly of big power plants of particularly important socio-economic, defense or security significance" [Article 4]. The existing structure of the EVN could be transformed smoothly in this organization. However, a major disadvantage of this structure is that non-EVN generators and investors may not believe the single buyer will be fair and unbiased to in procuring the power supply. Thus, more stringent regulations and strict enforcement to ensure fair and efficient competition is required.

By strict definition, a single buyer may be created from existing utilities by divestiture, or they may be new producers who enter the market when a new plant is needed. However, with this model, it is possible to have competition in generation, with a single buyer purchasing wholesale electricity. In the case of Vietnam's electricity industry, it goes further to establish a pool for electricity trading. The pool in the first stage is operated internally within the EVN. Although operated on a trial basis, the internal market can provide a good training for both market participants. The regulator has enough time to prepare and check efficient market rules. The newly established generators, single buyer, transmission company, system operator, and distribution companies can be trained and possess adequate commercial management capability to negotiate and execute commercial contracts such as power purchase agreements, transmission service agreements, bulk power sales contracts, system operation contracts, etc. The generators, who should be targeted for training, have strengthened their financial and technical management capacity and are financially viable. Moreover, to serve the pool and create perquisites for the next stage, they must be equipped with information technology infrastructure as well as metering systems.

In terms of the operation mechanism, several schemes can be considered for this stage, mostly including 'pay as bid' or 'pay as marginal'. The authors suggests the use of a 'pay as marginal' mandatory pool with, initially, 'regulator imposed' vesting contracts for differences (CfDs[26]) that can be changed to negotiated bilateral CfDs at the multiple buyer stage. 'Pay as bid' is not a satisfactory approach. It is contrary to the economic principle to set market price based on the incremental cost of supplying power and also causes volatile dispatch.

The level of vesting contracts initially imposed by the regulator depends on many factors, mostly on the total demand, in which suppliers cannot be changed. In practice, the level of CfDs in the initial period of market development is 80–90% of the total energy traded, as is the case in the Australian market.

The vesting quantities are effectively load profiles based on average historical load curves that are scaled up to reflect an assumed load growth. Variations in load growth (economy) and in day to day demand (weather effects) will cause the effective vesting level to vary on each day. According to an analysis of EVN system load, the hourly difference of the averaged day type curve in a month relative to the actual demand indicates variations ranging from -12.2% to $+13.3\%$ over a calendar year (2004).

If the reserve plant margin is low then there will be a greater chance of price volatility. Accordingly the vesting should be set as high as possible and not reduced until rpm levels are adequate. Given that market generators will only supply about 60% of the EVN system load, a vesting level of >85% for market generators would be acceptable.

Therefore, the internal generation market includes: (i) 90% long term vesting contracts settled at contract prices in order to prevent price vibrations, (ii) a day-ahead market settled at the day-ahead price and with a relatively small portion of 10% of total energy traded Balancing of the market is not proposed in this stage due to its high requirement for on-line services which is impossible to mobilize in the initial period market development. In the internal market, the three hydro-power plants that belong to the EVN could assume the role of balancing generators under normal conditions and even be able to act as a system intervention in case system price is raised due to a lack of reserve margin, (iii) IPPs have a choice to sell electricity to EVN according to existing PPAs or sell electricity to large customers. In the later option, IPPs need to pay transmission fee and other fees to EVN. A single buyer, a transmission company, and a market/system operator company are all

---

26 CfDs are financial "swaps", which is one form of an agreement for the exchange of a contract for the right to settle a specific quantity of electrical energy at a fixed price for a contract for the right to settle the same quantity of electrical energy at a floating price.

separated for accounting purposes, so fee regulation should be established for transparency reasons. In Figure 13. 2 a, the dark and light arrows respectively represent flows of electricity and payment in the market.

A single buyer, in turn, sells electricity wholesale to distribution companies according to wholesale pricing framework regulated by the regulator. Distribution companies, including several joint-stock companies, are regional monopolies and selling electricity to customers complied with end-use tariffs regulated by government.

− Essential issues

The most considerable issue in establishing an internal generation market might be its relationship to the regulatory requirement. Requirements for the regulatory framework include promulgating the Electricity Law and by-law documents, and creating a regulatory authority. Such tasks have been discussed in Chapter 12. Another important task for the regulator is the development of market regulations, which include least-cost resource planning and bidding code, grid code, tariff regulations and especially anti-competitive regulations.

One of the most important tasks for Vietnam's power sector reform is to attract new investments to the power sector so system planning is also very important. This task also discussed in Chapter 12.

− Prerequisites for moving to a fully competitive generation market

The key preconditions for moving from the internal EVN market to a fully competitive generation market summarized as follows: (i) Organization: complete the restructuring process of EVN, (ii) Capacity building: market players have been adequately trained in terms of technical, commercial, and financial management capability, (iii) Technical requirements for the power system: Adequate generation capacity − international experience has shown that a spot power market with 25% or more in surplus generation capacity and adequate fuel supply at reasonable prices would have a better chance for success in improving economic efficiency and lowering power supply price, has no major transmission constraints, and has an adequate number of generators in the market. (iv) Technical requirements for information system support: Software and hardware systems for managing power trading, determining clearing prices, scheduling, metering, billing, settlement, and other activities have been commissioned and readied for market operation (v) Regulatory framework: An independent regulatory authority with qualified commissioners and staff are in place to give advice on market transformation, establish market rules and codes, review and approve unbundled electricity tariffs, monitor market performance, enforce regulations and codes, mitigate anti-competitive behavior, and resolve market disputes. In particular, affiliate and anti-competitive regulations have been established for the single buyer, system operator and transmission companies to prevent unfair and discriminatory practices against non EVN-owned generation plants, (vi) Tariff system: Transparent criteria, methodologies, and procedures for developing and approving tariffs for generation, transmission, and distribution in the single buyer market have also been established.

Fully Competitive Generation Market
− Structure and operation (Figure 13. 2 b)

After a certain period[27] of implementation of the internal EVN market, the meeting of prerequisites for a fully competitive generation market mentioned above can be examined, IPPs will participate in the power pool. Power plants are gradually separated for the EVN to establish separate generation companies. The transmission system is opened to all generators and should be independent from the EVN.

Operation schemes of this structure are not so different from the last stage. The difference might be the number of market players, due to the involvement of IPPs. The time-frame for the single buyer model should be based on the time required to fulfill prerequisites for the next step of market development, such as the reserve margin, the availability of market participants, etc. According to the fifth master plan of electric power development, a stable reserve margin will be available after 2010.

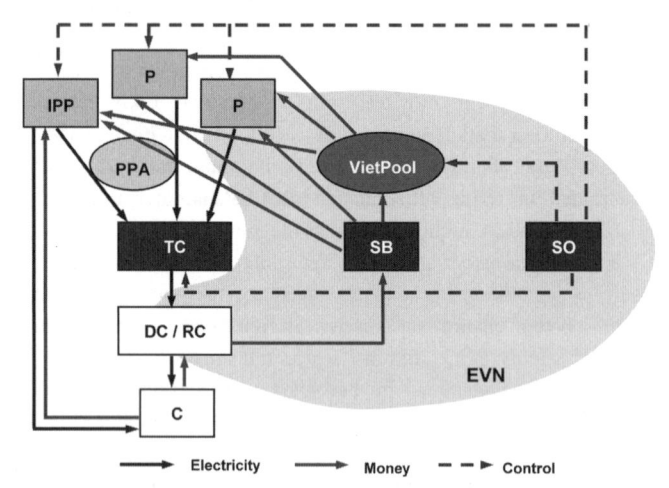

IPP = Independent power producer   PEVN= EVN power plant VietPool = Power exchange
TC = Transmission company   SB = Single buyer   SO = System/market operator
DC = Distribution company   RC = Retailer   C = Customer

Figure 13. 2 b Competitive generation market step 2: a fully competitive generation market

---

27 5 years as provided in the Prime Minister's decision No 26/2006/QD-TTg.

Therefore, from the authors' point of view, the period for the single buyer model is from 2005 to 2014. This milestone conforms to decision No 26/2006/QD-TTg.
– Essential issues
Two important issues emerge in this stage. First is the establishment of regulations on stranded costs of long term PPA's. Clear and fair regulations need to be established to compensate for potential stranded costs of the approved long term PPA's that the single buyer holds. These may include all the long term PPA's that the single buyer has signed or taken over from the EVN in the single buyer market phase. It is essential that the single buyer continue to fulfill its obligations of these long-term PPA's to maintain investors' confidence in Vietnam's power sector. If the single buyer uses Least-Cost Resource Planning and Bidding Code in procuring new resources, the prices of the signed long-term PPA's should be very close to the economic prices. In this case, the stranded costs should not be substantial. As noted previously, there are different options for managing these long term PPA's in the transition period. If a long-term PPA is declared a must run unit, regulations need to be established regarding how the costs of the long-term PPA's can be spread among end-use consumers. The second way to deal with PPA's stranded costs should be developed for compensating the stranded costs if other options, such as buy-out and/or remarketing of PPA's are to be implemented in the wholesale competition market. One more possible solution is persuading producers transforming PPA into CfD contracts. The last solution is preferable to the incorporation of IPPs in a competitive market. In Vietnam, there are fortunately not many fixed and long- term PPA's signed. Most PPAs, except for the Phu My 2.2 and Phu My 3 BOT contracts, include the provision for power market participation so that the stranded costs issue is not too difficult to solve.

The second issue deals with the establishment of an appropriate number of generation companies. In this stage, a number of power plants are separated from the EVN an turned into independent generation companies. It is recommended that an optimal number of generation companies should be determined to ensure fair and efficient competition in Vietnam's power sector. The key principle is that no single generation company should have market domination. The international experience shows a rule of thumb that no individual generation company can have more than 25 to 30% of the market share in a power grid. In other words, there should ideally be at least 3 to 4 generation companies of similar size in terms of generation capacity in the power grid to prevent market domination. If the study finds that it may not be feasible to regulate market share of all generation companies, for example, the largest hydropower plant, Hoa Binh hydropower plant already accounts for over 20% of the total installed capacity in 2003, then strict anti-competitive regulations will have to be established and enforced. On the other hand, consideration should also be

given to the mix of generation types and locations to ensure individual generation companies have adequate asset bases, stable revenues throughout the year, and adequate financial viability.
  – Prerequisites for moving to a trial competitive wholesale market
The preconditions for moving from a single buyer market to a competitive wholesale market are summarized below. It should be noted that the preconditions shown below are in addition to those identified for the single buyer market structure. In terms of organization, first, buyers must have a spot market or power exchange and a forward market. Second, a competitive wholesale market requires a sufficient number of unaffiliated suppliers and active customers because it will be introduced on a trial basis so the requirement of separating distribution companies may be limited in a particular area. The national transmission company has been transformed from an EVN subsidiary to an independent company. In terms of technical matters, an economically reliable transmission network is required. In terms of regulatory framework, a credible, effective, fast-acting regulatory mechanism to deal with flaws in market design and encourage efficient behavior by market participants is needed, market rules for wholesale competition market operations are established and market participants have been trained, tariff regulations have to be updated, especially adjustment of the national uniform tariff. License requirements and procedures for bulk power traders are established.

13. 3   Competitive Electricity Wholesale Market

Model Selection and Arguments
The single buyer market is not an optimal market structure. After implementing a single buyer market for several years, adequate generation capacity should be procured by the single buyer, and when the medium-term PPA's between the single buyer and generators are about to expire, the power sector will be ready to move to the next phase, the wholesale competition market. In the wholesale competition market, there will be bilateral a PPA market between power suppliers and distribution companies, plus a spot power market for the residual and imbalanced energy market. A financial market would also be established during this stage to allow power companies to enter into financial contracts with generators of their choice and enable them to hedge the risks of volatile electricity wholesale prices. The wholesale competition market will increase competitive pressure on generators to cut costs and reduce overall power supply prices. A wholesale competitive market is obviously more competitive and dynamic than a single buyer market. However, several prerequisites, mentioned above, must be met for wholesale electricity markets to succeed. In order to make a smooth transition, the progress of wholesale market development also consists of two steps: a trial market and a fully competitive

market, both of which will be explained below.
Wholesale Electricity Market: Trial Step
- Structure and operation (Figure 13. 3 a)

During the trial step in the transformation from a single buyer model to a wholesale competitive market, the regulator will select a particular area, which is able to meet the prerequisites, to introduce the wholesale market. The objective of this period is to test the wholesale mechanism in a limited scope, in structure, operation, rules, capacity building, technical infrastructure, etc. In a wholesale market trial, selected distribution companies, which are separated from the EVN, have the right to sign bilateral physical and financial contracts with generators or buy electricity from a single buyer through the "Vietpool". The EVN still exists and functions as a single buyer and system operator, and operates several power plants. Vesting contracts also remain. The structure and operational scheme of the trial wholesale electricity market is illustrated in Figure 13. 3 a.

In terms of electricity pricing, generation prices are set as follows: (i) in the bilateral market, prices are negotiated between buyers and sellers. These are private contracts and considered confidential, (ii) in the spot market, power prices are set by the balance of supply and demand. Generators make competing offers in the "pool" and the highest-priced dispatched generator offer sets the spot price. The spot price is used by sellers and buyers as a reference for negotiating contract prices. For retail pricing: (i) large customers directly negotiate with generators or distributors, and (ii)

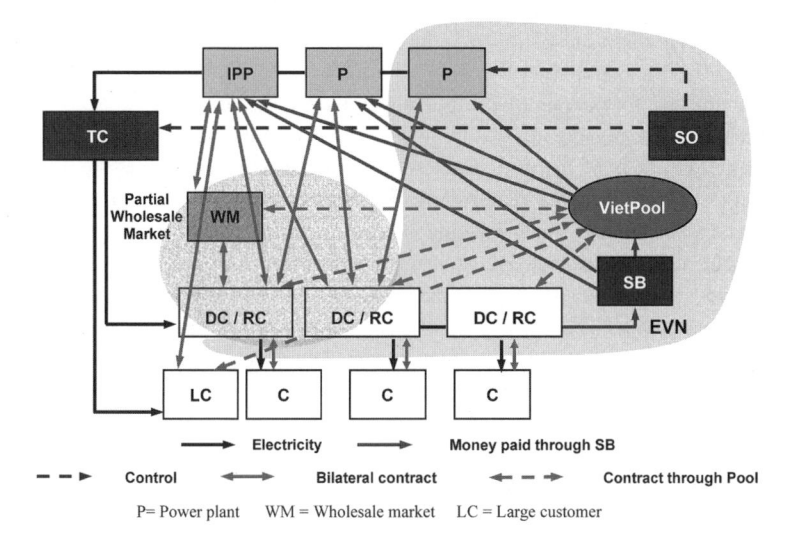

Figure 13. 3 a Wholesale electricity market: trial step

small customers apply the tariff regulated by government.
- Essential issues
A wholesale competitive market requires a sufficient number of unaffiliated suppliers and active customers. The establishment of generation companies has been solved to some extent following the single buyer period. In this stage, the attention should be focused on distribution companies. To develop the optimal number of distribution companies, the following should be considered: (i) the economy of scale and optimal geographical coverage for efficiency and commercial value of fixed assets and (ii) the optimal customer mix, load profiles and diversity, and opportunity for load growth.

In this stage of market development, the number of market players and amount of trade has increased. New market players now participate in electricity trading, the bulk power traders. International experience shows that bulk power traders can play an important role in the wholesale competition market. Bulk power traders can act as an agent or as a trader to buy and sell power for distribution companies and/or generation companies. A trader also can offer a variety of financial hedges, such as contracts for difference, to the sellers and buyers for market risk management.

The regulatory functions in the single buyer model continue, however, market rules and tariff regulations need to be updated. In this period, the competitive supervision function becomes more complicated, while the function of tariff regulation is reduced. For this segment of the wholesale market, the regulator no longer sets tariffs, prices are set through bilateral negotiation between distribution and generation companies. However, the regulator needs to complete the regulation provision for the transmission fee.
- Prerequisites for moving to a fully competitive wholesale market
Preconditions for moving to a fully wholesale competition market are not much different from those of the trial stage. Additional prerequisites include the scope of the market players involved. The right to choose electricity sellers is given to all of distribution companies. Thus, meeting the requirement of a sufficient number of unaffiliated suppliers and active customers becomes hard. The single buyer should be transformed into a special company the remarkets and manages the portfolio of the PPA's it has signed. The national transmission company has been transformed from an EVN subsidiary to an independent company. The national system operator of the EVN's subsidiary also becomes an independent national company. The regulatory framework, including the establishment of a credible and effective regulator and market rules for wholesale competition market operations need to be completed.

Fully Competitive Wholesale Market
- Structure and operation (Figure 13. 3 b)
Upon the successful implementation of the trial stage, a competitive wholesale

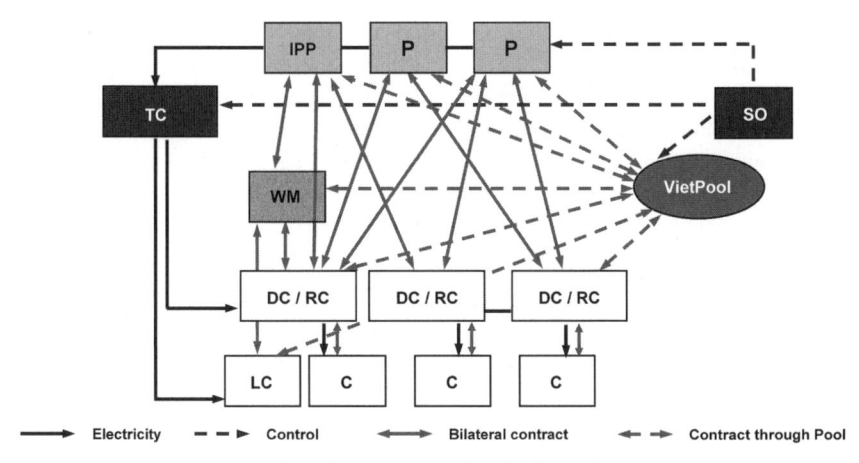

Figure 13. 3 b Fully competitive wholesale market

market will be extended nationwide. All distribution companies are separated from the EVN and electricity can be purchased from many available sellers. Under a fully competitive wholesale market, the EVN is no longer a single buyer. The number of power traders increases and take an important role in the financial market, which should be accompanied by a physical power market.
- Essential issues

The trend of reducing tariff regulation while increasing competitive supervision continues. The regulator is now responsible for end-use tariffs only, no more bulk is supplied at all. While functions of the single buyer model are continued, market rules and tariff regulations need to be updated. In this period, the function of competitive supervision becomes more complicated, while the function of tariff regulation is reduced. For this segment of the wholesale market, the regulator will no longer sets tariffs, the price are set by bilateral negotiation between distribution and generation companies. However, the regulator needs to complete the regulation provision for the transmission fee.

One challenge that emerges in this stage comes from the clearance of the single buyer. All issues related to existing PPAs need to be solved. The regulator has to prepare for new incentives to meet the requirement of system planning and investment after single buyer no longer exists. This topic was discussed in Chapter 12.
- Prerequisites for moving to trial competitive retail market

Regarding organization, to prepare for the next step of the trial competitive retail market, distribution companies in selected areas need to be unbundled into the distribution network (wheeling) service, and retail power supply service, providing retail power, metering, billing, and collection services. Regulation requirements are to: (i) establish a plan for phased retail competition, (ii) establish rules and procedures for retail competition, (iii) update tariff regulations, in particular the regulatory authority will need to develop new tariff regulations for the retail power supply, distribution network (wheeling) service, and retail services such as metering, billing, and collection, (iv) update grid code and establish distribution code to provide all the users an open and non-discriminatory access to the distribution network, and ensure the reliability of the distribution system.

13. 4   Competitive Electricity Retail Market

Model Selection and Arguments
Moving to a retail competition market will increase competitive pressure on both generators and distribution companies to lower their power supply prices as consumers are allowed to shop in the market for the lowest possible power supplies. This stage should involve the gradual opening up of the retail electricity market and a distinction between the energy market retail and distribution network access. Customers can freely choose an electricity retailer. Electricity prices are negotiated between suppliers and customers, except in the case of some specially targeted customer groups. The retail competitive market is also divided into two steps: a trial step and a fully competitive market as illustrated below.

Trial Competitive Retail Electricity Market
- Structure and operation (Figure 13. 4 a)
During the trial step, only a certain number of retailers in specific regions are allowed to sign contracts with generators through bilateral negotiations, or buy electricity from the "Vietpool", or even buy electricity from distribution companies. In other distribution power companies, however, the functions of distributing and selling electricity still remain integrated.
- Essential issues
The retail companies are separated from distribution companies, however, the number of retail companies need to be considered. There should not be too many companies to avoid a loss in social resources. In the area of retail competition, there is no requirement for tariff regulation, even for end-use. However, for some special classes of customers, tariffs still need to be subsidized and regulated, for example people under economic hardship. Of course regulated tariff will remain available in other areas and end-use customers can choose to buy electricity with regulated or negotiated prices.

To facilitate a retail market, transmission and especially distribution network

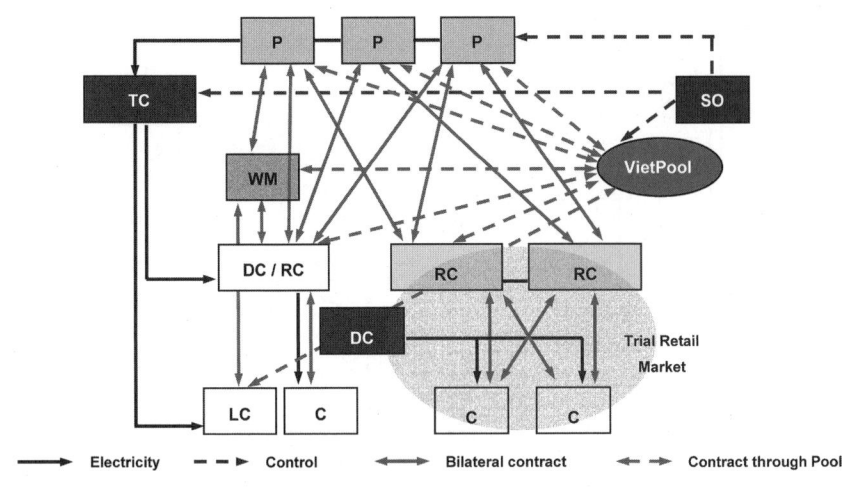

Figure 13. 4 a Retail electricity market: trial step

need to allow open and non-discriminatory access, at least in the areas where retail competition is introduced.
- Prerequisites for moving to a fully competitive retail market

A fully competitive retail market has a large impact on society the regulated tariff is abolished. Accompanied risks for customers may increase. The cost of implementing the competitive retail market is very expensive due to the huge requirement of the metering system, information technology infrastructure, etc. Thus, it is far from the objective of market development in Vietnam's electricity industry. Prerequisites for this stage are the political commitment, comprehensive and powerful regulatory framework, well-educated to consumers, and cooperation from unions and the public media. Technical requirements include the maturation of the power system with regard to generation, transmission, distribution and metering services, and the advanced information technology infrastructure, specifically advanced hardware and software. The organization requirement is fully unbundled distribution and retail functions.

Fully Competitive Retail Electricity Market
- Structure and operation (Figure 13. 4 b)

This structure gains the highest level of competition. In principle, customers are free to choose their supplier. Distribution and retail energy services are completely separated nationwide. However, even at this step, a phased retail competition plan is needed. The costs of retail competition are usually very high, therefore initially, the

Figure 13. 4 b Fully competitive retail market

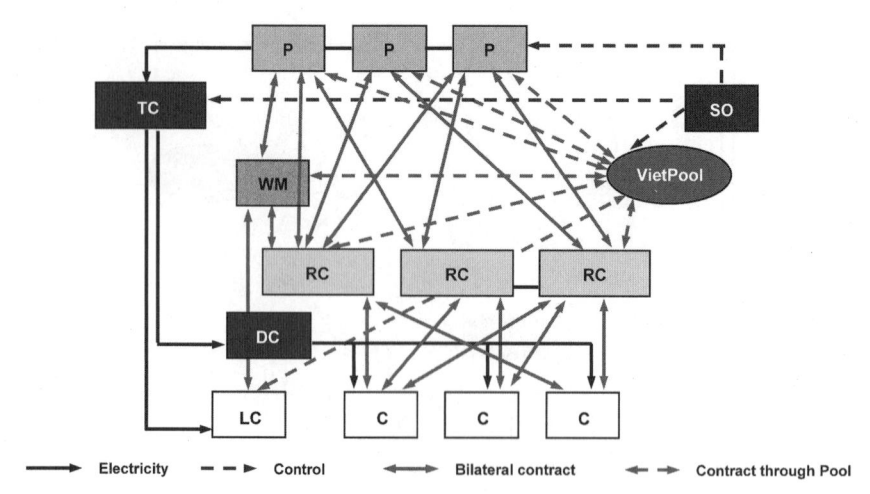

VietPool = Power exchange WM = Wholesale market RC = Retailer C = Customer LC = Large customer

retail market may be opened up to large consumers only, allowing smaller consumers to enter the market in later phases or using a common load profile for each category of customers.

As mentioned above, the risks accompanied with this industry structure are considerable. The role of regulation is still the linchpin that minimizes these risks. Tariff regulation may be reduced with the exception of the subsidized tariff, however supervision of competition regulation needs to extend market-wide with more concentration given to system planning and new investment incentive mechanisms to ensure reliability of the electricity supply.

In conclusion, it should be noted that the process of market development requires a time, resources, and government commitment and leadership. It needs to be implemented with care phase by phase and step by step. As stipulated by the Prime Minister's decision, a fully competitive retail market can be considered after 2022.

CHAPTER 14
IMPLEMENTING PARTIAL PRIVATIZATION IN VIETNAM'S ELECTRICITY INDUSTRY

The Vietnamese government's policy on electricity industry privatization has outlined in the Electricity law: "... to attract all economic sectors to participate in activities of electricity generation, electricity distribution, electricity wholesaling, electricity retailing and/or specialized electricity consultancy. The State holds monopoly in activities of transmission, national electric system regulation, construction and operation of big power plants of particularly important socio-economic, defense or security significance." [Article 4 Chapter 1]. Privatization is considered as one essential aspect for electricity industry reform in Vietnam. As stated in Chapter 11, privatization strategy consists of two branches (i) attracting new investment in the power sector and (ii) privatizing power companies. The former is related to Greenfield projects, which have been used historically and the latter is a part of the national program of "equitization"[28] of state owned enterprises (SOEs), which has recently become attractive. The privatization program in Vietnam's EI should be conducted carefully and slowly to avoid loosing public money to private pockets. Therefore, the principles of privatization here are partial privatization.

This chapter will try to shape the future of Vietnam's electricity industry under the partial privatization program. Several privatization plans are considered and a cost benefit analysis will be used to propose the best option. The author also includes supplementary mechanisms to ensure the success of the privatization program.

14.1 Private Investment Promotion in Electricity Industry

Greenfield power projects have been used in Vietnam since 1998. Vietnam began to implement this lesser-known form of privatization as a provision of new the economic infrastructure. Sometimes called a 'soft option' policy, Build-Operate-Transfer, or BOT, projects are recognized by their proponents as useful mechanisms by which new private sector investment can be mobilized into sectors that remain in

---

28 Refer to Section 14.3 for an explanation of "equitization".

public hands for political reasons [World Bank, 1994]. They are implemented as discrete projects and therefore do not seem to threaten the overall structure of ownership in the industry.

EVN's Annual Report 2004 noted that non-EVN installed generation capacity as of December 2004 was 2518 MW, and that 2004 generation was 6,363 GWh. The share of IPP installed capacity was 22% and 13.0% in generation.

According to the first draft of Master Plan on electricity power development for period 2006–2020, extended to 2025 (Master Plan VI), the investment capital from non-EVN sources mobilized for period 2006–2010 accounts for 34% of the total investment requirement. The list of IPP projects registered in the Ministry of Industry by 2010 consists of 86 projects with a total capacity of 5,159 MW.

The master plan has estimated the investment requirement for other five-year periods to average USD 5–6 billion per period. This amount of investment can not be covered without significant non-EVN contributions. As proposed by the master plan, the EVN should be responsible for all transmission network investments and only a portion of the investment required for the generation segment covering multi-

Table 14.1 Investment requirements for electricity industry development in the period of 2006–2025 (Scenario: mobilize non-EVN capital sources)   Unit: billion USD

Item	2005	2006–2010	2011–2015	2016–2020	2021–2025	2006–2025
I Investment requirements for generation	1.117	17.863	15.950	20.327	13.467	67.607
A Net investments	1.005	16.280	15.274	19.083	12.918	63.555
Thermal power plants	0.327	8.560	11.432	17.426	10.566	47.984
Hydro power plants	0.678	7.720	3.843	1.657	2.352	15.572
1 EVN sources	0.766	8.418	4.026	6.187	1.815	20.446
Thermal power plants	0.221	3.104	2.994	6.187	1.815	14.100
Hydro power plants	0.545	5.314	1.032	0.000	0.000	6.346
2 Non-EVN sources	0.240	6.458	11.249	12.897	11.103	41.707
Thermal power plants	0.107	4.053	8.438	11.239	8.751	32.481
Hydro power plants	0.133	2.406	2.811	1.657	2.352	9.226
B IDC	0.112	1.583	0.676	1.243	0.550	4.052
II Investment requirements for network expansion	0.726	4.025	7.481	10.223	10.541	32.271
A Net investments	0.651	3.173	6.506	9.113	9.765	28.557
Transmission system	0.300	1.264	2.269	2.832	2.042	8.407
Distribution system	0.352	1.909	4.237	6.281	7.723	20.150
B IDC	0.075	0.853	0.974	1.111	0.776	3.713
Total	1.843	21.888	23.431	30.550	24.009	99.877
Net investments	1.656	19.452	21.781	28.196	22.683	92.113
IDC	0.187	2.436	1.650	2.354	1.326	7.765

Source: [Institute of Energy, 2006]

propose hydro power plants, nuclear power plants, and several essential projects. Non-EVN sources will invest in 46% of the total investment required for other power plants for the period of 2006–2025.

Therefore, the role of private investment Vietnam's power development seems to be indispensable. IPPs, as already mentioned in Chapter 3, are not the only answer to power sector development needs, however handled correctly, they can undoubtedly make a major contribution and are an attractive option for financing future investments in power infrastructure. The issue that needs to be addressed is how to attract such a huge amount of money from private investors?

Chapter 3 Part I addressed several IPP issues under the context of reform prevailing in the electricity industry, for example, the problem of stranded costs, the inflexibility of PPAs, the requirement of government guarantee, etc. From the investors' perspective, they normally want to sign fixed PPAs in order to deal with price fluctuation and unpredicted risk in the electricity market. In order to attract further investment, the most challenging aspect is how to deal with the issues of finding the appropriate approaches to reconciling existing IPP contracts with emerging power markets and designing new IPP contracts to better to facilitate subsequent integration into electricity markets.

Fortunately, the problem with existing PPAs in Vietnam does not seem too critical. Except for the Phu My 2.2 and Phu My 3 BOT contracts, other PPAs include the provision of power market participation so that the stranded costs issue can be resolved. The signed PPAs could also be dealt with by persuading producers transforming PPA into CfD contracts. For the new potential IPPs, the EVN should use Least-Cost Resource Planning and Bidding Code in procuring new resources so that the prices of the signed long term PPA's should be very close to the economic prices and then the stranded costs should not be substantial. In addition, more options for promoting private investments in the new environment of industry structure may include merchant plants, and joint-venture or joint-stock companies through which independent power producers are offered the right to move from power purchase arrangements to merchant operation as part of the introduction of wholesale competition.

14. 2   Equitization and Electricity Industry Reform

- **Privatization and Equitization of State-owned Enterprises in Vietnam**

What does Equitization Mean in Vietnam?
Equitization of state-owned enterprises (SOEs) is a major policy of the Vietnamese government which aims to create a major evolution in enhancing the production and business efficiency of SOEs, improve the effectiveness and competitiveness of the economy, promoting the transition to the market economy mechanism and Vietnam's

integration into the international economy, consist with the world's overall trend.

SOE equitization is, in essence, the conversion of SOEs with the sole owner as the State into share holding companies with multiple owners as shareholders, in order to restructure the SOE sector for better efficiency, to commercialize the enterprise activities. Equitization can take one of the following four forms: (i) keeping state shares intact and issuing new shares; (ii) selling part of the existing state shares; (iii) detaching and then selling parts of an SOE (a method mostly applied to state government corporation (SGCs)); and (iv) selling off all state shares to workers and private shareholders, a method mostly applied to loss-making SOEs [Anh, 2005]. Compared to the widely accepted definition of privatization, which is addressed in Chapter 3 Part I – "all initiatives designed to increase the role of private enterprises in using society's resources and producing goods and services by reducing or restricting the roles that government or public authorities play in such matters"– there is not much difference between Vietnam's equitization and privatization. However, for official documents the term of "equitization" should be utilized.

The major objectives of equitization are (i) to transform SOEs which are not necessarily owned 100% by the State into enterprises with various owners; to mobilize capital from local and foreign individuals, economic organizations, social organizations to enhance financial capability, renovate technology, renovate management modes with the aim to improve efficiency and competitiveness of the economy; (ii) to harmonize interests of the State, enterprises, investors and employees; and (iii) to ensure the publicity and transparency in line with the market principles, to promote the development of the capital market and stock market [http://www.nscerd.org.vn/].

Recent Obstacles to Equitization

Equitization is no easy task due to the lack of experience and knowledge of the policy making process, necessary techniques and legal formalities. The following are several major obstacles to equitization in Vietnam:

- Policies on equitization aren't consistent and detailed enough. Many problems, such as equitization procedures, methods of evaluating company assets, preferential treatment of equitized companies, and the use of proceeds from the sale of shares, aren't clarified. That is why equitization seems unwelcome to many managers and was unnecessarily prolonged.
- Argument over equitization continues. Many people even thought that equitization would harm socialism. Most laborers in the public sector are afraid equitization will deprive them of jobs, interests, and position.
- Certain conditions needed for equitization do not meet the requirement of equitization, such as methods of revaluating assets, stock market, etc. The

revaluation of assets is based only on account books and no attention is paid to intangible assets. It's hard to trade in stocks and shares when the capability of the stock market is limited. This obstacle will be discussed in more detail in the power sector experiences that follow. Financial information isn't disclosed regularly enough to attract investing public.
- Certain policies relating to the equitization aren't suitable for the market mechanism, such as interest rates on savings accounts, limits on bonuses, treatment of workers, etc.
- Companies to be equitized struggle with existing bad debt and dead stocks.

- **Three Initial EVN Equitization Experiences**

Under the government policy of equitization, from the EVN's perspective, an important goal of equitization is to renovate the management mode in order to improve the operational efficiency of enterprises after being equitized to enhance competitiveness, prevent monopolies, and develop the power market. At the same time, equitization is one of solution for the EVN to supplement capital for investment in power projects to meet the power demand from 88 to 93 billion kWh in 2010.

In execution of Decision No. 219/2003/QD-TTg issued by the Prime Minister on 28 th October 2003 on approval of the Master Study for Rearrangement and Renovation of State Own Enterprises of EVN through 2005, up to now the EVN has finalized the equitization of sixteen subsidiaries, in which fourteen were sold with 100% share. The two remaining units, which belong to the main production chain, are the Vinhson-Songhinh hydro power plant and the Khanhhoa electricity provincial department. The EVN sold off a 5% share of the Vinhson-Songhinh power plant through employee buy-outs and a 20% share (43.6 million shares) through a public offer on the stock market. The percentage of the Khanhhoa provincial electricity department that was sold off was even higher: 35% by employee buy-outs and 14% (2.1 million shares) through the stock market. Preliminary examinations show that the successful IPOs, via auction of Vinh Son – Song Hinh on 10th March 2005 and Khanh Hoa on 17 th March 2005, were considered a remarkable contribution to the equitization process of the EVN. All of shares on offer have been auctioned off and EVN collected VND 490.5 billion, or VND 31.8 billion higher than the offered face value of share through these auctions. The average price of a Vinhson's share is VND 10,576 per share compared to the initial price of VND 10,200 per share. The average price of a Khanhhoa's share is VND 13,109 per share compared to the initial price of VND 10,500 per share.

Another experience is the initial public offering of a minority equity stake in the Pha Lai coal-fired power plant was completed on November 25, with the official

Vietnam News Agency reporting that "almost all of its 73,000 shares" on offer had been auctioned off. The Pha Lai generation plant is the third electricity enterprise to be partly privatized by being unbundled from the state-owned power utility (EVN) and corporatized, with part of the resultant equity then being sold off.

The successful equitization of enterprises and the business performance of those enterprises afterward proved to be the proper government policy, as well as help the EVN draw useful lessons for further equitization activities.

Several valuable lessons have been learned through such experiences. First, capability of the newly-established Vietnam stock market should be considered. Vietnam's stock market only lists 28 stocks after more than four years in operation. Value of the stock traded on the market accounts for less than USD 10 million as a daily average. Most companies listed on the stock market are small and average-size companies. Equitized SOE were allowed to sell shares as unofficial listed companies since the end of 2004 [Decree No 187/2004/ND-CP and Circulation No 126/2004/TT-BTC]. All three of EVN's equitized companies are under such kind of SOEs. The government considered it as an approach to push for further measures to expand the capital market. However, due to the new establishment and weaknesses of the stock market in Vietnam shown by the number of potential investors and the number of commodities sold in market, it is difficult to avoid the power of the market. Negotiations with several important investors could reduce the share price. The rate of return (RoR) for power companies in Vietnam is not really attractive with investors, the RoR of EVN is only in 4–5%. The announced RoR of equitized companies was set at 10% since the authorities (MoI) did not expected the price of power company share to gain a high price on the stock market similar to other famous stocks like Vinamilk, etc. However, with the exception of Khanh hoa company's stock selling 25% higher than the initial offering price, the average selling price for the other two companies, Vinhson-Songhinh and Phalai, was only 3% higher than the initial offering price. The average prices gained seem to be inversely proportional to amount of stock sold. The Phalai power company could not even sell off all 72 million shares. The average price of the initial public offering do not reflect the actual value of power companies because at the beginning year of 2006, the value of these stocks on the secondary stock market reach 20,000 VND per share, nearly double the initial price. Therefore, before conducting a privatized deal, the proportion of assets to be sold and the appropriate privatization methods, which may not be via public offering but through the private sale of shares, needs to be carefully considered.

14. 3   Proposed Equitization Plan

- **Comments on EVN's Existing Equitization Program**

The Vietnamese government's policy of electricity industry privatization as outlined in the Electricity Law is: "... to attract all economic sectors to participate in activities of electricity generation, electricity distribution, electricity wholesaling, electricity retailing and/or specialized electricity consultancy. The State holds monopoly in activities of transmission, national electric system regulation, construction and operation of big power plants of particularly important socio-economic, defense or security significance" [Article 4 Chapter 1].

At the moment, the EVN has an ambitious plan to equitize its subsidiaries. The sale of the Pha Lai shares will be followed by the sale of a minority equity stake in the Thac Ba hydroelectric plant, according to Dang Phan Tuong, the head of EVN's Equitization and Securities Division. The tender is scheduled to take place in early 2006. More power plants are scheduled for equitization in 2006. The EVN said that the initial public offerings for 5 of remaining power plants, including the 100 MW Ninh Binh coal-fired plant, 380 MW Ba Ria gas-fired facility and 410 MW Uong Bi coal-fired plant, and the 150 MW Thac Mo, 160 MW Da Nhim, 300 MW Ham Thuan and 175 MW Da Mi hydroelectric plants, are planned from 2006 to 2008.

Other generation plants will be offered in 2008, a number of which will have only been operation for only one year. Equity stakes in 17 provincial electricity companies, 4 power engineering consulting companies, and 3 other supporting units will be tendered over the next few years. The EVN currently envisages the competitive electricity market will begin operating from 2010, with only the transmission system remaining as a state monopoly.

Several comments should be made regarding EVN's equitization plan. It seems to be ambitious. EVN does not put adequate attention on the analysis the initial equitization experiences. To plan for equitization, the following aspects should be taken into consideration: (i) the demand side with the stock market capacity or purchasing capability of potential investors, and (ii) EVN's capital requirement for new investment and normal operation. To clearly understand the purchasing capability of potential investors, comprehensive analysis from financial experts is needed, however, it is apart from this book, except for several preliminary ideas mentioned in the section 14.2. The author's objective is to address the equitization program based on the EVN point of view. The author will try to point out the weaknesses of EVN's program and propose methods for overcoming these weaknesses.

In terms of methodology, the equitization program is set without the base of cost-benefit analysis. This may lead to misjudgment in terms of setting the time to equitize subsidiaries, as well as the amount of stocks being sold. Equitization is considered a means of accumulating capital for new investments. At the same time, equitization creates huge opportunities for corruption, thus increasing the risk economic inequality in society. Therefore, cost-benefit analysis should be utilized. The author proposes use of cost-benefit analysis for evaluating the strategy of equitization. The argument will be presented in the following paragraph. In addition to the introduction of the cost-benefit method as a general methodology, the author also considers the method for evaluating the value of assets.

Regarding the pace of implementation of the equitization program, the author believes that the EVN's plan is quite accelerated. Regarding the plan of equitizing generation, most of existing potential power plants will be sold before 2008 and the EVN will further equitize newly operational power plants. First, the author proposes that the EVN consider selling only existing power plants, and then delay the sale of new power plants until 2010. The author will try to examine the effect of this program on EVN's financial status and compare it with a new proposal. Regarding the distribution equitization plan of 17 provincial electricity companies, it should be noted that forming these more autonomous companies is complicated by the policy of the national uniform retail electricity tariff. The costs of providing power services are inevitably different for each locality. Under the national uniform tariff policy, tariffs for the same class of customers have to be the same throughout the nation. Furthermore, the current national uniform electricity tariffs for rural domestic customers are heavily subsidized. Under such a situation, it would be very difficult for any single-share or joint-stock limited liability company to achieve a healthy financial performance. Therefore, in the author proposes equitizing distribution companies on a trial basis, with wide-scale equitization of provincial power companies after 2010, when the government plans reconsider its national uniform retail tariff policy in order to facilitate the power sector reform.

- **Evaluation Methods of SOE**

According to Circular No. 79/2002/TT-BTC on guidelines on enterprise evaluation issued on September 12, 2002 by the Ministry of Finance, there are two methods for SOE valuation under the equitization process:

Method 1 is referred to as the Net Asset Valuation (NAV) method. According to this methodology, an enterprise will be valuated based upon its real value of the total intangible and tangible assets at the time of valuation. SOEs and their dependent units operating in commercial and production activities, except for those subject to the Discounted Cash Flow (DCF) method as mentioned below, shall be subject to the NAV method.

Method 2 is referred to as the Discounted Cash Flow method. According to this framework, an enterprise will be valuated based on its profit-making potential only, regardless of the enterprise's asset value. The DCF method is applicable to SOEs operation in commercial services, consulting services, construction, design, financial and auditing services, and information technology. These SOEs have an average after-tax profit margin per owner's equity in 5 consecutive years before equitization higher than the 10-year government bond rate at the nearest time before the enterprise valuation time.

In the case of Electricity of Vietnam, the method proposed is DCF. Moreover, it should be linked between the value of the equitized company and the financial statement of the EVN because the appropriate equitized program should strike a balance between the benefits from cash raised by proceeds of share sold in advance and the costs for the higher price of electricity purchased from power plants paying over a long-term period. Without the proper balance, equitization could result in losses for the EVN. Thus, evaluation of equitized companies needs to be incorporated as a module of the cost-benefit analysis.

- **Cost-benefit Analysis**

We use EVN's financial projection model with a sub-module of the re-evaluated companies' market value to assess the cost and benefit of equitization program. Model forecast variables such as electricity prices, electricity sales and revenues, costs of generation, transmission and distribution, and wages and employment are under an assumed market structure. Figure 14.1 illustrates the flow chart of the cost-benefit model.

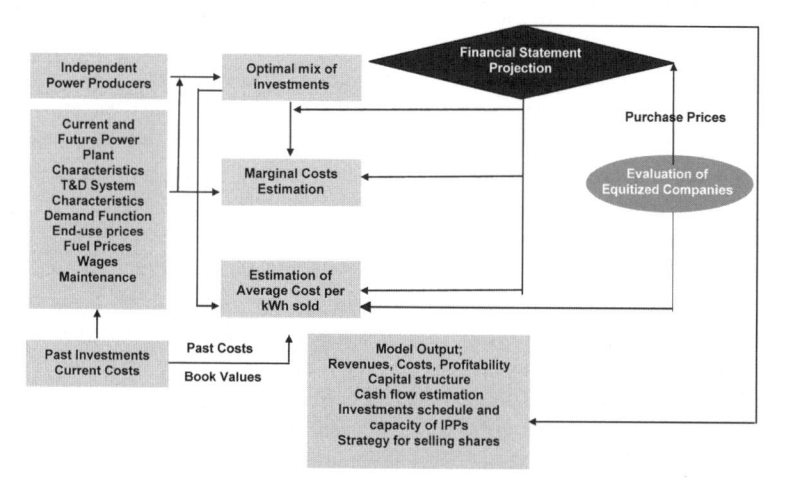

Figure 14. 1 A flow chart of the cost-benefit model

Due to data limitations, analysis extends up to 2010 only with the electricity industry structure remaining in single buyer model. Further analysis may be necessary to include the entire reform process until 2025. The analysis should also be extended to examine the cost-benefit analysis not only for equitization, but competition as well.

14.4 Applying Cost-benefit Analysis into the Proposed Equitization Plan

Financial Performance Requirements
Using the financial projection model, the equilibrium should be gained through financial performance requirements imposed on the EVN by international financial institutions. Under such conditions: (i) EVN would be reviewed every year to ensure that funds generated from internal sources would be equivalent to no less than 30% of the annual three-year average of their capital expenditures. It means the self-financing ratio requires to be over or at least equal to 30% (SFR $\geq$ 30%). This covenant seems to be hard for the EVN considering that the same condition applies to EGAT with 15% of SFR or further considering the normal banking practice with 15% of SFR for project financing arrangement. EVN is trying to negotiate with WB to loosen the covenant. Unofficially, a level of 25% of SFR is acceptable; (ii) EVN shall maintain net revenues at a level 2 times their total estimated debt service requirements for normal projects and 1.5 times for rural electrification projects. In that case, the debt service coverage ratio must be at least equal to 1.5.

Analyzed Scenarios
Our cost-benefit analysis considers four scenarios: (i) no further equitization, (ii) the accelerated program proposed by EVN, and (iii) the moderate program proposed by the author. This scenario is divided into Scenario 3_1 and Scenario 3_2, in which the former is original and the latter considers the increase of market value of stock due to the impact of appropriate schedule that can capture the maximum benefit from the demand-supply relationship. Although the actual effects may be greater, a 10% increase in the market value of stock will be considered for Scenario 3_2.

Model Solutions
Figure 14.2 presents the projection of EVN's self-financing ratio in different scenarios. Obviously, under Scenario 1 there is no equitization, EVN is unable to meet the covenant, and the SFR drops to less than 25% for the entire period. Under Scenario 2 an accelerated equitization program is proposed by the EVN, the SFR is improved in the period 2005-2008 with the SFR over 30%, however the SFR drastically drops to less than 20%, which is even lower than the Scenario 1. Under Scenario 3, both 3_1 and 3_2, the EVN generally meets the loan covenant with the range of SFR of 25-30% for the entire period of 2005-2010. Thus, the effects of

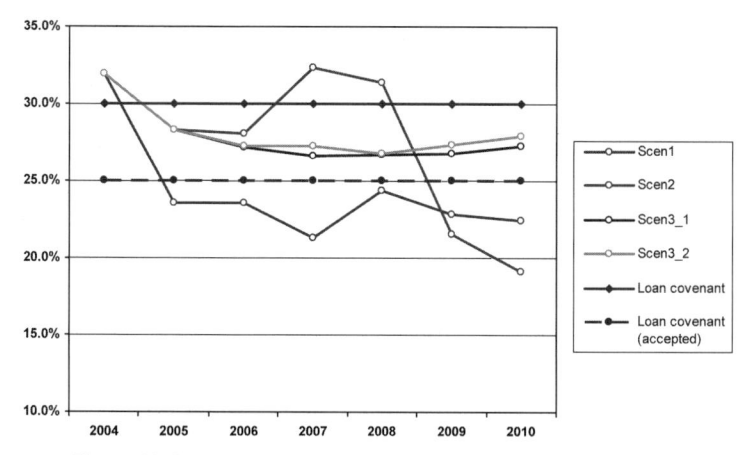

Figure 14. 2 Self-financing ratio of EVN in the period of 2004–2010

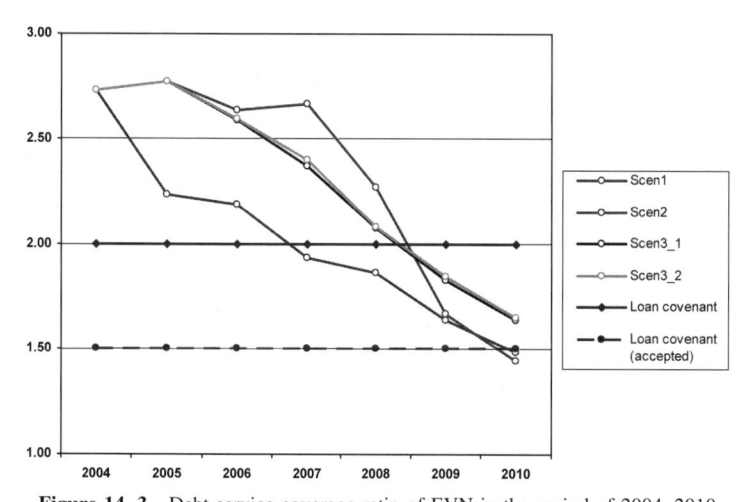

Figure 14. 3 Debt service coverage ratio of EVN in the period of 2004–2010

equitization program are greatly influenced by the pace of the program and the time taken to sell shares.

In terms of the covenant on DSCR, in all scenarios, there is decrease in the DSCR trend. However, the amount of reduction is differs between scenarios. Under the Scenario 1 and 2, the DSCR drops to the red line 1.5 times at the end of period. Meanwhile, under Scenario 3_1 and 3_2, DSCR remains in the safety area with the

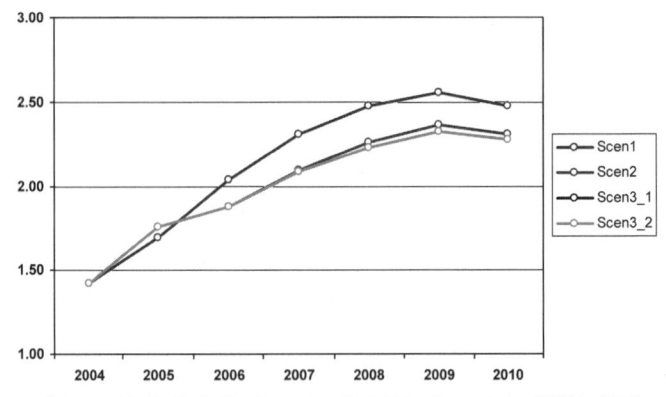

Figure 14. 4 Debt/Equity ratio of EVN in the period of 2004–2010

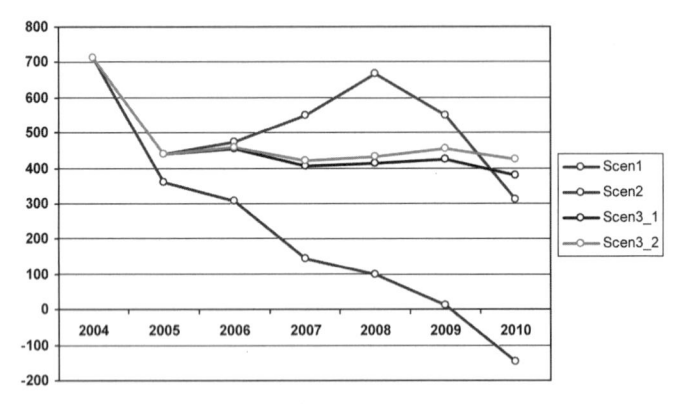

Figure 14. 5 Cash balance of EVN in the period of 2004–2010

lowest value of 1.65 times in the year of 2010 (Figure 14. 3).

The loan covenants require for only two financial ratios mentioned above. However, we go further to see the impact of the equitization program on two more essential financial corporation ratios: debt/equity ratio and cash balance (Figure 14. 4 and 14. 5). Similar conclusions are drawn from both financial indicators. The equitization program can improve the financial status of the EVN.

Thus, the cost-benefit analysis confirms the author's comments regarding the EVN equitization strategy. The equitization program may have positive impacts on EVN's financial performance if it is scheduled appropriately. Then, considering both EVN's capital requirements and the capability of potential investors for the period of 2005–2010, only the equitization program for the existing power plants should be implemented. It is not necessary to extend the program to new-operated power plants. This option may be considered after the 2010. Similarly, equitization of large scale provincial distribution companies should be postponed beyond the 2010. The period of 2005–2010 should be given to trial equitization and the conversion of government policies, especially regarding changes in tariff regulation.

CONCLUSIONS, RECOMMENDATIONS AND FURTHER RESEARCH

Electric utility companies were traditionally owned and operated by the state until the end of 1970's. During the 1980's and 1990's, the period known as the "liberalization tide", electricity industry reforms occurred worldwide. However, at the end of the 1990's and especially at beginning of 2000's, the perspective on industry reform changed. After a series of financial crises, corporate scandals, the California electricity crisis, and a series of major blackouts in North America and other areas of the world, the need to discover how EI reform should be carried out arose. Several countries re-examined their reform strategies, while others even temporarily delayed the privatization process.

This book endeavors to support policymakers in addressing the question of how EI reform should be carried out by introducing a multidimensional and strategic analysis of electricity industry reform. This book follows two paths of logic. The first addresses the multidimensionality of electricity reform. Reform can be seen as a comprehensive process with three dimensions of competition, privatization, and regulation. Regulation is defined here as the creation and enforcement of rules that promote the efficiency and optimal operation of markets as well as the devising mechanisms to address specific social and environmental issues. The role of the regulator is to maximize social benefits while taking into account social values and the need to maintain financial viability and promote economic efficiency of the sector in the long run. Thus, electricity industry reform could be examined from a multidimensional point of view. The second path of logic in the book is systematic of analysis. Throughout the entire analysis, the scope of examined objects is reduced from a worldwide scale to Asian countries and at finally to the case of Vietnam. In contrast, the depth of analysis increases stage by stage from the standard prescription to empirical modeling and policy implication.

The multidimensionality of reform has been throughout the book, while the systematic analysis is addressed in respective parts. In Part I, a broad and extensive analysis, which can be applied to any country regarding reform in the electricity industry according to the standard prescription is carried out. Except the background chapter, various issues related to reform are grouped into three chapters, from Chapter 2 to Chapter 4, which respectively covers three sides of reform such as ownership, competition, and regulation. Moreover, Chapter 5 of this part carries out a worldwide comparison of electricity industry reform. Valuable lessons should be drawn from the successes and failures. Then, discussion proceeds to electricity industry reform from the viewpoint of several main stakeholders, who have important roles in the reform process.

The second part of the book endeavored to carry out a comprehensive analysis of the EI reform picture in Asian countries. Chapter 6 includes facts and evaluations of EI reform. In the chapter, the author tried to construct a set of core indicators for EI reform. Such indicators follow the main line of this book and will be divided into three categories: regulation, competition, and privatization. The results of this part were essential for the empirical analysis of industry performance in the following chapter. Chapter 6 also extended to classify country groups based on the level of advancements in electricity industry reform and also on the political and economic conditions within the three dimensional matrix of restructuring, privatization, and regulation. Based on this classification, a representative of each country group was selected for further analysis in Chapter 8. Chapter 7 will carry out an econometrics analysis, which allows policy makers to evaluate the impact of each reform policy option on the EI performance. On that basis, this book intended to derive several predictions about the future and shape of change in the electricity industry reform of several representative countries. The scope of this analysis opened further to include nuclear power policy under the process of EI reform in Asian countries.

Part III of the book presented the most pointed and deep analysis of policy implication for the case of the multidimensional reform in Vietnam's electricity industry. This part at first attempted to examine the reform of the industry to date, and then applied policy implication findings to establish comprehensive strategies for the electricity industry reform in Vietnam, which might cover most of reform aspects from establishing an efficient regulatory framework, restructuring the Electricity of Vietnam, introducing a competitive market, implementing privatization to address regulation mechanisms for several sensitive social-benefit issues. Such challenging tasks were solved respectively in Chapters 10 to 14.

On the whole, the book contributed with a number of essential findings. First, the book, based on a cautious review of existing analyses in the field of electricity industry reform, provides its readers with a standard prescription for multidimensional reform, which covers competition, privatization, and regulation, as well as the requirement of sustainable sector development. Second, the book emphasized electricity industry reform in Asian countries which until now has not been adequately addressed by research. The book also collected and created a compact set of data which could be served not only the author's research objectives, also further analyses as well. The data of reform in eighteen Asian countries is presented in the Appendix 6_1.

The third contribution of the book is given by a clear country group classification for Asian countries. Based on a set of reform indicators, which consists of eleven indicators reflecting different reform policies, and thanks to a multiple correspondence analysis principle, four country groups were recognized. These groups reflect the different nature of electricity industry reform. The first group, including Nepal, Sri Lanka, Laos, Cambodia, and Bangladesh, is characterized by both a low level of competition, privatization, and regulation. The second group, including Thailand, Indonesia, Pakistan, India, Taiwan, and Vietnam, has initiative reform experience, mostly related to establishment of legal framework, corporatization and the introduction of new generation entrants, and pre-planning new reform steps with a regulatory body, introducing a power pool and launching privatization. The group 3, which includes Japan and Hong Kong, is characterized by a very high level of privatization against the low level of competition and regulation. The final group, including Singapore, Malaysia, Korea, the Philippines, and China, is represented by a relatively advanced level in the introduction of competition, and establishing regulatory authority.

The fourth valuable contribution of this book comes from an effort to carry out an econometrics analysis, which allows policymakers to address the impact of each reform policy option on the EI performance. Despite data availability limitations, the regression may offer several interesting results. It should be noted that the most influential policy increases the non-governmental share in electricity generation as a measure of privatization, and the second position is held by the occurrence of legal framework for industry reform. One more notation is that while most reform policy options, in general, have positive impacts to electricity generation per capita and installed capacity per capita, they have different effects on the share of nuclear power generation. The option of corporatization and establishment of an independent regulator have a positive impact, while other options related to more competition and privatization has a negative impact. Therefore, the more electricity industry reform implemented, the more attention that should be paid to regulation in order to maintain the sustainable energy policy, especially for countries that import energy, such as Japan or Korea.

Based on the fourth contribution, the book tried to draw several reform strategies for Asian countries. Obviously, empirical analysis itself serves only as a foundation rather than a competed argument for policy implication. Four case studies of respective country group representatives will be utilized as a verification and supplementation analysis. This directly contributes to the fifth contribution in this book.

Last, but not least, the sixth contribution of the book is provided by policy implications for the pointed and deep analysis of the Vietnamese case. Starting later than many reformed countries, Vietnam gains the advantages of those countries' experiences. Making use of these experiences may allow Vietnam to carry out EI reform at a quicker pace. Among many valuable lessons learned, the two most principal lessons are: (i) step-by-step policy is the best practice for Vietnam. "Slow and steady wins the race". This policy guideline needs to be obeyed in all aspects of proposed reform design, and (ii) multidimensionality of the reform policy. Balance should be maintained between economic efficiency and social benefit.

However, as is the nature of any research, this book contains many limitations, which are acknowledged by the author and enumerated upon here for further analysis. To begin with, the sample is composed of most Asian countries, except Central Asia, which covers Kazakhstan, Uzbekistan, etc. There will be a sample selection bias if the countries making this data available have differing results for the dependent variables than those that do not make data available. If the time allowed, the scope of this analysis should have been opened to the reform processes in those countries. Another limitation faced by the analysis is the quality of data. The results of the empirical analysis should be considered with caution since, as discussed above, the performance measures are imperfectly represented by net generation per capita, installed capacity, the capacity utilization rate, loss ratio and the nuclear power share. Moreover, there are issues of cross-country comparability of data. The IEA performance data is based on submissions from national administrations. Different countries have different classifications and reporting conventions, so observations in a given performance data series may not have the same meaning unilaterally.

An additional possible source of bias is that several indicators serve only as crude proxies of the policy options they are meant to measure. For example, the TPA indicator is based on the formal approach to network access rather than actual use of Third Party Access; there may be no formal barriers to TPA, however, simultaneously a monopolist may be the only producer in a market, making TPA rules moot. The legal TPA may not result in actual entry and especially if the incumbent retains practical control of the market.

In terms of modeling, the following drawbacks should be recognized. The most bothersome come from employing dummy variables to measure the regulation and competition policy options. Dummy variables can help avoid problems related to subjectivity and judgment when constructing quantitative indicators from qualitative information. However, the worst problem with the use of a dummy variable, as recognized by the author, is the capability to measure actual development, since there are only two levels, 0 or 1, except for two other variables, R 1 and C 1, which represented as 2 dummies. The numeric measurements to policy options should be applied for analysis and might be the objective of further analysis. Moreover, the model employed here also has a problem with variable lag. Several variables used in the model measure policy, which should impact to performance indicators for more than one year. To deal with that employment of the dynamic panel data technique may be needed, which should be left for further analysis.

Also due to the unavailability of cross-country data, empirical analysis in this book is unable to conduct many useful examinations which may include the impact of sequenced reform policies, the impacts of reform option on prices, the role of international institutions on reform, etc. These empirical analyses will be able to partly solve the debate of how efficient reform policy in the electricity industry is. Finally, the analysis has not explored the social and long-term developmental effects resulting from privatization and market liberalization in the electricity sector which remains the subject of future research.

Apart from the limitations mentioned above and regardless of the implication of applying policy to a particular country, such as Vietnam, it should be noted that such implications are drawn from cross-section empirical analysis so it may not be completely suitable for the country.

REFERENCES

ADB (2001), *Special Evaluation Study on the Privatization of Public Sector Enterprise: Lessons for Developing Member Countries.*
ADB (2003), *TA 3763 VIE: Vietnam Roadmap for Power Sector Reform,* PA Consulting, Hanoi, Vietnam.
ADB (2005), *ADB Key Indicators 2005,* www.adb.org/statistics.
ADB-JBIC-WB (2005), *Connecting East Asia: A New Framework for Infrastructure.*
Anh, V. T. T. (2005), "Vietnam – The Long March to Equitization", Policy Brief No. 33, The William Davidson Institute, University of Michigan.
APEC Energy Working Group (1997), *Manual of Best Practice Principles for Independent Power Producers.*
APEC Energy Working Group (2003), *Electricity Reform in APEC Economies – The Way Ahead,* Peter Smiles and Associates, Resources Law International and Country Energy.
Armstrong, M. and D. Sappington (2005), "Recent Developments in the Theory of Regulation", in Armstrong, M. and R. Porter (eds.), *Handbook of Industrial Organization,* forthcoming.
Asia Pacific Energy Research Centre (2000), *Electricity Sector Deregulation in the APEC Region.*
Asia Pacific Energy Research Centre (2004 a), *New and Renewable Energy in the APEC Region: "Prospects for Electricity Generation".*
Asia Pacific Energy Research Centre (2004 b), *Nuclear Power Generation in the APEC Region.*
Asia Power Sector Reforms Workshop (2002), *Electricity Sector Reforms in Asia: Experiences and Strategies,* Chulalongkorn University Campus, Bangkok, Thailand, 7–10 October.
Bacon, R. (1999), "Global Energy Sector Reform in Developing Countries: A Scorecard", Report No. 219-99, UNDP/World Bank.
Bacon, R. W. and J. Besant-Jones (2001), "Global Electric Power Reform: Privatization and Liberalization of the Electric Power Industry in Developing Countries", *Annual Reviews of Energy and the Environment,* Vol. 26.
Bacon, R. W. and J. Besant-Jones (2002), "Global Electric Power Reform, Privatization and Liberalization of the Electric Power Industry in Developing Countries", Paper No. 2, The Energy and Mining Sector Board, World Bank.
Balce, G. R. (2002), "Overview of the Privatization and Deregulation Initiatives in the ASEAN Power Sector", Presented at the Metering Asia-Pacific 2002–Singapore, 9–11 April.
Baltagi, B. H. (1995), *Econometric Analysis of Panel Data,* John Wiley & Sons.
Balzhiser, R. (1996), "Technology – It's Only Begun to Make a Difference", *Electricity Journal,* Vol. 9, No. 4.
Baron, D. (1989), "Design of Regulatory Mechanisms and Institutions", in Schmalensee R. and

R. D. Willig (eds.), *The Handbook of Industrial Organization,* Elsevier North-Holland.
Bayless, C. (1994), "Less Is More: Why Gas Turbines Will Transform Electric Utilities", *Public Utilities Fortnightly,* Vol. 132, No. 22.
Bernow, S., W. Dougherty, M. Duckworth and M. Brower (1998), "An Integrated Approach to Climate Change Policy in the US Electric Power Sector", *Energy Policy,* Vol. 26, No. 5.
Besant-Jones, J. E. (ed.) (1993), *Reforming the Policies for Electric Power in Developing Countries,* World Bank.
Bjornsson, H., R. Crow and H. Huntington (2004), "International Comparisons of Electricity Restructuring: Considerations for Japan", CIFE Technical Report No. 150, Stanford University.
Borenstein, S. (2000), "Understanding Competitive Pricing and Market Power in Wholesale Electricity Markets", *Electricity Journal,* Vol. 13, No. 6.
Bortolotti, B. and D. Siniscalco (2004), *The Challenges of Privatization, an International Analysis,* Oxford University Press.
Breyer, S. (1982), *Regulation and Its Reform,* Harvard University Press.
Byrne, J. and Y. M. Mun (2003), "Rethinking Reform in the Electricity Sector: Power Liberalisation or Energy Transformation?", in Wamukonya, N. (ed.), *Electricity Reform: Social and Environmental Challenges,* Pitney Bowes Management Services, Denmark.
Byrne, J., L. Glover, H. Lee, Y. D. Wang and J. M. Yu (2004), "Electricity Reform at a Crossroads: Problems in South Korea's Power Liberalization Strategy", *Pacific Affair,* Vol. 77, No. 3.
Casten, T. R. (1995), "Electricity Generation: Smaller is Better", *Electricity Journal,* Vol. 8, No. 10.
Choynowski, P. (2004), "Restructuring and Regulatory Reform in the Power Sector: Review of Experience and Issues", ERD Working Paper No. 52, Asian Development Bank.
Colley, P. (1997), *Reforming Energy: Sustainable Futures and Global Labour,* Pluto Press.
Covarrubias, A. J. (1996), *Lending for Electric Power in Sub-Saharan Africa,* World Bank.
Cubbin, J. and J. Stern (2005), "Regulatory Effectiveness and the Impact of Variations in Regulatory Governance: Electricity Industry Capacity and Efficiency in Developing Countries", World Bank Policy Research Working Paper 3535.
David, A. K. and F. S. Wen (2001), "Transmission Open Access", in Lai L. L. (ed.), *Power System Restructuring and Deregulation,* John Wiley & Sons.
Deloitte Touche Tomatsu Emerging Markets (2004), "Sustainable Power Sector Reform in Emerging Markets – Financial Issues and Options", Joint World Bank/USAID Policy Paper.
Drillisch, J. and C. Riechmann (1998), "Liberalisation of the Electricity Supply Industry – Evaluation of Reform Policies", EWI Working Paper No. 98/5, Cologne/ Tokyo.
Dubash, N. K. (2003), "Revisiting Electricity Reform: the Case for a Sustainable Development Approach", *Utilities Policy,* Vol. 11, No. 3.
Dung, P. H. (1996), "The Political Process and the Private Sector's Role in Vietnam", *International Journal of Health Planning and Management,* Vol. 11.
Eberhard, A. (2003), "Electricity Sector Reforms: Timing, Sequencing, Pre-conditions and

Advancing Public Benefits", Graduate School of Business University of Cape Town.

Energy Data and Modeling Center (2004), *Handbook of Energy and Economics – Statistics in Japan,* Energy Conservation Center, Japan.

Energy Information Administration (2002), *International Energy Outlook* 2002.

Energy Information Administration (2003), *International Energy Outlook* 2003.

Energy Information Administration (2004), *International Energy Outlook* 2004.

Energy Market Authority (2004), *Singapore Law and Regulations on Energy Security and the Role of the Private Sector in Ensuring Continued Supply of Oil and Gas.*

Energy Sector Management Assistance Programme (1999), "Global Energy Sector Reform in Developing Countries: A Scorecard", Joint UNDP/World Bank Energy Sector Management Assistance Programme.

Ergas, H. and J. Small (2001), "Price Caps and Rate of Return Regulation", Network Economics Consulting Group, Kingston, Australia.

Estache, A. and D. Martimort (1999), "Politics, Transaction Costs, and the Design of Regulatory Institutions", Policy Research Working Paper 2073, World Bank.

Estache, A., P. M. Rodriguez, J. M. Rodriguez and G. Sember (2003), "An Introduction to Financial and Economic Modeling for Utility Regulators", World Bank Policy Research Working Paper 3001.

Eto, J., C. Goldman and S. Nadel (1998), "Ratepayer-Funded Energy-Efficiency Programs in a Restructured Electricity Industry: Issues and Options for Regulators and Legislators", http://eetd.lbl.gov/EA/EMP/.

EVN (2004), *Project of Sustainable Development of Electricity of Vietnam,* Hanoi, Vietnam.

Feler, L. (2001), "Electricity Privatization in Argentina", Mimeo, World Bank.

Glachant, J. M. (2003), "The Making of Competitive Electricity Markets in Europe: No Single Way and No 'Single Market'", in Glachant, J. M. and D. Finon, (eds.), *Competition in European Electricity Markets: A Cross-Country Comparison,* Edward Elgar.

Gujarati, D. (2003), *Basic Econometrics,* 4th ed., McGraw-Hill.

Gupta, N. (2002), "Partial Privatization and Firm Performance", FEEM Working Paper No. 110.

Gutierrez, L. H. (2003), "Regulatory Governance in the Latin American Telecommunications Sector", *Utilities Policy,* Vol. 11, No. 4.

Hall, D., S. Thomas and K. Bayliss (2002), "Resistance and Alternatives to Energy Privatization", http://www.psiru.org/reports/2002-12-E-Resist.

Hattori, T. (2004), "Making Electricity Trading Work in India: Lessons from the Experiences in Developed Countries", JBIC Working Paper No. 16.

Hattori, T. and M. Tsutsui (2004), "Economic Impact of Regulatory Reforms in the Electricity Supply Industry: A Panel Data Analysis for OECD Countries", *Energy Policy,* Vol. 32.

Hogan, W. W. (1998), "Competitive Electricity Market Design: A Wholesale Primer", John F. Kennedy School of Government, Harvard University.

Hogan, W. W. (2005), "Market Design Failures and Electricity Restructuring", CAEM Conference, Washington, D. C.

Hunt, S. and G. Shuttleworth (1996), *Competition and Choice in Electricity,* John Wiley &

Sons.

IADB (2001), *Competitiveness: The Business of Growth, Economic and Social Progress in Latin America,* Research Department, Inter-American Development Bank, Washington, D. C.

IAEA (2004 a), "Nuclear Power: An Evolving Scenario", *IAEA Bulletin,* Vol. 46, No. 1.

IAEA (2004 b), *Nuclear Fuel Cycle Information System, International Atomic Energy Agency.*

IAEA (2004 c), *Nuclear Power's Changing Future: Fastest Growth in Asia.*

IAEA (2004 d), *Power Reactor Information System Database, International Atomic Energy Agency.*

IAEA (2004 e), *Research Reactor Database.*

IEA (1999), "Impacts of Electricity Market Liberalisation on Generation Cost", in *Electricity Reform: Power Generation Costs and Investment.*

IEA (2002), *Electricity in India.*

IEA (2003), *Power Generation Investment in Electricity Markets.*

IEA (2005), *Key World Energy Statistics.*

Iimi, A. (2003), "An Empirical Note: Privatization Transaction and Macroeconomy in Developing Countries", Working Paper No. 10, JBIC Institute.

Institute of Energy (2006), *Master Plan on Electricity Power Development – The Period of 2006–2015 (Master Plan No. VI),* Hanoi, Vietnam.

Institute of Strategy (2004), *Energy Outlook and National Energy Policies by 2010 and Extended to 2020,* Ministry of Industry, Hanoi, Vietnam.

Jackson, I. (2005), "Nuclear Energy: Economics vs Pragmatics – A Viable Option for the Future of Liberalised Energy Supply in Europe? ", 1st Annual European Energy Policy Conference, Brussels.

Jacobs, S. (2002), "Private Investment in Infrastructure in Asia: Rethinking Regulatory Governance", Jacobs and Associates, APEC Privatization Forum Regional Roundtable Hanoi, 22–24 May.

Jacobs, S. (2004), "Governance of Asian Utilities: New Regulators Struggle in Difficult Environments", *The Governance Brief,* Issue 10, ADB.

JAEC (2005), *Framework for Nuclear Energy Policy.*

Jamasb, T. (2002), "Reform and Regulation of the Electricity Sectors in Developing Countries", Cambridge Working Papers in Economics No. 0226.

Jamasb, T. and M. Pollitt (2000)[29], "Benchmarking and Regulation of Electricity Transmission and Distribution Utilities: Lessons from International Experience", Cambridge Working Papers in Economics No. 0101.

Jamasb, T., R. Mota, D. Newbery and M. Pollitt (2004)[30], "Electricity Sector Reform in Developing Countries: A Survey of Empirical Evidence on Determinants and Performance", CMI Working Paper 47.

Jamasb, T., D. Newbery, M. Pollitt and R. Mota (2005 a)[31], "Electricity Sector Reform in

---

29 Allowed by the author to be cited in this book.
30 Allowed by the author to be cited in this book.
31 Allowed by the author to be cited in this book.

Developing Countries: A Survey of Empirical Evidence on Determinants and Performance", World Bank Policy Research Working Paper 3549.

Jamasb, T., D. Newbery and M. Pollitt (2005 b)[32], "Core Indicators for Determinants and Performance of the Electricity Sector in Developing Countries", World Bank Policy Research Working Paper 3599.

Jamasb, T. and M. Pollitt (2005), "Electricity Market Reform in the European Union: Review of Progress toward Liberalisation and Integration", *Energy Journal*, Vol. 26, Special Issue.

Japan Atomic Industrial Forum (2004), "Nuclear Power Plants in the World".

JEPCJ (2005), *Electricity Review Japan 2004–2005*.

Joskow, P. L. (1998 a), "Regulatory Priorities for Reforming Infrastructure Sectors in Developing Countries", Paper presented at the Annual World Bank Conference on Development Economics, Washington, D. C., 20–21 April.

Joskow, P. L. (1998 b), "Electricity Systems in Transition", *Energy Journal*, Vol. 19, No. 2.

Joskow, P. L. (2003), "Electricity Sector Restructuring and Competition: Lessons Learned", 03 –014 WP, A Joint Center of the Department of Economics, Laboratory for Energy and the Environment and Sloan School Management.

Joskow, P. L. and R. Schmalensee (1985), *Market for Power – An Analysis of Electricity Utility Deregulation*, MIT Press.

Kahn, E., C. Robinson and H. Kibune (1995), "International Comparison of Privatization and Deregulation among the USA, the UK and Japan – Volume II: Electricity", *The Economic Analysis*, Vol. 142.

Kennedy, R. M. and L. P. Jones (2003), "Reforming State-Owned Enterprises: Lessons of International Experience, especially for the Least Developed Countries", Working Paper No. 11, United Nations Industrial Development Organization.

Kessides, I. (1997), "Regulation of the Argentine Network Utilities: Issues and Options for the National Government", Economic Notes No 16, World Bank.

Kessides, I. N. (2003), *Infrastructure Regulation Promises, Perils and Principles*, AEI-Brooking Joint Center for Regulatory Studies.

Kessides, I. N. (2004), *Reforming Infrastructure – Privatization, Regulation, and Competition*, World Bank and Oxford University Press.

Kikeri, S. and A. F. Kolo (2005), "Privatization Trends and Recent Development", World Bank Policy Research Working Paper 3765.

King, M., C. K. Woo, A. Tishler and L. C. H. Chow (2006), "Costs of Electricity Deregulation", *Energy*, Vol. 31.

Kirkpatrick, C. and D. Parker (2004), "Infrastructure Regulation: Models for Developing Asia", Research Paper Series No. 60, ADB Institute.

Kirkpatrick, C. and D. Parker (2005), "Towards Better Regulation? Assessing the Impact of Regulatory Reform in Developing Countries", Prepared for Presentation at the CRC Workshop, 22–24 June, University of Manchester, UK.

KPX (2004), Power Statistics website, www.kpx.or.kr.

---

32 Allowed by the author to be cited in this book.

Laffont, J. J. (1996), "Industrial Policy and Politics", *International Journal of Industrial Organization*, Vol. 14.

Laffont, J. J. and J. Tirole (1993), *A Theory of Incentives in Procurement and Regulation*, MIT Press.

Laffont, J. J. and J. Tirole (2000), *Competition in Telecommunications*, MIT Press.

Lammers, G. (2003), "A Third Way for the Electricity Industry, Electricité de France African Energy Commission", Presentation in Workshop for African Energy Experts: Operationalising the NEPAD Energy Initiative, 2–4 June, Dakar, Senegal.

Lee, D. W. (2004), *Intermediary Report on Comparative Analysis of Electricity Reform in OECD Pacific Countries*, IEA.

Lefevre, T. and J. L. Todoc (2002), "Energy Deregulation in Asia: Status, Trends and Implications on the Environment", Center for Energy-Environment Research and Development, Asian Institute of Technology.

Lovei, L. (2003), "The Single-Buyer Model, a Dangerous Path toward Competitive Electricity Markets, Viewpoint", View Point 22403, World Bank.

Lucas, N. (2001), "Reform of the Legal and Institutional Energy Sector Framework: Experience of Asian Countries", Frankfurt Electricity Reform Workshop.

Maddala, G. S. (2002), *Introduction to Econometrics*, 3rd ed., John Wiley & Sons.

Mahmood, I. H. (2002), *Power Sector Reforms in Bangladesh*, Shanghai, China.

Mai, N. T. N. and M. Nomura (2005), "Electricity Industry Reform in Vietnam, an Extended View on Asian Countries", *Journal of Public Utility Economics*, Vol. 57, No. 3.

Marcovitch, J. (1999), "Privatization, Restructuring and Economic Democracy: Synthesis Report", ILO Action Programme, Geneva.

Martinot, E. (2002), "Power Sector Restructuring and Environment: Trends, Policies and GEF Experience", UNEP/IEA Brainstorming Meeting on Power Sector Reform and Sustainable Development, May 21–22, Paris.

Martinot, E. and K. Reiche (2000), "Regulatory Approaches to Rural Electrification and Renewable Energy: Case Studies from Six Developing Countries", Working Paper, World Bank.

Matsuo, S. (2005), *Liberalization of the Electric Power Market in Japan*, Kyushu Electric Power Co.

Mayer, C. (2001), "The Design of Regulatory Institutions", Conference Paper, Competition and Regulation: The Energy Sector in Brazil and the UK/EU, Centre for Brazilian Studies, University of Oxford, 4–5 June.

McGoverna, T. and C. Hicks (2004), "Deregulation and Restructuring of the Global Electricity Supply Industry and Its Impact upon Power Plant Suppliers", *International Journal of Production Economics*, Vol. 89.

Megginson, W. L. and J. M. Netter (2001), "From State to Market: A Survey of Empirical Studies on Privatization", *Journal of Economic Literature*, Vol. 39.

Millan, J., E. Lora and A. Micco (2001), "Sustainability of the Electricity Sector Reforms in Latin America", Prepared for the Annual Meetings of the Board of Governors, Inter-American Investment Corporation, Santiago, Chile.

MIT (2003), *The Future of Nuclear Power.*
Nagayama, H. (2004), *Risks and Its Countermeasures for Power Sector Reform in Asian Developing Countries,* Mitsubishi Research Institute.
Newbery, D. (2001 a), "Problems of Liberalizing the Electricity Industry", Paper for Presentation at the EEA Meetings in Lausanne, 1 Sep.
Newbery, D. (2001 b)[33], "Issues and Options for Restructuring Electricity Supply Industries", DAE Working Paper WP 0210.
Newbery, D. (2002), "Regulatory Challenges to European Electricity Liberalisation", *Swedish Economic Policy Review,* Vol. 9.
Newbery, D. (2003), "Privatisation Experiences in the EU – Privatising Network Industries", Paper prepared for the CESifo Conference on Privatisation, Cadenabbia, 31 Oct–3 Nov.
Ni, C. C. (2005), "Analysis of Applicable Liberalization Models in China's Electric Power Market", *International Public Economy Studies,* Vol. 16.
Nilsson, L. J., A. Arvidson and A. Eberhard (2003), "Public Benefits and Power Sector Reform", Report from an International Workshop held in Stockholm, 12–13 May.
Noll, R. (2000), "Telecommunications Reform in Developing Countries", In Krueger A. (ed.), *Economic Policy Reform: The Second Stage,* University of Chicago Press.
Nuntavorakarn, S. (2002), "The Reform of the Electricity Sector in Thailand: A Civil Society Perspective", Asia Power Sector Reforms Workshop.
Ocana, C. (2002), *Regulatory Reform in the Electricity Supply Industry, An Overview,* IEA.
OECD (1999), "Cross-Country Patterns of Product Market Regulation", in OECD, *Economic Outlook 66.*
OECD (2000), *Regulatory Reform in Korea – Regulatory Reform in the Electricity Sector.*
OECD (2003), *Decommissioning Nuclear Power Plants: Policies, Strategies and Costs.*
Oren, S. S. (2003), "Ensuring Generation Adequacy in Competitive Electricity Markets", University of California.
Outhred, H. (2004), "Global Trends in Electricity Markets", Conference on Australian Energy Reform, Duxton Hotel, Melbourne, 29 June–1 July.
Outhred, H. (2005), "Electricity Industry Restructuring: Issues for Asian Countries", Presented at the EPRI International Conference on Global Electricity Industry Restructuring, 11–12 May.
Petrazzini, B. (1997), "Regulating Communications Services in Developing Countries", in Melody W. (ed.), *Telecom Reform: Principles, Policies and Regulatory Practices,* Technical University of Denmark.
Pinto Junior, H. (2002), "Institutional Designs and Regulatory Reforms in the Energy Industries: An International Comparatitve Analysis and Lessons for Brazil", Research Paper No 1, Center for Brazilian Studies, University of Oxford.
Pollitt, M. (2003), "Electricity Reform in Chile and Argentina: Lessons for Developing Countries", http://www.econ.cam.ac.uk/electricity/news/autumn 03/pollitt.pdf.
Public-Private Infrastructure Advisory Facility (2001), *Toolkit: A Guide for Hiring and*

---

33 Allowed by the author to be cited in this book.

Managing Advisors for Private Participation in Infrastructure – What is PPI and How Can Advisors Help?

Rector, J. (2005), "The IPP Investment Experience in Malaysia", Working Paper No. 46, Program on Energy and Sustainable Development at Stanford University.

Regulatory Assistance Project: Gardiner, M. and V. Montpelier (2000), *Best Practices Guide: Implementing Power Sector Reform,* The Energy Group, Institute of International Education, Washington, D. C.

Rothwell, G. and T. Gomez (2003), *Electricity Economics: Regulation and Deregulation,* IEEE.

Rudnick, H. (1998), "The Electric Market Restructuring in South America: Successes and Failures on Market Design", Harvard Electricity Policy Group, San Diego.

Rufin, C. (2003), *The Political Economy of Institutional Change in the Electricity Supply Industry,* Edward Elgar.

Sappington, D. and D. Weisman (1996), *Designing Incentive Regulation for the Telecommunications Industry,* MIT Press.

Sharma, D., S. E. Madamba and R. L. Chan (2004), "Electricity Industry Reforms in the Philippines", *Energy Policy,* Vol. 32.

Shehadi, K. S. (2002), *Lessons in Privatization: Considerations for Arab States,* United Nations Development Programme.

Shin, J. S. (2005), "Korean Electricity Industry: Overview, Restructuring and Its Future, Global Electricity Industry Restructuring", 11–12 May, San Francisco, California.

Shleifer, A. and R. Vishny (1994), "Politicians and Firms", *Quarterly Journal of Economics,* Vol. 46.

Shrestha, R. M., S. Kumar, M. J. Todoc and S. Sharma (2004), "Institutional Reforms and Their Impact on Rural Electrification: Case Study in South and Southeast Asia – Sub-Regional "Energy Access" Study of South and Southeast Asia", Global Network on Energy for Sustainable Development.

Shrestha, R. M., S. Kumar, S. Sharma and M. J. Todoc (2004), "Institutional Reforms and Electricity Access: Lessons from Bangladesh and Thailand", Energy Field of Study, School of Environment, Resources and Development, Asian Institute of Technology.

Sidak, J. G. and D. F. Spulber (1998), *Deregulatory Takings and the Regulatory Contracts,* Cambridge University Press.

Sihag, A. R., N. Misra and V. Sharma (2003), "Impact of Power Sector Reform on the Poor: Case Studies of South and South-East Asia", The Energy and Resources Institute, India.

Situmeang, H. H. (2005), *The Challenge of Financing Power Projects,* Jakarta, Indonesia.

Smith, W. (1997), "Utility Regulators – The Independence Debate", *Public Policy for the Private Sector Note 127,* World Bank.

Son, L. D. (2005), "Electricity Industry in Vietnam – Current Situation and Development Plan", Workshop of Electricity Power Development – the Cooperation between China and Vietnam.

Steiner, F. (2001), "Regulation, Industry Structure and Performance in the Electricity Supply Industry", *OECD Economic Studies,* No. 32.

Stern, J. (2000 a), "Electricity and Telecommunication Regulatory Institutions in Small and Developing Countries", *Utilities Policy,* Vol. 9.

Stern, J. (2000 b), "Styles of Regulation: The Choice of Approach to Utility Regulation in Central and Eastern Europe", in Coen, D. and M. Thatcher (eds.), *Utilities Reform in Europe,* Nova Science Publishers.

Stern, J. (2004), "Regulatory Governance and Its Relationship to Infrastructure Industry Outcomes in Developing Economies", New Directions in Regulation Seminar, Kennedy School of Government, Harvard University.

Stoft, S. (2002), *Power System Economics: Designing Markets for Electricity,* John Wiley & Sons.

Studenmund, A. H. and H. J. Cassidy (1987), *Using Econometrics: A Practical Guide,* Little, Brown and Company.

Suzuki, T. (2005), "Japan's Nuclear Power Program –Trends and Issues–", Asia Energy Security Workshop, Beijing, 12–17 May.

TAIPOWER (2004), *2003 Annual Report.*

TEPCO (2004), *Electricity Market in Japan.*

Thomas, S. (2004), "Electricity Liberalization: the Beginning of the End", Public Services International Research Unit.

Thomas, S., D. Hall and V. Corral (2004), "Electricity Privatisation and Restructuring in Asia-Pacific", Asia-Pacific Meeting in Changmai, Thailand.

Thomas, S., D. Hall and V. Corral (2006), "Electricity Privatization and Restructuring in Asia-Pacific", Public Services International Research Unit.

Tumiwa, F. (2003), "Power Sector Restructuring in Indonesia: Viable Solution or Recipe for Disaster?", Asia Power Sector Reforms Workshop 2002.

United Nations (2004), "Competition, Competitiveness and Development: Lessons from Developing Countries", United Nations Conference on Trade and Development, Geneva.

United States Agency for International Development (2000), *Best Practices Guide: Implementing Power Sector Reform.*

United States Agency for International Development and the United States Energy Association (2000), *Power Sector Privatization in Central/Eastern Europe and Eurasia: A Summary Report.*

University of Chicago (2004), *The Economic Future of Nuclear Power.*

Ure, J. (2002), "Regulatory Issues and Privatization, Asian Forum on ICT Policies and e-Strategies", 20–22 October, Kuala Lumpur: UNDP-APDIP.

Vazquez, C., M. Rivier and J. I. Perez-Arriaga (2002), "A Market Approach to Long-term Security of Supply", *IEEE Transactions on Power Systems,* Vol. 17, No. 2.

Vickers, J. and G. Yarrow (1988), *Privatization: An Economic Analysis,* MIT Press.

Victor, D. G. (2004), *Electricity Market Reform in Developing Countries: Results from a Five Country Study,* State Level Electricity Reforms in India, Delhi.

Vietnam National Assembly (2004), *Electricity Law,* Vietnam.

Villot, X. (1996), "Market Instruments and the Control of Acid Rain Damage", *Energy Policy,* Vol. 24, No. 9.

Wallsten, S. J. (2000), "An Econometric Analysis of Telecommunications Competition, Privatization and Regulation in Africa and Latin America", *Journal of Industrial Economics*, Vol. 49, No. 1.

Wallsten, S., G. Clarke, L. Haggarty, R. Kaneshiro, R. Noll, M. Shirley and L. C. Xu (2004), "New Tools for Studying Network Industry Reforms in Developing Countries: The Telecommunications and Electricity Regulation Database", *Review of Network Economics*, Vol. 3, Issue 3.

Wamukonya, N. (2003 a), "Power Sector Reform in Developing Countries: Mismatched Agenda", *Energy Policy*, Vol. 31, No. 12.

Wamukonya, N. (2003 b), *Electricity Reform: Social and Environmental Challenges*, United Nation Environment Programme, Pitney Bowes Management Services, Denmark.

Wardle, D. and N. Towle (1996), "Global Privatization", in Campbell, D. (ed.), *International Privatization*, Kluwer Law International.

Weinmann, J. and D. Bunn (2004), "Resource Endowment and Electricity Sector Reform", Paper presented at the 6th Annual Conference of the International Society for New Institutional Economics and the 22nd North American Conference of the United States Association for Energy Economics.

Weisman, D. (2002), "Is There 'Hope' for Price Cap Regulation?", *Information Economics and Policy*, Vol. 14.

Weisman, D. (2005), "Market Concentration, Multi-market Participation and Merger in Network Industries", *Review of Network Economics*, Vol. 4, Issue 2.

Weizsacker, E. U. von, O. R. Young and M. Finger (2005), *Limits to Privatization*, Earthscan.

Woo, C. K., M. King, A. Tishler and L. C. H. Chow (2005), "Costs of Electricity Deregulation", *Energy*, Vol. 31, Issues 6-7.

Woo, P. Y. (2005 a), "China's Electric Power Market: The Rise and Fall of IPPs", Working Paper No. 45, The Program on Energy and Sustainable Development at Stanford University.

Woo, P. Y. (2005 b), "Independent Power Producers in Thailand", Working Paper No. 51, The Program on Energy and Sustainable Development at Stanford University.

Woodhouse, E. J. (2005), "The IPP Experience in the Philippines", Working Paper No. 37, The Program on Energy and Sustainable Development at Stanford University.

Wooldridge, J. M. (2002), *Econometric Analysis of Cross Section and Panel Data*, MIT Press.

Woolf, F. and J. Halpern (2001), "Integrating Independent Power Producers into Emerging Wholesale Power Markets", Policy Research Working Paper No. 2703, World Bank.

World Bank (1993), *The World Bank's Role in the Electric Power Sector, Policies for Effective Institutional, Regulatory and Financial Reform.*

World Bank (1994), "Power and Energy Efficiency Status Report on the Bank's Policy and IFC's Activities", Joint World Bank/IFC Seminar Report.

World Bank (2002), *Private Sector Development Strategy – Diretions for the World Bank Group*.

World Bank (2003), *Statistical Annex: Infrastructure Indicators, Connecting East Asia: A New Framework for Infrastructure.*

World Energy Council (2001), *Electricity Market Design and Creation in Asia Pacific*, London.
Yajima, M. (1997), *Deregulatory Reforms of the Electricity Supply Industry*, Greenwood Publishing Group.
Yanase, T. (2005), *The Challenges and Directions for Nuclear Energy Policy in Japan*, Ministry of Economy, Trade and Industry.
Yergin, D. and J. Stanislaw (1998), *The Commanding Heights: The Battle Between Government and the Marketplace That Is Remaking the Modern World*, Simon & Schuster.
Yoon, Y. T. and M. D. Ilic (2000), "Transmission Expansion in the New Environment", Energy Laboratory Publication, MIT EL 00-002.
Zhang, C. (2003), "Reform of Chinese Electric Power Market: Economics and Institutions", Program on Energy and Sustainable Development, Stanford, Working Paper No. 3.
Zhang, C. (2004), *China's Electricity Industry Reform: Economics and Institutions*, New Delhi, India.
Zhang, Y. F., D. Parker and C. Kirkpatrick (2002)[34], "Electricity Sector Reform in Developing Countries: An Econometric Assessment of the Effects of Privatisation, Competition and Regulation", Working Paper No. 31, University of Manchester[35].
Zhang, Y. F., D. Parker and C. Kirkpatrick (2005), "Competition, Regulation and Privatisation of Electricity Generation in Developing Countries: Does the Sequencing of the Reforms Matter?", *Quarterly Review of Economics and Finance*, Vol. 45, Issues 2-3.
Zhang, X. (2005), "Can competitive reforms solve the long run investment problem (incentive and coordination) in electricity market?", http://www.grjm.net/documents/xinying/wp-wei_2005.pdf.
Zhou, S., T. Grasso and G. Niu (2003), "Comparison of Market Designs", Project 26376, Rulemaking Proceeding on Wholesale Market Design Issues in the Electric Reliability Council of Texas, Public Utility Commission of Texas.

---

34 Allowed by the author to be cited in this book. The paper will be published soon in *Journal of Regulatory Economics*.
35 Updated version of the paper was issued in 2005.

Index

ancillary service   27, 46–7, 231, 299
Asian Development Bank (ADB)   104, 193

back-end fund   257–8
bilateral trades   49

Clean Development Mechanism   90, 258, 293
combined cycle gas turbines (CCGT)   23
congestion management   44, 47–8, 231
correspondence analysis   119–20, 326
cost-plus regulation   82, 84

decommissioning   252, 257–9
demand-side-management (DSM)   24
Discounted Cash Flow (DCF)   318

Electric Power System Council of Japan (ESCJ)   212
equitization   5, 178–9, 267, 274, 311, 313–23, 329
European Bank for Reconstruction and Development (EBRD)   106
European Commission   95

golden share   68, 70–1

Independent Power Producers (IPPs)   34, 167–8, 210
Independent System Operator (ISO)   39, 224

Japan Electric Power Exchange (JEPX)   151, 212–3
Japanese Bank of International Cooperation (JBIC)   106

management buy-out (MBO)   57
mandatory pool   48, 300
multidimensionality   15–6, 277–8, 324, 327

National Load Dispatch Center   299
Net Asset Valuation (NAV)   318
nodal pricing   43–4
nuclear power   6, 8, 18, 22, 97, 111, 193, 195, 197–8, 203–4, 207, 213, 234, 236–52, 254–6, 258–9, 297, 313, 325–7

one-member limited liability companies   270

partial privatization   68, 70, 205, 208, 230, 278, 311
performance indicators   186–7, 192, 198, 204–5, 328
postage stamp pricing   44

Power Exchange (PX)   39
Power Purchasing Agreement (PPA)   34
price cap   6, 52, 81–5, 94, 285, 331, 338
privatization   3, 5, 7, 15, 17–8, 28, 30, 54–71, 73–4, 92–100, 104–6, 109, 111–2, 114–8, 121, 123, 125–6, 130–2, 134–9, 142, 144–6, 148, 150–4, 157–60, 162, 165, 167, 170, 172–6, 180, 182, 184–7, 190, 192–5, 197–8, 200, 203–5, 207–8, 214, 216, 220–1, 223–5, 229–30, 232, 234, 261, 268, 272, 277–8, 311, 313–4, 316–7, 324–6, 328–34, 336–8
public offerings   55, 57, 317

Rate of Return (ROR)   13, 84–5, 331
rural electrification   33, 103, 113, 127–9, 168, 172, 174, 184, 205, 225, 232, 264, 275, 291–2, 320, 334, 336

Scheduling Coordinators (SCs)   39
Single Buyer Model (SBM)   34

Third Party Access (TPA)   32

unbundling   16, 32, 40, 47, 75, 92–3, 95, 104–5, 116, 121, 150, 153, 164, 167–9, 171–2, 175, 188, 194, 205, 219, 226, 229, 269

vertical integration   7, 25, 33, 42, 76, 188, 214
Vietpool   305, 308

World Bank (WB)   104

zonal pricing   44, 47

NGUYEN THI NGOC MAI　グエン・ティ・グォック・マイ

1974年	ベトナム生まれ
1990年9月〜1995年6月	国立ハノイ工科大学（Hanoi National University of Technology, Hanoi, Vietnam）卒業
1998年1月〜1999年8月	アジア工科大学大学院（Asian Institute of Technology, Bangkok, Thailand）修士号取得
2004年9月〜2006年3月	関西学院大学大学院（経済学研究科博士課程後期課程）在籍
2001年6月〜現在に至る	ベトナム電力公社（Electricity of Vietnam）勤務

The Multidimensionality of Electricity Industry Reform:
A Strategic Analysis for the Case of Asian Countries and Vietnam

2007 年 2 月 20 日初版第一刷発行

著　者	Nguyen Thi Ngoc Mai
発行者	山本　栄一
発行所	関西学院大学出版会
所在地	〒662-0891　兵庫県西宮市上ケ原一番町 1-155
電　話	0798-53-5233
印　刷	協和印刷株式会社

© Nguyen Thi Ngoc Mai 2007
Printed in Japan by Kwansei Gakuin University Press
ISBN: 978-4-86283-007-4
乱丁・落丁本はお取り替えいたします。
http://www.kwansei.ac.jp/press/